AF606425

VOLUME FIVE HUNDRED AND FORTY EIGHT

METHODS IN ENZYMOLOGY

Protein Kinase Inhibitors in Research and Medicine

METHODS IN ENZYMOLOGY

VOLUME FIVE HUNDRED AND FORTY EIGHT

METHODS IN ENZYMOLOGY

Protein Kinase Inhibitors in Research and Medicine

Edited by

KEVAN M. SHOKAT

Howard Hughes Medical Institute and Department of Cellular & Molecular Pharmacology, University of California, San Francisco, California, USA

AMSTERDAM • BOSTON • HEIDELBERG • LONDON
NEW YORK • OXFORD • PARIS • SAN DIEGO
SAN FRANCISCO • SINGAPORE • SYDNEY • TOKYO
Academic Press is an imprint of Elsevier

Academic Press is an imprint of Elsevier
225 Wyman Street, Waltham, MA 02451, USA
525 B Street, Suite 1800, San Diego, CA 92101-4495, USA
32 Jamestown Road, London NW1 7BY, UK
The Boulevard, Langford Lane, Kidlington, Oxford OX5 1GB, UK

First edition 2014

ISBN: 978-0-12-397918-6
ISSN: 0076-6879

For information on all Academic Press publications visit our website at store.elsevier.com

CONTENTS

CONTRIBUTORS

Veselin I. Andreev

Fiona P. Bailey
Department of Biochemistry, Institute of Integrative Biology, University of Liverpool, Liverpool, United Kingdom

Valerio Berdini
Astex Pharmaceuticals, Cambridge, United Kingdom

Philip A. Cole
Department of Pharmacology and Molecular Sciences, Johns Hopkins School of Medicine, Baltimore, Maryland, USA

Patrick A. Eyers
Department of Biochemistry, Institute of Integrative Biology, University of Liverpool, Liverpool, United Kingdom

Doriano Fabbro
PIQUR Therapeutics AG, Hochbergerstrasse 60C, Basel, Switzerland

Nathanael S. Gray
Department of Biological Chemistry and Molecular Pharmacology, Harvard Medical School, Department of Cancer Biology, Dana-Farber Cancer Institute, Boston MA, USA

Joseph I. Kliegman
Howard Hughes Medical Institute and Department of Cellular & Molecular Pharmacology, University of California, San Francisco, California, USA

John Kuriyan
Department of Molecular and Cell Biology; California Institute for Quantitative Biosciences; Howard Hughes Medical Institute; Department of Chemistry, University of California, and Physical Biosciences Division, Lawrence Berkeley National Laboratory, Berkeley, California, USA

Michael S. Lopez
Howard Hughes Medical Institute and Department of Cellular & Molecular Pharmacology, University of California, San Francisco, California, USA

Rand M. Miller
Chemistry and Chemical Biology Graduate Program, University of California, San Francisco, California, USA

Paul N. Mortenson
Astex Pharmaceuticals, Cambridge, United Kingdom

Marc O'Reilly
Astex Pharmaceuticals, Cambridge, United Kingdom

Daniel Rauh
Chemical Genomics Centre of the Max Planck Society, and Fakultät Chemie, Chemische Biologie, Technische Universität Dortmund, Dortmund, Germany

Kevan M. Shokat
Howard Hughes Medical Institute and Department of Cellular & Molecular Pharmacology, University of California, San Francisco, California, USA

Jeffrey R. Simard*
Chemical Genomics Centre of the Max Planck Society, Dortmund, Germany

Jack Taunton
Howard Hughes Medical Institute and Department of Cellular and Molecular Pharmacology, University of California, San Francisco, California, USA

Qi Wang
Department of Molecular and Cell Biology, and California Institute for Quantitative Biosciences, University of California, Berkeley, California, USA

Zhihong Wang
Department of Chemistry and Biochemistry, University of the Sciences, Philadelphia, Pennsylvania, USA

Julie A. Zorn
Department of Molecular and Cell Biology, and California Institute for Quantitative Biosciences, University of California, Berkeley, California, USA

*Current address: Amgen Inc., 360 Binney Street, Cambridge, Massachusetts, USA

PREFACE

The field of protein kinase inhibitors has exploded in the past 20 years. This pace of discovery has been driven by the diversity of biological processes controlled by protein kinases and the ability to develop small-molecule inhibitors capable of potently inhibiting kinases. This volume of *Methods in Enzymology* begins with a framework for understanding how protein kinases carry out the regulated transfer of phosphate to substrate proteins (Chapter 1). Protein kinase inhibitors were first viewed simply as ATP-competitive agents which occlude ATP binding and thereby block catalysis. As a result of the great number of structural studies, we now appreciate that certain inhibitors can bind to kinases in a conformation-specific manner or even outside the ATP pocket (Chapter 2). Following these mechanistic and structural chapters, we move to the discovery of inhibitors. There have been relatively few transformations in drug discovery since the implementation of structure-based design. One particularly noteworthy advance has been the appreciation of ligand efficiency and the implementation of this guiding principle at the earliest stages of kinase inhibitor discovery, termed fragment-based inhibitor discovery (Chapter 3). A particularly important clinical aspect of kinase inhibitors has been the need to inhibit oncogenic kinases potently such that the pathways in which they reside are inhibited at the level of 80% or greater (Vemurafenib clinical experience). The challenge of achieving such potent and durable inhibition while maintaining target specificity is a challenge the entire field faces. Covalent inhibitors which target nonconserved cysteines solve the potency and selectivity challenge simultaneously (Chapter 4). The main liability of covalent inhibitors, that of irreversible binding to off-targets, has been creatively addressed by the development of covalent-reversible inhibitors (Chapter 4). Although the core catalytic residues of kinases are conserved across the 538 family members, there are many residues in the active site which tolerate mutations resulting in drug resistance. The clinical emergence of these mutations and their creation and use *in vitro* to understand the mechanism of action of drugs is described in Chapter 5. The importance of identifying conformation-specific inhibitors and alternatives to traditional ATP site blockers has led to new tailored fluorescent screens (Chapter 6) and phenotypic screens (Chapter 7) for kinase inhibitors. Methods for discovering kinase inhibitors have improved greatly in the last 20 years, yet the kinase family is enormous with 538 human kinases. Genetics offers a means to inactivate each family member individually, but the adaptability of kinase networks often masks the effect of knocking out or knocking down a specific kinase. In Chapter 8, the use of chemical genetics to develop an inhibitor of each protein kinase is described which combines the advantages of genetics and pharmacology in a widely used system across multiple organisms.

The field of kinase inhibitors is vast. Many important advances and techniques that were not covered in this volume will be covered in future editions.

Kevan M. Shokat

CHAPTER ONE

Catalytic Mechanisms and Regulation of Protein Kinases

Zhihong Wang*, Philip A. Cole[†,1]

*Department of Chemistry and Biochemistry, University of the Sciences, Philadelphia, Pennsylvania, USA
[†]Department of Pharmacology and Molecular Sciences, Johns Hopkins School of Medicine, Baltimore, Maryland, USA
[1]Corresponding author: e-mail address: pcole@jhmi.edu

Contents

Abstract

Protein kinases transfer a phosphoryl group from ATP onto target proteins and play a critical role in signal transduction and other cellular processes. Here, we review the kinase kinetic and chemical mechanisms and their application in understanding kinase structure and function. Aberrant kinase activity has been implicated in many human diseases, in particular cancer. We highlight applications of technologies and concepts derived from kinase mechanistic studies that have helped illuminate how kinases are regulated and contribute to pathophysiology.

1. INTRODUCTION

The discovery of protein kinases in the 1950s led to a massive influence on clarifying biological pathways and disease mechanisms and developing therapies over the subsequent six decades (Hunter, 2000; Krebs & Beavo, 1979). Eukaryotic protein kinases are enzymes that catalyze phosphoryl transfer from MgATP to Ser/Thr and Tyr side chains in proteins. Their

Methods in Enzymology, Volume 548
ISSN 0076-6879
http://dx.doi.org/10.1016/B978-0-12-397918-6.00001-X

importance is in part evidenced by their frequency in eukaryotic genomes, typically representing 2–3% of the genes, including in human where 518 protein kinases have been annotated (Manning, Whyte, Martinez, Hunter, & Sudarsanam, 2002). While each specific kinase is thought to have a specialized function, there are many conserved features among kinases regarding their structures and catalytic mechanisms (Hanks, Quinn, & Hunter, 1988). This protein kinase chapter is written from an enzymology perspective and will cover the kinetic and chemical mechanisms of kinases and how an understanding of these features has been used to explore the structure, function, and regulation of these important catalysts.

2. KINETIC MECHANISM

Protein kinases operate on two substrates, proteins, and MgATP and produce phosphoproteins and MgADP (Adams, 2001; Taylor & Kornev, 2011). While it is sometimes the case that free ATP rather than Mg-bound ATP is thought of as the phosphoryl-donor substrate, the affinity of Mg for ATP is high enough that there is only a low concentration of non-Mg-bound ATP in cells. Thus with one apparent exception (Mukherjee et al., 2008), protein kinases require at least one divalent ion, Mg or Mn, for catalysis. Two substrate group transfer enzymes like kinases can be classified into two general types, those that follow ternary complex mechanisms and those that follow ping-pong mechanisms (Segel, 1993). Ternary complex mechanisms typically involve direct reaction between the two substrates to afford the two products, whereas ping-pong mechanisms proceed through a covalent enzyme intermediate, which in the case of kinases would be a phosphoenzyme species.

Classical two substrate steady-state kinetics experiments revealing an intersecting line pattern in double reciprocal plots (Segel, 1993) as well as more technically sophisticated stereochemical studies showing inversion at the phosphoryl group (Knowles, 1980) helped define protein kinase A (PKA) as following a ternary complex mechanism. Subsequently, two substrate kinetic studies on a variety of Ser/Thr and Tyr kinases and many X-ray structures of these enzymes in complex with substrate analogs have confirmed this to be a general feature of the kinase superfamily (Zheng et al., 1993). However, recently, an X-ray crystal structure of an atypical kinase showed the surprising finding that an active site aspartate was phosphorylated (Ferreira-Cerca et al., 2012). This phosphoAsp was proposed to correspond to a phosphoenzyme intermediate that could deliver the phosphoryl group to a protein substrate, though further experiments will be needed to establish this mechanism. Of note, nucleoside diphosphokinase does proceed through a phosphohistidine intermediate so there is enzymatic precedence for a small-molecule kinase using a related mechanism (Admiraal et al., 1999).

For the vast majority of protein kinases that involve direct phosphoryl transfer through a ternary complex, other kinetic mechanism issues that have been addressed are whether there is a preference for MgATP or protein substrate to bind first and what step(s) is rate-limiting for catalysis? These features have been analyzed for a variety of protein kinases and the results are somewhat enzyme and reaction condition dependent. For example, PKA displays a clear preference for MgATP binding prior to peptide substrate whereas Csk kinase shows no apparent-binding preference between nucleotide or peptide substrates (Cole, Burn, Takacs, & Walsh, 1994; Qamar, Yoon, & Cook, 1992; Zheng et al., 1993). Interestingly, experiments on p38 MAP kinase have led to contradictory models. Models in which protein substrate binds first, MgATP binds first, or random order binding have all been proposed for p38 MAP kinase (LoGrasso et al., 1997; Szafranska & Dalby, 2005). While these different models could be traced to the distinct methods used for measurement, they also highlight the limitation of steady-state kinetic approaches to provide unambiguous mechanistic portraits. The most comprehensive studies on p38 MAP kinase that include complementary methods including calorimetry and structural considerations point to a random order of substrate binding for this enzyme (Szafranska & Dalby, 2005). One potential practical application that emanates from such models relates to the development of specific kinase inhibitors that target the ATP pocket (Noble, Endicott, & Johnson, 2004). If the protein substrate binds to the kinase in the absence of MgATP, there may be an influence on drug affinity.

Regarding rate-limiting steps, a combination of viscosity effects, presteady-state kinetic techniques, and alternate substrates have been employed with various kinases to define the microscopic rate constants. To some extent, the kinetic models not only depend on the conditions of the kinase assay conditions (salt concentration, divalent ion (Mg vs. Mn), peptide, or protein substrate), but they also show differences among the kinases themselves. With PKA, MgADP release is fully rate determining (Adams & Taylor, 1992; Qamar & Cook, 1993), whereas for Csk phosphoryl transfer is partially or fully rate limiting depending on whether MgATP or MnATP is used as the substrate (Grace, Walsh, & Cole, 1997). When Mg is used with Csk, product release is fast and chemistry is rate determining, whereas when Mn is employed, product release slows down, presumably because of metal–enzyme interactions. Increasing the ionic strength of the buffer can also speed product release, possibly by weakening the interactions between nucleotide and enzyme. Some kinases such as Ser–Arg protein kinase or Src protein tyrosine kinase can show processive phosphorylation of its protein substrate, effectively indicating that protein substrate/product release is the slow step in turnover (Aubol et al., 2003; Pellicena & Miller, 2001). Furthermore, many protein kinases like the insulin receptor tyrosine kinase (IRK) are regulated by accessory domains, phosphorylation, or allosteric ligands which can dramatically impact the nature of the rate-limiting steps (Ablooglu, Frankel, Rusinova, Ross, & Kohanski, 2001; Hubbard & Miller, 2007).

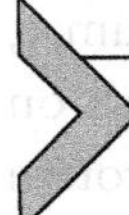

3. CHEMICAL MECHANISM OF KINASE PHOSPHORYL TRANSFER

Despite the apparent simplicity of the reaction chemistry, there has been significant effort to understand the details of how the phosphoryl group moves from ATP to the protein substrate hydroxy group in the kinase active site. This interest stems from several considerations. One is the fundamental challenge in defining the catalytic mechanism of an important family of enzymes. A second factor relates to our fascination with how kinase enzymes interconvert between more active and less active forms. Kinase regulation by ligands, phosphorylation, as well as mutation can alter the alignment of active site residues, which ultimately translates to effects on the chemistry of phosphoryl transfer. A third reason for interest in the chemical mechanism is to aid in the design of synthetic compounds which can artificially switch kinase activity on or off. Such mechanism-inspired chemical biology approaches can and have shed light on the biological functions of kinases in cellular signaling.

A central issue in defining the kinase mechanism is clarifying the nature of the phosphoryl transfer transition state. In the study of nonenzymatic phosphoryl transfer mechanisms, research dating back to the 1960s showed that phosphate monoesters like phenol phosphates display "dissociative" transition states (Kirby & Jencks, 1965). A dissociative transition state is one in which the bond between the phosphorus atom and the leaving group is largely broken prior to significant bond formation between the incoming nucleophile and the phosphorus, involving a metaphosphate-like intermediate (Admiraal & Herschlag, 1995; Mildvan, 1997; Fig. 1.1A). In contrast,

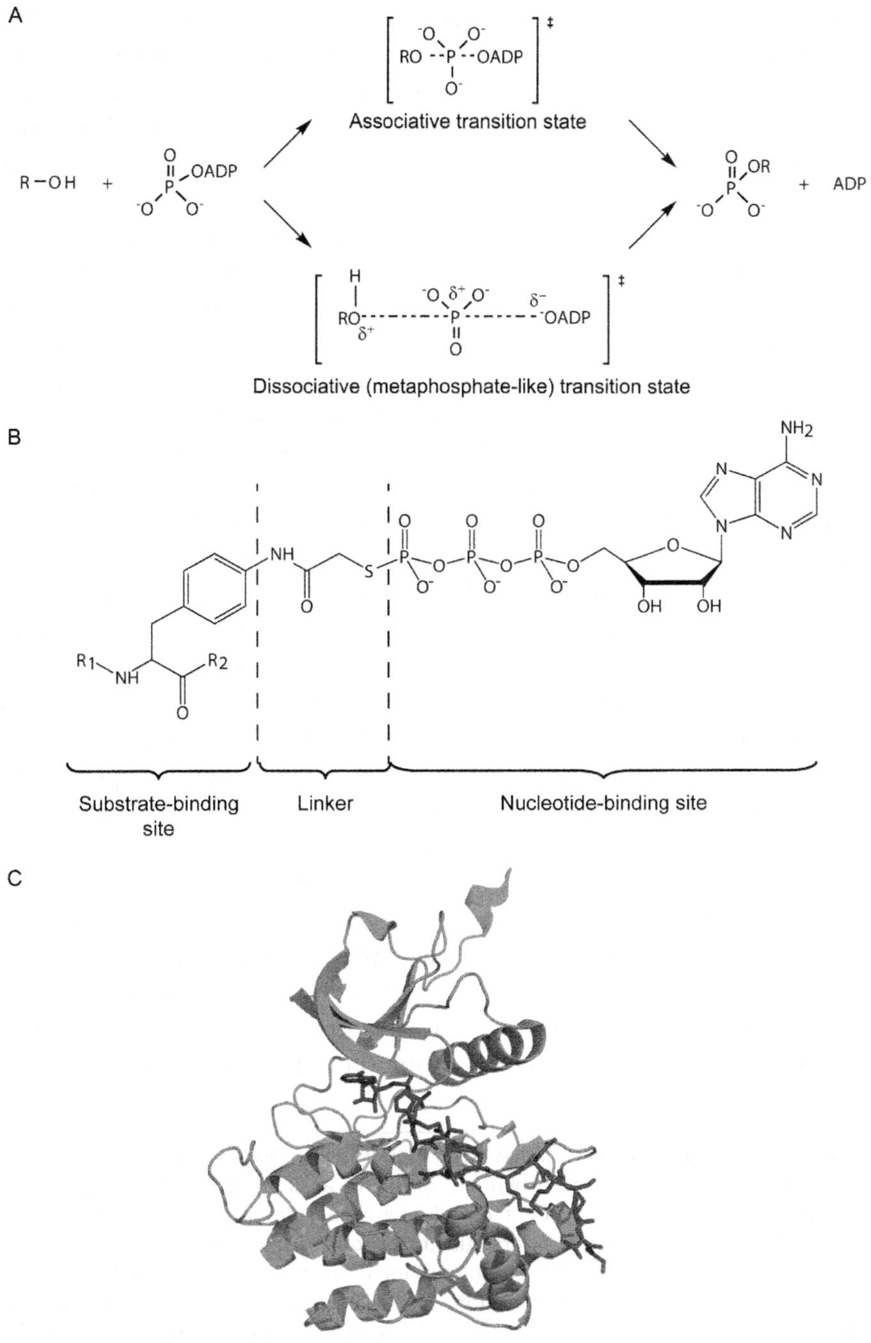

A

Associative transition state

Dissociative (metaphosphate-like) transition state

B

Substrate-binding site

Linker

Nucleotide-binding site

C

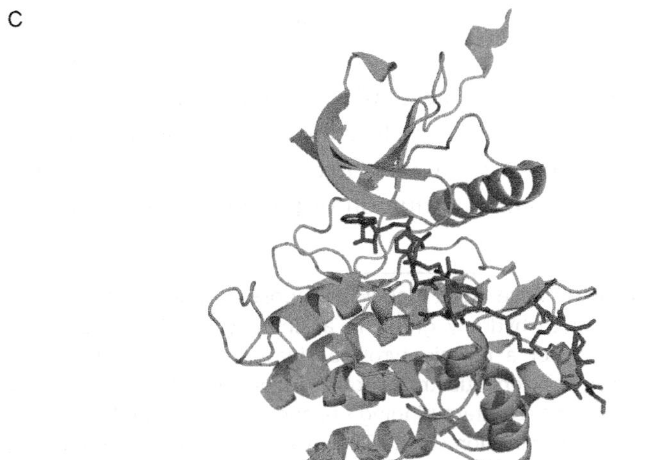

Figure 1.1 Protein kinase mechanism and bisubstrate analog inhibitors. (A) Mechanistic scheme of associative and dissociative transition states of phosphoryl transfer.

(Continued)

nonenzymatic phosphate triester reactions typically show an associative transition state where the nucleophile forms a substantial bond with the phosphorus prior to leaving group departure. This dissociative character of the monoester is rationalized based on its negatively charged nonbridging oxygens repelling nucleophiles and stabilizing the partial positive charge on the metaphosphate-like intermediate.

Relative to their nonenzymatic counterparts, transition states for enzyme reactions are much harder to analyze because of the complexity of the large protein catalyst and the more limited repertoire of methods available. Based on the early X-ray crystal structures of PKA, a chemical phosphoryl transfer mechanism with associative character was implied (Knighton et al., 1991). However, analysis of a static view of an enzyme complexed with substrate analogs can be an imprecise way for probing the dynamics of rapid chemical processes. Linear-free energy studies designed to measure the Bronsted nucleophile coefficient (βnuc) gave a different interpretation of the transition state. By analyzing the tyrosine kinase activity dependence with a series of tyrosine analogs substituted with fluorine atoms on the phenol ring, it was established that phosphoryl transfer rate is nearly independent of the nucleophile pK_a, that is a near zero value for βnuc (Kim & Cole, 1997, 1998). In addition, the unexpected observation that the neutral phenol, rather than the phenoxide anion, was the obligate functional substrate form based on pH-rate studies further underscored the minimal role, the nucleophile likely plays in the transition state (Kim & Cole, 1998).

Other experiments have corroborated that tyrosine kinase transition states are likely dissociative, analogous to their nonenzymatic counterparts (Ablooglu et al., 2000; Parang et al., 2001; Sondhi, Xu, Songyang, Eck, & Cole, 1998; Williams & Cole, 2002). Interestingly, rate reductions of kinases with MgATPγS were initially considered as evidence against kinase-dissociative transition states (Cole, Grace, Phillips, Burn, & Walsh, 1995), since nonbridging sulfur substitution of phosphate monoesters can

Figure 1.1—Cont'd A dissociative state model is proposed for the transfer of the γ-phosphoryl group of ATP to hydroxy group in a kinase–substrate. (B) Designed bisubstrate analog inhibitor for a protein tyrosine kinase. R1 and R2 amino acid sequences are derived from efficient substrate motifs. The linker is predicted to result in a 5–5.7 Å spacer between the substrate-binding site and the nucleotide-binding site, compatible with a dissociative transition state. ATPγS was used as the ATP-mimic analog. (C) The crystal structure of IRK (green ribbon) in complex with the above peptide–ATP conjugate (red stick), determined at 2.7 Å resolution (PDB code: 1GAG; Parang et al., 2001). (See the color plate).

accelerate nonenzymatic phosphoryl transfer through stabilization of a metaphosphate-like species. However, follow-up measurements with various divalent ions showed that thiophilic metals like Ni and Co could complement the assays with ATPγS (Grace et al., 1997). Of note, it should be mentioned that kinase-mediated thiophosphate installation has been elegantly used for protein substrate identification in chemoselective labeling (Allen, Lazerwith, & Shokat, 2005).

The totality of the solution-phase evidence strongly supports a dissociative transition state for the protein kinase reactions that have been studied. Perhaps this result was to be expected since it would be energetically costly to force the phosphoryl transfer down a mechanistic pathway that traverses a steeper energetic barrier. Furthermore, based on the conserved features of all ~500 human kinases, it is reasonable to assert that dissociative character is a general property of these reactions.

This model does raise the somewhat untidy question of how a protein kinase would catalyze a dissociative transition state based on the architecture and electrostatics of the kinase active site. Since the enzyme is organizing a ternary complex, the templating of the ATP and protein substrate in position for reaction could be thought most compatible with a more associative reaction, as the crystallographic interpretations appeared to suggest. Perhaps an overly simple explanation for what kinases need to do to facilitate dissociative reactions is that they must focus positive charge on the departing ADP-anionic oxygens to sever the gamma-phosphoryl bond. Indeed, positive charges including a Lys side chain and Mg ion are pointed toward this β-phosphate and are observed in many kinase crystal structures in an arrangement that could stabilize the buildup of negative charge in the leaving group. It is also true that many kinases show substantial ATP-hydrolytic activity, which is consistent with generation of a metaphosphate-like species that can be intercepted by water in the absence of precise positioning of a protein hydroxyl (Rominger et al., 2007).

A number of computational studies have been performed to understand the kinase phosphoryl transfer mechanism (Cheng, Zhang, & McCammon, 2005; Valiev, Yang, Adams, Taylor, & Weare, 2007). Many of these simulations have analyzed the contributions of specific active site interactions, as well as geometry and charge states as a function of phosphoryl transfer progress. Such calculations generally are consistent with a dissociative transition-state mechanism. While these studies are generally not readily testable with precise experiments, site-directed mutagenesis and substrate analog experiments generally support their plausibility. Most fully active states of kinases are defined crystallographically by macroscopic and microscopic features that seem to be crucial for phosphoryl transfer. There is a particular alignment of the N-terminal and C-terminal lobes, generating a hydrophobic spine, that correspond to active kinase conformations (Kornev, Haste, Taylor, & Eyck, 2006). The formation of a salt bridge between conserved catalytic site Glu and Lys correlates well with active enzymes (Huse & Kuriyan, 2002). Availability of the catalytic base (Asp residue) for hydrogen bonding to the substrate hydroxyl also seems to be an important feature from structural studies.

Small molecule tools have also been used to decipher the nature of kinase catalytic mechanisms. The phosphate mimic aluminum fluoride has been crystallized with PKA in complex with ADP and a peptide substrate analog (Madhusudan, Akamine, Xuong, & Taylor, 2002). Despite the weak affinity of AlF_3 for ADP in the absence of enzyme, the fact that the crystal structure captures an arrangement expected for catalysis suggests that the enzyme has selected a transition-state conformation. The distance between the substrate Ser nucleophile and the aluminum ion in this structure is sufficiently long to suggest a dissociative transition state in the phosphoryl transfer reaction (Mildvan, 1997).

Bisubstrate analogs for kinases have been designed based on a dissociative transition state. For enzymes that obey ternary complex mechanisms, covalent linkage of the two substrates can generate high-affinity inhibitors, if the linkers are installed to approximate the geometry of the reaction coordinate. In favorable cases, bisubstrate analog-binding energies to enzymes can equal or exceed the sum of the energies of the individual substrates, since the entropic penalty of assembling the three components of the reaction is reduced. Early work on PKA bisubstrate analog inhibitors involved synthetic compounds in which the peptide Ser oxygen was covalently linked to ADP, ATP, and adenosine-tetraphosphate via a phosphodiester with the terminal phosphate (Medzihradszky, Chen, Kenyon, & Gibson, 1994). While none of these compounds was a stronger PKA inhibitor than ADP alone, the tetraphosphate-containing bisubstrate analog was more potent than the shorter phosphate strings. This suggests that an extended spacer between the two substrates better approximates the reaction coordinate distance of kinase reaction. A theoretical analysis by Mildvan of the preferred reaction coordinate distance for a dissociative transition state, which is the optimal distance between the entering oxygen and the attacked phosphorus, should be 4.5–6 Å (Mildvan, 1997). The tetraphosphate compound is in this range.

Another bisubstrate analog approach that has been applied to several kinases employs a thioacetyl-bridge and replaces the entering oxygen with a nitrogen atom (Cheng et al., 2006; Levinson et al., 2006; Parang et al., 2001; Zhang, Gureasko, Shen, Cole, & Kuriyan, 2006). This spacer places the nitrogen and ATP gamma phosphorus about 5 Å apart (Fig. 1.1B). This design has produced some very potent kinase inhibitors, some with K_is in the low nanomolar range. Thus, it appears that bisubstrate analogs with geometries matching the Mildvan reaction coordinate model for a dissociative mechanism are most effective.

4. APPLICATIONS OF MECHANISTIC STUDIES IN UNDERSTANDING KINASE FUNCTION AND REGULATION

4.1. Bisubstrate analogs

As discussed, peptide–ATP conjugates inspired by a dissociative transition state have been developed for several protein kinases. A principal motivation for developing these analogs is for their use in structural biology studies to clarify the basis of protein substrate recognition by kinases. The field is still lacking a comprehensive understanding of how particular protein substrates are selected by kinases. Various bioinformatic and peptide library approaches have been applied to deduce specificity but they have enjoyed a limited impact. Despite the thousands of X-ray crystal structures of protein kinases, there are only a small number with kinases complexed with protein or peptide substrates. As a result, for the vast majority of kinases, we have little insight into the molecular basis of substrate recognition. The difficulty in obtaining kinase–substrate crystal complexes is thought to emanate from the relatively low affinity of most kinase–substrate binding interactions. From an efficiency perspective, it is logical that enzyme–substrate complexes not be too tight since this could limit catalytic turnover.

Bisubstrate analogs facilitate the anchoring of the peptide moiety to the kinase surface involved in substrate recognition, often allowing for enhanced cocrystallization (Fig. 1.1C). To date, this strategy has been used to capture four kinase–substrate crystal structures: insulin receptor, Abl, EGFR, and cyclin-dependent kinase (Bose, Holbert, Pickin, & Cole, 2006; Levinson et al., 2006; Parang et al., 2001; Zhang et al., 2007). These structures have revealed several of the contacts that give rise to substrate selectivity for these enzymes.

In the case of the Abl and EGFR tyrosine kinases, these cocrystal structures provided helpful clues to activity regulation. One of the pharmacologically important features of Abl is its propensity to populate a conformation which can bind tightly to the drug gleevec. This conformation, known as DFG (Asp–Phe–Gly) out, leads to inactivation of the kinase. A crystal structure of the Abl-kinase domain in complex with the peptide–ATP bisubstrate analog captured Abl in a novel conformation, a catalytically inactive state reminiscent of the downregulated Src tyrosine kinase (Levinson et al., 2006). Based on this unexpected structure, a new framework was proposed for how kinases shuttle between various activation states.

The cocrystal structure of EGFR kinase domain complexed with a peptide–ATP bisubstrate analog showed EGFR kinase in its likely active conformational state (Zhang et al., 2006). This structure also showed high-quality electron density for an apparent asymmetric dimer between two neighboring EGFR kinase molecules. The dimeric interactions are somewhat akin to how a cyclin binds to a cyclin-dependent kinase where it allosterically activates the enzyme. Using a combination of mutagenesis and vesicle-binding studies, it was revealed that the asymmetric EGFR kinase dimer is critical for the mechanisms of EGF ligand-induced activation of EGFR.

4.2. Oncogenic kinase mutants

Enzyme mechanism studies on kinases have also been pursued to understand how mutations can stimulate kinase activity and altered drug sensitivity. A battery of structural, chemical, and kinetic experiments has been channeled to address the functional effects of mutation. Two protein kinases that have received considerable attention in this regard are EGFR tyrosine kinase and B-RAF Ser/Thr kinase, and the experimental progress in these areas is highlighted later.

EGFR mutation has been shown to drive oncogenesis, with L858R point mutation and Δ(746–750) deletion the most prevalent EGFR mutations in nonsmall cell lung cancer (Lynch et al., 2004; Paez et al., 2004; Pao et al., 2004). Patients with tumors carrying these mutations are particularly responsive to small-molecule tyrosine kinase inhibitors erlotinib and gefitinib, which bind the ATP pocket. The ATP site-targeting drug lapatinib inhibits WT EGFR and Her2/Neu and is approved for the treatment of breast cancer (Cameron & Stein, 2008; Wood et al., 2004). *In vitro* studies on isolated EGFR kinase domains showed that the L858R EGFR kinase domain is about ~50-fold more active relative to the wild-type EGFR kinase domain (Yun et al., 2007; Zhang et al., 2006). In addition, L858R EGFR kinase domain has a dramatically increased ATP K_m for ATP and a lower gefitinib IC_{50} relative to the corresponding parameters for wild type. More recent experiments on near full-length EGFRs (tEGFRs), which lack part of the C-terminal tail, investigated the mechanistic basis for oncogenic activation by mutation (Qiu et al., 2009; Wang et al., 2011). These tEGFR studies revealed that L858R and Δ(746–750) are about as active as wild type in the presence of EGF, but retained their full activity when EGF was replaced by the competitive antibody antagonist, cetuximab, which blocks EGF from binding to the EGFR and prevents ectodomain dimerization. In contrast, wild-type tEGFR in complex with cetuximab showed <1% activity relative to its EGF-bound form.

Interestingly, in the presence of EGF, wild-type and oncogenic mutant tEGFRs were equally susceptible to strong erlotinib inhibition but the oncogenic mutant tEGFRs were quite resistant to lapatinib (Wang et al., 2011). Wild-type tEGFR was about as potently inhibited by lapatinib whether in complex with EGF or cetuximab (Wang et al., 2011). Since lapatinib is thought to have a strong binding preference for the inactive conformation of EGFR (Wood et al., 2004), while erlotinib selectively binds the active conformation of EGFR (Stamos, Sliwkowski, & Eigenbrot, 2002), this was puzzling initially. However, a model to explain this data hypothesizes that the active conformation of wild-type EGF/tEGFR complex is roughly isoenergetic with the inactive kinase conformation (Fig. 1.2A; Wang et al., 2011). In contrast, the oncogenic tEGFR mutants greatly favor their active over inactive kinase conformations, regardless of the engagement with EGF (Fig. 1.2B). Separate studies involving negative staining electron microscopy and computational analysis are consistent with this model (Mi, Lu, Nishida, Walz, & Springer, 2011; Shan et al., 2012). tEGFR enzyme kinetic experiments also helped to illuminate why both L858R and Δ(746–750) have very high ATP K_ms, even though these mutations are somewhat remote from the adenine pocket. It appears that the reduced binding affinity of the oncogenic tEGFRs for the nucleotide *per se* is only a minor part of the answer. The greater impact appears to be related to a faster chemical (phosphoryl transfer) step with these oncogenic mutants. By increasing the chemical step rate, the K_m will be elevated, independent of the dissociation constant of EGFR for ATP.

In principle, oncogenic mutation could drive activation in a dimer-dependent or dimer-independent fashion. The tEGFR oncogenic mutants

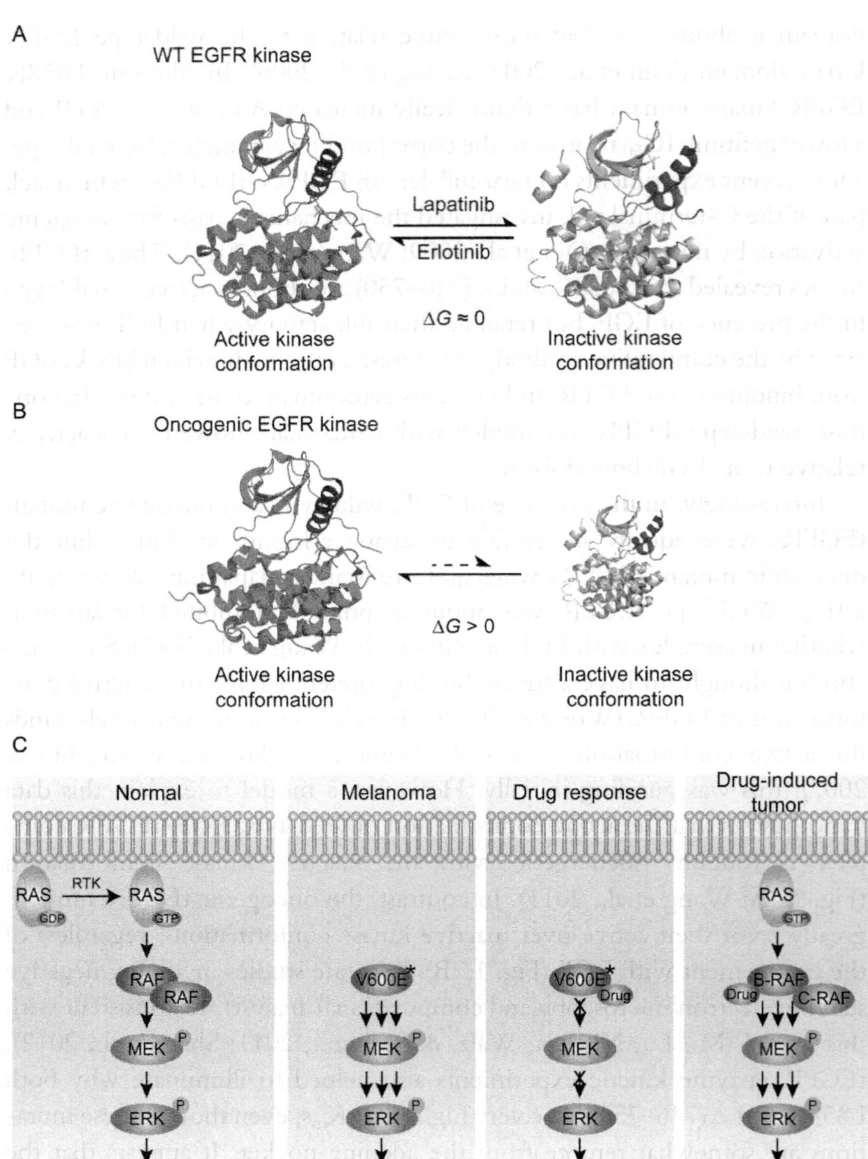

Figure 1.2 Oncogenic kinase mutant mechanisms. (A) The crystal structures of active and inactive EGFR kinase conformations are shown in ribbon representation. Two key structural elements within the kinase domain, the activation loop and helix, are colored orange and red. For WT EGF-bound EGFR, the active kinase conformation (green ribbon, PDB code: 1M17) (Stamos et al., 2002) is nearly isoenergetic to the inactive kinase conformation (gray ribbon, PDB code: 1XKK). Erlotinib selectively binds and stabilizes the active conformation; therefore, the active conformation is dominant in the

were resistant to the protein inhibitor Mig6, which blocks dimerization by coating the EGFR kinase domain C-lobe surface (Zhang et al., 2007). These findings suggested that oncogenic kinase domain dimerization could be dispensable for activation. Alternatively, resistance to Mig6 could suggest that oncogenic kinase domain asymmetric dimerization could be very tight. The fact that Mig6 is a potent inhibitor of the isolated L858R EGFR kinase domain, whereas tEGFR L858R is resistant to Mig6 suggests that the isolated kinase domain could be misleading as a model system compared with the full-length EGFR (Wang et al., 2011). In fact, N-lobe and C-lobe mutations that disrupt the asymmetric kinase dimer interface caused steep reductions in kinase activity for both L858R and Δ(746–750) tEGFRs (Wang et al., 2011). These experiments establish that oncogenic mutants drive activation through the asymmetric kinase dimer, and that the dimer is likely too tight to be effectively blocked by Mig6.

The most frequently mutated protein Ser/Thr kinase in human cancer, B-RAF, has been studied intensively over the past decade (Dhomen & Marais, 2007; Greenman et al., 2007; Wellbrock, Karasarides, & Marais, 2004). B-RAF and its homologs, A-RAF and C-RAF (also called Raf-1), as well as a related kinase KSR1 are involved in the RAS/RAF/MEK/ERK signaling cascade which is activated by growth factors, hormones, and cytokines (Matallanas et al., 2011; Mercer & Pritchard, 2003; Roskoski, 2010; Stephens et al., 1992; Fig. 1.2C). B-RAF mutations have been found in 7% of human cancers and in 66% of malignant melanomas, among which the hyperactivating mutation V600E is the predominant form (Davies et al., 2002). Oncogenic B-RAF mutations are within the catalytic domain, and most enhance B-RAF's ability to phosphorylate MEK, the only

presence of erlotinib. Similarly, the inactive conformation is dominant in the presence of lapatinib (Wood et al., 2004). Erlotinib and lapatinib are both potent inhibitors against WT EGFR. (B) Oncogenic mutations alter the energy landscape to favor the active conformation. As a result, lapatinib has little access to the majority of oncogenic EGFR mutants, which are locked into the active conformation. The size of each structure has been schematized to depict the relative population of active and inactive conformations. (C) In normal cells, activated RAS recruits B-RAF from the cytosol to the plasma membrane for dimerization and activation. Once activated, B-RAF phosphorylates MEK, which in turn phosphorylates and activates ERK. Oncogenic forms of B-RAF (marked with V600E*) highly activate the ERK pathway in a RAS-independent manner. In B-RAF mutant cancer cells, such as melanoma, B-RAF inhibitor efficiently blocks ERK activation. In RAS mutant cancer cells, the same ATP-competitive inhibitor unexpectedly activates the ERK pathway by promoting RAF dimerization. (See the color plate).

well-established substrate of B-RAF (Wan et al., 2004). Cell-based assays showed that the presence of V600E B-RAF results in constitutive ERK activation (Davies et al., 2002). Whereas wild-type B-RAF seems to require dimerization of the kinase domain for activation, a series of elegant cellular experiments suggest that the "gain-of-function" of V600E B-RAF is independent of RAS and dimerization (Davies et al., 2002; Freeman, Ritt, & Morrison, 2013; Poulikakos et al., 2011). The crystal structures of the catalytic domains of WT and V600E B-RAFs verified that replacing the hydrophobic Val600 with a charged residue destabilizes the interactions that maintain the DFG motif in an inactive conformation and favors conversion of B-RAF into an active, closed conformation (Wan et al., 2004). This "gain-of-function" mutation presumably mimics activation loop phosphorylation.

In addition to mutation affording enhanced kinase activity, several studies also suggest that protein–protein interaction functions of B-RAF may confer cell transformation. Some oncogenic B-RAF mutants have decreased or undetectable kinase activity but retain the ability to activate the ERK pathway (Dhomen & Marais, 2007). These mutations have been suggested to stimulate C-RAF activity through the formation of B-RAF/C-RAF heterodimers (Garnett, Rana, Paterson, Barford, & Marais, 2005; Moretti et al., 2009; Rushworth, Hindley, O'Neill, & Kolch, 2006). Various RAF inhibitors have been developed and are in different stages of preclinical and clinical developments. Treatment with vemurafenib (PLX-4032), a small-molecule B-RAF inhibitor, was shown to be beneficial in melanoma patients harboring the V600E B-RAF mutant and recently was approved for the treatment of metastatic melanoma (Bollag et al., 2010; Flaherty et al., 2010; Joseph et al., 2010; Yang et al., 2010). Unexpectedly, B-RAF inhibitors caused adverse effects in tumor cells carrying wild-type B-RAF. Follow-up studies demonstrated that inhibitor binding to RAF promotes formation of dimeric RAF complexes and activation of the ERK pathway (Hatzivassiliou et al., 2010; Heidorn et al., 2010; Poulikakos, Zhang, Bollag, Shokat, & Rosen, 2010). It remains unresolved precisely how various drug resistance mutants evade kinase inhibition and whether RAF dimerization is thermodynamically coupled to activation loop phosphorylation.

4.3. Chemical rescue of tyrosine kinases

A crystal structure of the IRK domain in its active conformation in complex with a peptide substrate and ATP analog revealed a well-defined triangle of hydrogen bonds with vertices involving the side chains of the catalytic base

Asp, the catalytic loop Arg, and the substrate phenol (Parang et al., 2001). Site-directed mutagenesis on this Arg in several tyrosine kinases showed that it is critical for activity, with 100-fold or greater rate reductions associated with its mutation to Ala (Muratore et al., 2009; Qiao, Molina, Pandey, Zhang, & Cole, 2006; Williams, Wang, & Cole, 2000). It is also interesting that this Arg, while present in the sequences in all canonical tyrosine kinases, is flexibly located in two spots in the catalytic loop, either +2 or +4 upstream of the catalytic base Asp (Manning, Plowman, Hunter, & Sudarsanam, 2002). It has been shown previously with various mutant enzymes that loss of activity can sometimes be rescued by the addition of small molecules that complement the missing residue. To assess the specific role of the Arg side chain in the tyrosine kinase reaction, a series of nitrogen-containing small molecules were studied to see if they could complement the Arg to Ala mutant. In fact, several diamino and triamino compounds could at least partially rescue activity. Interestingly, the small-molecule imidazole, the functional group of the side chain of histidine, is the most efficient rescue agent for the tyrosine kinases Csk, Src, and Abl (Muratore et al., 2009; Qiao et al., 2006; Williams et al., 2000). A combination of pH-rate measurements and structure–activity relationship studies indicated that it is the imidazolium (protonated imidazole) state rather than the neutral imidazole form that is necessary for chemical rescue. The positively charged imidazolium would be predicted to be the stronger hydrogen bond donor and acceptor and a better mimic of the Arg guanidinium, which has a higher pK_a than imidazole.

It is noteworthy that replacement of the conserved catalytic Arg by His fails to confer enzymatic activity, despite the fact that imidazole is more efficacious in chemical rescue than guanidinium, the functional group of the Arg side chain (Muratore et al., 2009). It is hypothesized that the histidine side chain is simply too short to extend into hydrogen-bonding position when located at the Arg position. Nevertheless, X-ray crystallographic studies of the Src Arg/Ala mutant reveal that a distantly located histidine side chain from an adjacent molecule in the crystal lattice can occupy the cavity associated with Arg removal (Muratore et al., 2009).

Interestingly, where investigated, the chemical rescue of Arg/Ala tyrosine kinases preserves the natural substrate specificity and regulation of the enzymatic activity. Because imidazole is relatively nontoxic (at less than 30 mM) and is cell permeable (Iguchi, Usui, Ishida, & Hirano, 2002), it was tested in cells stably expressing tyrosine kinases bearing the Arg/Ala replacements in place of the natural wild-type enzymes. The most complete studies have been carried out analyzing Arg/Ala cellular Src (c-Src) and v-Src in mouse embryonic fibroblasts lacking the three major Src tyrosine kinases, Src, Yes, and Fyn (Ferrando et al., 2012; Qiao et al., 2006). c-Src is the proto-oncoprotein that, after passage in Rous sarcoma virus, can become v-Src and thereby induce sarcoma in chickens. v-Src can no longer be downregulated by C-terminal phosphorylation and is a hyperactive tyrosine kinase. Chemical rescue studies of Arg/Ala c-Src and v-Src with imidazole treatment showed that both c-Src and v-Src Arg/Ala mutants could have their kinase activity robustly restored with chemical rescue (Ferrando et al., 2012; Qiao et al., 2006). The findings with c-Src complementation were somewhat unexpected because, in the absence of cell stimulation by growth factors, the basal activity of c-Src was believed to be very low. It must be remembered that cellular protein tyrosine phosphatase activity is robust and so at baseline, probably achieves an equilibrium with protein tyrosine kinase activity. Chemical rescue of the mutant kinase unmasks acutely this substantial enzymatic protein phosphorylation.

Chemical rescue in combination with mass spectrometry has been used to characterize previously unrecognized functions and targets of Src-mediated phosphorylation. Using stable isotope labeling in cell culture and mass spectrometry, chemical rescue can identify phosphorylation events kinetically linked to a specific nonreceptor kinase, bypassing the need for growth factor stimulation. In particular, chemical rescue of Arg/Ala c-Src has been used to suggest a role for Src-mediated tyrosine phosphorylation of the Rap1 guanine nucleotide exchange factor, C3G, in cellular adhesion regulation (Ferrando et al., 2012). In this fashion, chemical rescue is highly complementary to other powerful chemical genetic approaches that allow for specific inhibition of wild type or mutant kinases.

5. SUMMARY AND OUTLOOK

Advances in our understanding of the enzymology of protein kinases have led to new insights into their catalytic mechanisms in terms of kinetic steps, transition-state stabilization, substrate specificity, and basis of regulation. Yet, much of the effort in the analysis of kinases has centered, understandably, on a small proportion of the kinome that have been identified as important in biology and diseases. Hundreds of kinases have not yet been analyzed in enzymologic depth. In addition, there has been a limited work in exploring kinase mechanisms in the larger protein complexes in which they are often found or bearing the rich tapestry of posttranslational modifications (PTMs) that adorn many of them. Regarding the influence of PTMs', emerging methods to introduce these PTMs site specifically into proteins using unnatural amino acid mutagenesis (Park et al., 2011; Wang, Xie, & Schultz, 2006) or expressed protein ligation (Muir, Sondhi, & Cole, 1998) offers the potential to dissect their effects in greater detail. Thus, despite the major progress that has been made in the kinase field, there is far more work to do before a comprehensive understanding of these fascinating signaling enzymes is achieved.

REFERENCES

Ablooglu, A. J., Frankel, M., Rusinova, E., Ross, J. B., & Kohanski, R. A. (2001). Multiple activation loop conformations and their regulatory properties in the insulin receptor's kinase domain. *Journal of Biological Chemistry*, *276*, 46933–46940.

Ablooglu, A. J., Till, J. H., Kim, K., Parang, K., Cole, P. A., Hubbard, S. R., et al. (2000). Probing the catalytic mechanism of the insulin receptor kinase with a tetrafluorotyrosine-containing peptide substrate. *Journal of Biological Chemistry*, *275*, 30394–30398.

Adams, J. A. (2001). Kinetic and catalytic mechanisms of protein kinases. *Chemical Reviews*, *101*, 2271–2290.

Adams, J. A., & Taylor, S. S. (1992). Energetic limits of phosphotransfer in the catalytic subunit of cAMP-dependent protein kinase as measured by viscosity experiments. *Biochemistry*, *31*, 8516–8522.

Admiraal, S. J., & Herschlag, D. (1995). Mapping the transition state for ATP hydrolysis: Implications for enzymatic catalysis. *Chemistry & Biology*, *2*, 729–739.

Admiraal, S. J., Schneider, B., Meyer, P., Janin, J., Veron, M., Deville-Bonne, D., et al. (1999). Nucleophilic activation by positioning in phosphoryl transfer catalyzed by nucleoside diphosphate kinase. *Biochemistry*, *38*, 4701–4711.

Allen, J. J., Lazerwith, S. E., & Shokat, K. M. (2005). Bio-orthogonal affinity purification of direct kinase substrates. *Journal of the American Chemical Society*, *127*, 5288–5289.

Aubol, B. E., Chakrabarti, S., Ngo, J., Shaffer, J., Nolen, B., Fu, X. D., et al. (2003). Processive phosphorylation of alternative splicing factor/splicing factor 2. *Proceedings of the National Academy of Sciences of the United States of America*, *100*, 12601–12606.

Bollag, G., Hirth, P., Tsai, J., Zhang, J., Ibrahim, P. N., Cho, H., et al. (2010). Clinical efficacy of a RAF inhibitor needs broad target blockade in BRAF-mutant melanoma. *Nature*, *467*, 596–599.

Bose, R., Holbert, M. A., Pickin, K. A., & Cole, P. A. (2006). Protein tyrosine kinase–substrate interactions. *Current Opinion in Structural Biology*, *16*, 668–675.

Cameron, D. A., & Stein, S. (2008). Drug Insight: Intracellular inhibitors of HER2-clinical development of lapatinib in breast cancer. *Nature Clinical Practice. Oncology*, *5*, 512–520.

Cheng, K. Y., Noble, M. E., Skamnaki, V., Brown, N. R., Lowe, E. D., Kontogiannis, L., et al. (2006). The role of the phospho-CDK2/cyclin A recruitment site in substrate recognition. *Journal of Biological Chemistry*, *281*, 23167–23179.

Cheng, Y., Zhang, Y., & McCammon, J. A. (2005). How does the cAMP-dependent protein kinase catalyze the phosphorylation reaction: An ab initio QM/MM study. *Journal of the American Chemical Society*, *127*, 1553–1562.

Cole, P. A., Burn, P., Takacs, B., & Walsh, C. T. (1994). Evaluation of the catalytic mechanism of recombinant human Csk (C-terminal Src kinase) using nucleotide analogs and viscosity effects. *Journal of Biological Chemistry*, *269*, 30880–30887.

Cole, P. A., Grace, M. R., Phillips, R. S., Burn, P., & Walsh, C. T. (1995). The role of the catalytic base in the protein tyrosine kinase Csk. *Journal of Biological Chemistry, 270,* 22105–22108.

Davies, H., Bignell, G. R., Cox, C., Stephens, P., Edkins, S., Clegg, S., et al. (2002). Mutations of the BRAF gene in human cancer. *Nature, 417,* 949–954.

Dhomen, N., & Marais, R. (2007). New insight into BRAF mutations in cancer. *Current Opinion in Genetics & Development, 17,* 31–39.

Ferrando, I. M., Chaerkady, R., Zhong, J., Molina, H., Jacob, H. K., Herbst-Robinson, K., et al. (2012). Identification of targets of c-Src tyrosine kinase by chemical complementation and phosphoproteomics. *Molecular and Cellular Proteomics, 11,* 355–369.

Ferreira-Cerca, S., Sagar, V., Schafer, T., Diop, M., Wesseling, A. M., Lu, H., et al. (2012). ATPase-dependent role of the atypical kinase Rio2 on the evolving pre-40S ribosomal subunit. *Nature Structural and Molecular Biology, 19,* 1316–1323.

Flaherty, K. T., Puzanov, I., Kim, K. B., Ribas, A., McArthur, G. A., Sosman, J. A., et al. (2010). Inhibition of mutated, activated BRAF in metastatic melanoma. *New England Journal of Medicine, 363,* 809–819.

Freeman, A. K., Ritt, D. A., & Morrison, D. K. (2013). Effects of Raf dimerization and its inhibition on normal and disease-associated Raf signaling. *Molecular Cell, 49,* 751–758.

Garnett, M. J., Rana, S., Paterson, H., Barford, D., & Marais, R. (2005). Wild-type and mutant B-RAF activate C-RAF through distinct mechanisms involving heterodimerization. *Molecular Cell, 20,* 963–969.

Grace, M. R., Walsh, C. T., & Cole, P. A. (1997). Divalent ion effects and insights into the catalytic mechanism of protein tyrosine kinase Csk. *Biochemistry, 36,* 1874–1881.

Greenman, C., Stephens, P., Smith, R., Dalgliesh, G. L., Hunter, C., Bignell, G., et al. (2007). Patterns of somatic mutation in human cancer genomes. *Nature, 446,* 153–158.

Hanks, S. K., Quinn, A. M., & Hunter, T. (1988). The protein kinase family: Conserved features and deduced phylogeny of the catalytic domains. *Science, 241,* 42–52.

Hatzivassiliou, G., Song, K., Yen, I., Brandhuber, B. J., Anderson, D. J., Alvarado, R., et al. (2010). RAF inhibitors prime wild-type RAF to activate the MAPK pathway and enhance growth. *Nature, 464,* 431–435.

Heidorn, S. J., Milagre, C., Whittaker, S., Nourry, A., Niculescu-Duvas, I., Dhomen, N., et al. (2010). Kinase-dead BRAF and oncogenic RAS cooperate to drive tumor progression through CRAF. *Cell, 140,* 209–221.

Hubbard, S. R., & Miller, W. T. (2007). Receptor tyrosine kinases: Mechanisms of activation and signaling. *Current Opinion in Cell Biology, 19,* 117–123.

Hunter, T. (2000). Signaling—2000 and beyond. *Cell, 100,* 113–127.

Huse, M., & Kuriyan, J. (2002). The conformational plasticity of protein kinases. *Cell, 109,* 275–282.

Iguchi, K., Usui, S., Ishida, R., & Hirano, K. (2002). Imidazole-induced cell death, associated with intracellular acidification, caspase-3 activation, DFF-45 cleavage, but not oligonucleosomal DNA fragmentation. *Apoptosis,* 7, 519–525.

Joseph, E. W., Pratilas, C. A., Poulikakos, P. I., Tadi, M., Wang, W., Taylor, B. S., et al. (2010). The RAF inhibitor PLX4032 inhibits ERK signaling and tumor cell proliferation in a V600E BRAF-selective manner. *Proceedings of the National Academy of Sciences of the United States of America, 107,* 14903–14908.

Kim, K., & Cole, P. A. (1997). Measurement of a Brønsted nucleophile coefficient and insights into the transition state for a protein tyrosine kinase. *Journal of the American Chemical Society, 119,* 11096–11097.

Kim, K., & Cole, P. A. (1998). Kinetic analysis of a protein tyrosine kinase reaction transition state in the forward and reverse directions. *Journal of the American Chemical Society, 120,* 6851–6858.

Kirby, A. J. J., & Jencks, W. P. (1965). Reactivity of nucleophilic reagents toward p-nitrophenyl phosphate dianion. *Journal of the American Chemical Society*, *87*, 3209.
Knighton, D. R., Zheng, J. H., Ten Eyck, L. F., Xuong, N. H., Taylor, S. S., & Sowadski, J. M. (1991). Structure of a peptide inhibitor bound to the catalytic subunit of cyclic adenosine monophosphate-dependent protein kinase. *Science*, *253*, 414–420.
Knowles, J. R. (1980). Enzyme-catalyzed phosphoryl transfer reactions. *Annual Review of Biochemistry*, *49*, 877–919.
Kornev, A. P., Haste, N. M., Taylor, S. S., & Eyck, L. F. (2006). Surface comparison of active and inactive protein kinases identifies a conserved activation mechanism. *Proceedings of the National Academy of Sciences of the United States of America*, *103*, 17783–17788.
Krebs, E. G., & Beavo, J. A. (1979). Phosphorylation–dephosphorylation of enzymes. *Annual Review of Biochemistry*, *48*, 923–959.
Levinson, N. M., Kuchment, O., Shen, K., Young, M. A., Koldobskiy, M., Karplus, M., et al. (2006). A Src-like inactive conformation in the abl tyrosine kinase domain. *PLoS Biology*, *4*, e144.
LoGrasso, P. V., Frantz, B., Rolando, A. M., O'Keefe, S. J., Hermes, J. D., & O'Neill, E. A. (1997). Kinetic mechanism for p38 MAP kinase. *Biochemistry*, *36*, 10422–10427.
Lynch, T. J., Bell, D. W., Sordella, R., Gurubhagavatula, S., Okimoto, R. A., Brannigan, B. W., et al. (2004). Activating mutations in the epidermal growth factor receptor underlying responsiveness of non-small-cell lung cancer to gefitinib. *New England Journal of Medicine*, *350*, 2129–2139.
Madhusudan, S., Akamine, P., Xuong, N. H., & Taylor, S. S. (2002). Crystal structure of a transition state mimic of the catalytic subunit of cAMP-dependent protein kinase. *Nature Structural Biology*, *9*, 273–277.
Manning, G., Plowman, G. D., Hunter, T., & Sudarsanam, S. (2002). Evolution of protein kinase signaling from yeast to man. *Trends in Biochemical Sciences*, *27*, 514–520.
Manning, G., Whyte, D. B., Martinez, R., Hunter, T., & Sudarsanam, S. (2002). The protein kinase complement of the human genome. *Science*, *298*, 1912–1934.
Matallanas, D., Birtwistle, M., Romano, D., Zebisch, A., Rauch, J., von Kriegsheim, A., et al. (2011). Raf family kinases: Old dogs have learned new tricks. *Genes & Cancer*, *2*, 232–260.
Medzihradszky, D., Chen, S. L., Kenyon, G. L., & Gibson, B. W. (1994). Solid-phase synthesis of adenosine phosphopeptides as potential bisubstrate inhibitors of protein kinases. *Journal of the American Chemical Society*, *116*, 9413–9419.
Mercer, K. E., & Pritchard, C. A. (2003). Raf proteins and cancer: B-Raf is identified as a mutational target. *Biochimica et Biophysica Acta*, *1653*, 25–40.
Mi, L. Z., Lu, C., Nishida, N., Walz, T., & Springer, T. A. (2011). Simultaneous visualization of the extracellular and cytoplasmic domains of the epidermal growth factor receptor. *Nature Structural & Molecular Biology*, *18*, 984–989.
Mildvan, A. S. (1997). Mechanisms of signaling and related enzymes. *Proteins*, *29*, 401–416.
Moretti, S., De Falco, V., Tamburrino, A., Barbi, F., Tavano, M., Avenia, N., et al. (2009). Insights into the molecular function of the inactivating mutations of B-Raf involving the DFG motif. *Biochimica et Biophysica Acta*, *1793*, 1634–1645.
Muir, T. W., Sondhi, D., & Cole, P. A. (1998). Expressed protein ligation: A general method for protein engineering. *Proceedings of the National Academy of Sciences of the United States of America*, *95*, 6705–6710.
Mukherjee, K., Sharma, M., Urlaub, H., Bourenkov, G. P., Jahn, R., Sudhof, T. C., et al. (2008). CASK functions as a Mg^{2+}-independent neurexin kinase. *Cell*, *133*, 328–339.
Muratore, K. E., Seeliger, M. A., Wang, Z., Fomina, D., Neiswinger, J., Havranek, J. J., et al. (2009). Comparative analysis of mutant tyrosine kinase chemical rescue. *Biochemistry*, *48*, 3378–3386.

Noble, M. E., Endicott, J. A., & Johnson, L. N. (2004). Protein kinase inhibitors: Insights into drug design from structure. *Science*, *303*, 1800–1805.

Paez, J. G., Janne, P. A., Lee, J. C., Tracy, S., Greulich, H., Gabriel, S., et al. (2004). EGFR mutations in lung cancer: Correlation with clinical response to gefitinib therapy. *Science*, *304*, 1497–1500.

Pao, W., Miller, V., Zakowski, M., Doherty, J., Politi, K., Sarkaria, I., et al. (2004). EGF receptor gene mutations are common in lung cancers from "never smokers" and are associated with sensitivity of tumors to gefitinib and erlotinib. *Proceedings of the National Academy of Sciences of the United States of America*, *101*, 13306–13311.

Parang, K., Till, J. H., Ablooglu, A. J., Kohanski, R. A., Hubbard, S. R., & Cole, P. A. (2001). Mechanism-based design of a protein kinase inhibitor. *Nature Structural Biology*, *8*, 37–41.

Park, H. S., Hohn, M. J., Umehara, T., Guo, L. T., Osborne, E. M., Benner, J., et al. (2011). Expanding the genetic code of Escherichia coli with phosphoserine. *Science*, *333*, 1151–1154.

Pellicena, P., & Miller, W. T. (2001). Processive phosphorylation of p130Cas by Src depends on SH3-polyproline interactions. *Journal of Biological Chemistry*, *276*, 28190–28196.

Poulikakos, P. I., Persaud, Y., Janakiraman, M., Kong, X., Ng, C., Moriceau, G., et al. (2011). RAF inhibitor resistance is mediated by dimerization of aberrantly spliced BRAF (V600E). *Nature*, *480*, 387–390.

Poulikakos, P. I., Zhang, C., Bollag, G., Shokat, K. M., & Rosen, N. (2010). RAF inhibitors transactivate RAF dimers and ERK signalling in cells with wild-type BRAF. *Nature*, *464*, 427–430.

Qamar, R., & Cook, P. F. (1993). pH dependence of the kinetic mechanism of the adenosine 3′,5′-monophosphate dependent protein kinase catalytic subunit in the direction of magnesium adenosine 5′-diphosphate phosphorylation. *Biochemistry*, *32*, 6802–6806.

Qamar, R., Yoon, M. Y., & Cook, P. F. (1992). Kinetic mechanism of the adenosine 3′,5′-monophosphate dependent protein kinase catalytic subunit in the direction of magnesium adenosine 5′-diphosphate phosphorylation. *Biochemistry*, *31*, 9986–9992.

Qiao, Y., Molina, H., Pandey, A., Zhang, J., & Cole, P. A. (2006). Chemical rescue of a mutant enzyme in living cells. *Science*, *311*, 1293–1297.

Qiu, C., Tarrant, M. K., Boronina, T., Longo, P. A., Kavran, J. M., Cole, R. N., et al. (2009). In vitro enzymatic characterization of near full length EGFR in activated and inhibited states. *Biochemistry*, *48*, 6624–6632.

Rominger, C. M., Schaber, M. D., Yang, J., Gontarek, R. R., Weaver, K. L., Broderick, T., et al. (2007). An intrinsic ATPase activity of phospho-MEK-1 uncoupled from downstream ERK phosphorylation. *Archives of Biochemistry and Biophysics*, *464*, 130–137.

Roskoski, R., Jr. (2010). RAF protein-serine/threonine kinases: Structure and regulation. *Biochemical and Biophysical Research Communications*, *399*, 313–317.

Rushworth, L. K., Hindley, A. D., O'Neill, E., & Kolch, W. (2006). Regulation and role of Raf-1/B-Raf heterodimerization. *Molecular and Cellular Biology*, *26*, 2262–2272.

Segel, I. H. (1993). *Enzyme kinetics*. New York: Wiley.

Shan, Y., Eastwood, M. P., Zhang, X., Kim, E. T., Arkhipov, A., Dror, R. O., et al. (2012). Oncogenic mutations counteract intrinsic disorder in the EGFR kinase and promote receptor dimerization. *Cell*, *149*, 860–870.

Sondhi, D., Xu, W., Songyang, Z., Eck, M. J., & Cole, P. A. (1998). Peptide and protein phosphorylation by protein tyrosine kinase Csk: Insights into specificity and mechanism. *Biochemistry*, *37*, 165–172.

Stamos, J., Sliwkowski, M. X., & Eigenbrot, C. (2002). Structure of the epidermal growth factor receptor kinase domain alone and in complex with a 4-anilinoquinazoline inhibitor. *Journal of Biological Chemistry*, *277*, 46265–46272.

Stephens, R. M., Sithanandam, G., Copeland, T. D., Kaplan, D. R., Rapp, U. R., & Morrison, D. K. (1992). 95-Kilodalton B-Raf serine/threonine kinase: Identification of the protein and its major autophosphorylation site. *Molecular and Cellular Biology, 12*, 3733–3742.

Szafranska, A. E., & Dalby, K. N. (2005). Kinetic mechanism for p38 MAP kinase alpha. A partial rapid-equilibrium random-order ternary-complex mechanism for the phosphorylation of a protein substrate. *FEBS Journal, 272*, 4631–4645.

Taylor, S. S., & Kornev, A. P. (2011). Protein kinases: Evolution of dynamic regulatory proteins. *Trends in Biochemical Sciences, 36*, 65–77.

Valiev, M., Yang, J., Adams, J. A., Taylor, S. S., & Weare, J. H. (2007). Phosphorylation reaction in cAPK protein kinase-free energy quantum mechanical/molecular mechanics simulations. *Journal of Physical Chemistry B, 111*, 13455–13464.

Wan, P. T., Garnett, M. J., Roe, S. M., Lee, S., Niculescu-Duvaz, D., Good, V. M., et al. (2004). Mechanism of activation of the RAF–ERK signaling pathway by oncogenic mutations of B-RAF. *Cell, 116*, 855–867.

Wang, Z., Longo, P. A., Tarrant, M. K., Kim, K., Head, S., Leahy, D. J., et al. (2011). Mechanistic insights into the activation of oncogenic forms of EGF receptor. *Nature Structural and Molecular Biology, 18*, 1388–1393.

Wang, L., Xie, J., & Schultz, P. G. (2006). Expanding the genetic code. *Annual Review of Biophysics and Biomolecular Structure, 35*, 225–249.

Wellbrock, C., Karasarides, M., & Marais, R. (2004). The RAF proteins take centre stage. *Nature Reviews. Molecular Cell Biology, 5*, 875–885.

Williams, D. M., & Cole, P. A. (2002). Proton demand inversion in a mutant protein tyrosine kinase reaction. *Journal of the American Chemical Society, 124*, 5956–5957.

Williams, D. M., Wang, D., & Cole, P. A. (2000). Chemical rescue of a mutant protein-tyrosine kinase. *Journal of Biological Chemistry, 275*, 38127–38130.

Wood, E. R., Truesdale, A. T., McDonald, O. B., Yuan, D., Hassell, A., Dickerson, S. H., et al. (2004). A unique structure for epidermal growth factor receptor bound to GW572016 (Lapatinib): Relationships among protein conformation, inhibitor off-rate, and receptor activity in tumor cells. *Cancer Research, 64*, 6652–6659.

Yang, H., Higgins, B., Kolinsky, K., Packman, K., Go, Z., Iyer, R., et al. (2010). RG7204 (PLX4032), a selective BRAFV600E inhibitor, displays potent antitumor activity in preclinical melanoma models. *Cancer Research, 70*, 5518–5527.

Yun, C. H., Boggon, T. J., Li, Y., Woo, M. S., Greulich, H., Meyerson, M., et al. (2007). Structures of lung cancer-derived EGFR mutants and inhibitor complexes: Mechanism of activation and insights into differential inhibitor sensitivity. *Cancer Cell, 11*, 217–227.

Zhang, X., Gureasko, J., Shen, K., Cole, P. A., & Kuriyan, J. (2006). An allosteric mechanism for activation of the kinase domain of epidermal growth factor receptor. *Cell, 125*, 1137–1149.

Zhang, X., Pickin, K. A., Bose, R., Jura, N., Cole, P. A., & Kuriyan, J. (2007). Inhibition of the EGF receptor by binding of MIG6 to an activating kinase domain interface. *Nature, 450*, 741–744.

Zheng, J., Knighton, D. R., ten Eyck, L. F., Karlsson, R., Xuong, N., Taylor, S. S., et al. (1993). Crystal structure of the catalytic subunit of cAMP-dependent protein kinase complexed with MgATP and peptide inhibitor. *Biochemistry, 32*, 2154–2161.

CHAPTER TWO

A Structural Atlas of Kinases Inhibited by Clinically Approved Drugs

Qi Wang[*,†,1], **Julie A. Zorn**[*,†,1], **John Kuriyan**[*,†,‡,§,¶,2]
[*]Department of Molecular and Cell Biology, University of California, Berkeley, California, USA
[†]California Institute for Quantitative Biosciences, University of California, Berkeley, California, USA
[‡]Howard Hughes Medical Institute, University of California, Berkeley, California, USA
[§]Department of Chemistry, University of California, Berkeley, California, USA
[¶]Physical Biosciences Division, Lawrence Berkeley National Laboratory, Berkeley, California, USA
[1]These authors contributed equally.
[2]Corresponding author: e-mail address: kuriyan@berkeley.edu

Contents

Methods in Enzymology, Volume 548
ISSN 0076-6879
http://dx.doi.org/10.1016/B978-0-12-397918-6.00002-1

Abstract

The aberrant activation of protein kinases is associated with many human diseases, most notably cancer. Due to this link between kinase deregulation and disease progression, kinases are one of the most targeted protein families for small-molecule inhibition. Within the last 15 years, the U.S. Food and Drug Administration has approved over 20 small-molecule inhibitors of protein kinases for use in the clinic. These inhibitors target the kinase active site and represent the successful hurdling by medicinal chemists of the formidable challenge posed by the high similarity among the active sites of the approximately 500 human kinases. We review the conserved structural features of kinases that are important for inhibitor binding as well as for catalysis. Many clinically approved drugs elicit selectivity by exploiting subtle variation within the kinase active site. We highlight some of the crystallographic studies on the kinase-inhibitor complexes that have provided valuable guidance for the development of these drugs as well as for future drug design efforts.

1. INTRODUCTION

The drug discovery community has invested immense effort in the development of protein kinase inhibitors, with over 150 drugs currently in clinical trials and more than 20 already approved for cancer treatments (Cohen & Alessi, 2013). The link between protein kinases and cancer was first established in the late 1970s, when the first oncogene product, v-Src, was determined to be a tyrosine kinase (Collett & Erikson, 1978; Eckhart, Hutchinson, & Hunter, 1979; Levinson, Oppermann, Levintow, Varmus, & Bishop, 1978). It soon became widely appreciated that the deregulation of kinase activity can lead to cellular transformation and ultimately disease (Blume-Jensen & Hunter, 2001).

The activity of protein kinases is tightly regulated under normal cellular conditions (Hunter, 1995; Huse & Kuriyan, 2002; Nolen, Taylor, & Ghosh, 2004; Pawson & Hunter, 1994). When activated, these enzymes catalyze the transfer of a phosphate group from the γ-phosphate of adenosine triphosphate (ATP) to a serine, threonine, or tyrosine residue in their substrates (Adams, 2001; Manning, Whyte, Martinez, Hunter, & Sudarsanam, 2002; Schwartz & Murray, 2011). Protein phosphorylation can have drastic effects on cell signaling. It directly impacts diverse cellular events, including kinase localization, conformational changes, and substrate recruitment. Thus, mechanisms exist in the cell to restrict kinase activity, including inhibition by other proteins, autoinhibition by regulatory domains associated with the kinase, the control of kinase expression levels, and the restriction of kinase localization (Churchill, Qvit, & Mochly-Rosen, 2009; Hubbard, Mohammadi, & Schlessinger, 1998; Johnson, 2009a; Kuriyan & Eisenberg, 2007; Lemmon & Schlessinger, 2010; Pawson & Kofler, 2009; Pawson & Scott, 1997; Pellicena & Kuriyan, 2006; Shaw, Kornev, Hu, Ahuja, & Taylor, 2014; Wang et al., 2010). Mutations associated with malignancies often bypass these inhibitory mechanisms (Vogelstein & Kinzler, 2004).

An exciting development over the last 15 years has been the rapid validation in the clinic of the premise that small-molecule inhibitors can block the spurious activation of kinases that are associated with many cancers (Levitzki, 2003). There was, at first, considerable skepticism regarding this concept. The high cellular concentration of ATP, which is bound by all kinases, led many to question whether potent and selective inhibitors could ever be developed. These concerns were, however, assuaged rapidly (Knight & Shokat, 2005; Levitzki, 2003; Sawyers, 2003). High-affinity ligands that bound to the active site were identified in high-throughput screens, and these small molecules could outcompete ATP in a cellular setting (Cohen, 2002). Structural studies have revealed that kinases can adopt distinct inactive conformations that can be exploited to impart selectivity (Huse & Kuriyan, 2002; Nolen et al., 2004). Furthermore, selective small-molecule inhibitors that target even the highly conserved active kinase conformation have been developed (see, for example, dasatinib and vemurafenib, described later). Recent findings in the clinic also point to the fact that absolute specificity may not always be desirable. An argument can be made that the development of multitargeted inhibitors is an advantageous strategy in cancer (Knight, Lin, & Shokat, 2010). Most of the clinically approved kinase inhibitors block several kinases, yet exhibit relatively low toxicity profiles (Anastassiadis, Deacon, Devarajan, Ma, & Peterson, 2011; Davis et al., 2011; Fabian et al., 2005; Karaman et al., 2008).

The important developments in the discovery and design of protein kinase inhibitors have been highlighted in many reviews (Chiu et al., 2013; Cohen, 2002; Dar & Shokat, 2011; Johnson, 2009b; Levitzki & Mishani, 2006; van Linden, Kooistra, Leurs, de Esch, & de Graaf, 2014; Zhang, Yang, & Gray, 2009; Zhao et al., 2014). Here, we focus on the protein structural determinants for kinase-inhibitor specificity and restrict our discussion to clinically approved drugs. We examine the residues in the kinase active site that are important for ATP binding, substrate-peptide binding, and catalysis. We discuss how ATP-competitive inhibitors exploit these interactions and consider advances in our understanding of how to target kinases more selectively. Finally, we highlight a few recent strategies for inhibitor design that exploit additional "selectivity filters" within the kinase active site.

2. KINASE STRUCTURE AND CATALYTIC MECHANISM

At present, all of the clinically approved kinase inhibitors that bind directly to the kinase domain are ATP-competitive compounds. Some clinically approved drugs, such as rapamycin and related analogs, target additional domains or regions associated with the kinase domain. These additional domains can contribute to the regulation of kinase activity (Pawson & Kofler, 2009). Thus, small molecules that bind to these auxiliary sites represent a viable strategy for drug development (Barnett et al., 2005; Dar & Shokat, 2011; Engel et al., 2006; Gower, Chang, & Maly, 2014; Laudet et al., 2007; Rini, 2008). Compounds that compete for protein substrate binding, that bind to allosteric sites on the kinase domain, or that bind to other regulatory domains are also in development or in clinical trials, but these fall outside the scope of this review.

Susan Taylor and colleagues were the first to reveal the three-dimensional fold of a protein kinase, that of cyclic-AMP-dependent protein kinase A (PKA) (Knighton et al., 1991). The architecture of the catalytic domain, which is shared among all protein kinases, includes an N-terminal lobe (N lobe) and a C-terminal lobe (C lobe) (Fig. 2.1A). The N lobe consists of a five-stranded β-sheet with a single α-helix, the αC helix. In contrast, the C lobe is mainly α-helical.

The ATP-binding site lies in a large cleft between these two lobes, and ATP makes specific interactions with residues in both the N and C lobes as well as with conserved structural elements, including the hinge region, the phosphate-binding loop (P-loop), the activation loop, the catalytic loop, and the αC helix (Fig. 2.1A; Adams, 2001; Huse & Kuriyan, 2002; Schwartz & Murray, 2011). ATP maintains a similar binding mode in all kinases (Knight & Shokat, 2005). The key aspects of this interaction are illustrated in Fig. 2.1A, based on the structure of PKA (protein data bank (PDB) 1ATP). For tyrosine kinases, the active form of the insulin receptor

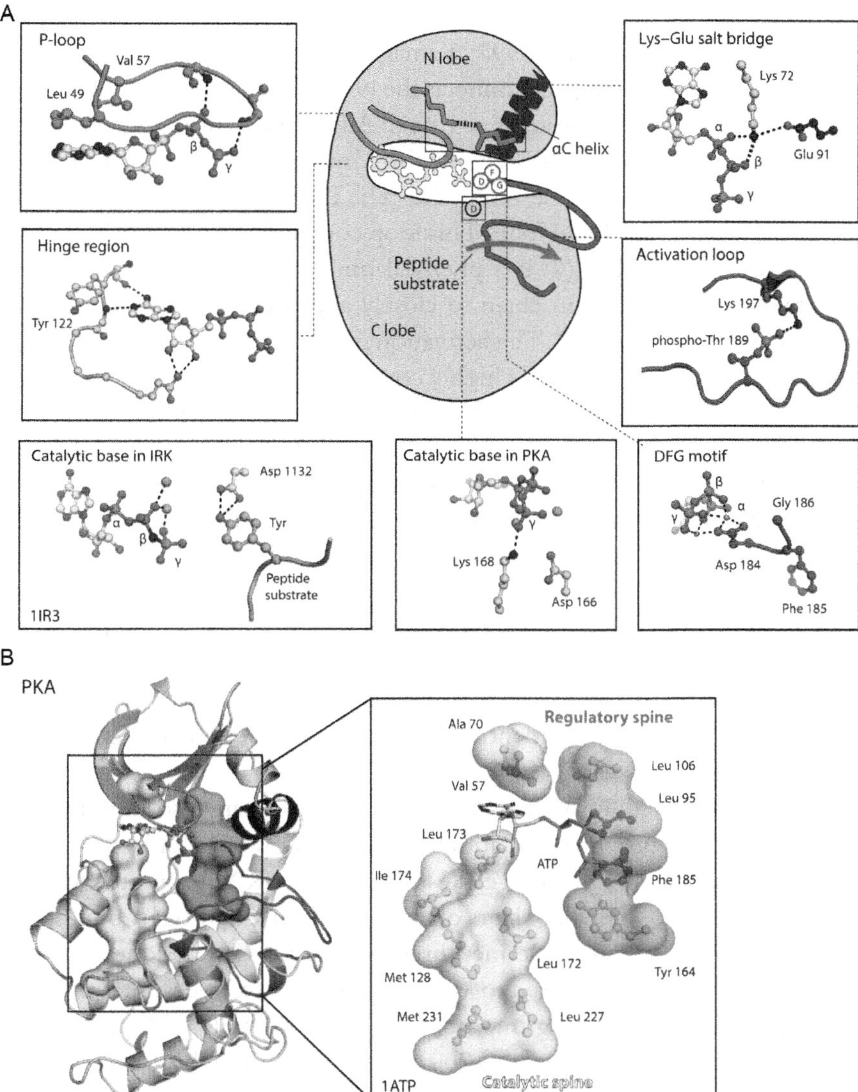

Figure 2.1 The key structural elements in a kinase domain. (A) An overview of catalytic and regulatory elements in a protein kinase. The active conformation is shown in light blue. The αC helix, P-loop, activation loop, and Lys–Glu ion pair are shown in stick representation. The bound ATP is illustrated in yellow. The Asp–Phe–Gly (DFG) motif is shown as red circles. The catalytic base is shown as a black circle. The zoomed-in windows display detailed atomic views of important regulatory elements, with hydrogen bonds indicated by dashed lines. Mg^{2+} or Mn^{2+} is shown as gray spheres. All structural figures are generated using PyMol. (B) The catalytic and regulatory spines of the kinase domain, as defined by Kornev, Taylor, and coworkers. The compact organization of the residues in these spines is disrupted in an inactive conformation of the kinase domain. (See the color plate.)

(IRK, PDB 1IR3) and the Src family kinase LCK (PDB 3LCK) are also good references (Hubbard, 1997; Yamaguchi & Hendrickson, 1996).

The hinge region, which connects the N lobe to the C lobe, makes backbone interactions with the adenine ring of ATP. The hydroxyl-groups at the 2′- and 3′-positions on the ribose ring form hydrogen bonds with residues in the C lobe at the base of the active site. The P-loop is situated between the β1 and β2 strands in the N lobe. This loop contains a characteristic glycine-rich motif, GXGXΦG (Φ is a phenylalanine or tyrosine residue), whose flexibility allows the main chain to closely approach and interact with the phosphate groups of ATP. The activation loop is positioned at the opening of the active site and contains a highly conserved Asp–Phe–Gly (DFG) motif at its N-terminal end (residues 184–186 in PKA); we shall use the residue numbering in the crystal structure of PKA (PDB 1ATP), unless otherwise stated. The aspartic acid residue of this motif, Asp 184, coordinates Mg^{2+}, which in turn acts as a counterion for the triphosphate moiety of ATP. In some kinases, a second divalent Mg^{2+} cation may interact with an asparagine residue (Asn 171 in PKA) in the catalytic loop at the base of the active site. Lys 72 in the β3 strand of the N lobe extends into the ATP-binding site to coordinate the α- and β-phosphate groups of ATP. Finally, the side chain of Glu 91 in the αC helix helps to orient Lys 72 in the active site. All of the residues that we have identified by number are essentially invariant in protein kinases.

Conserved residues within the kinase active site are also important for recognition and proper orientation of protein substrates. An aspartic acid residue at the base of the active site (Asp 166 in PKA) promotes substrate phosphorylation. The side chain of Asp 166 forms hydrogen bonds with the hydroxyl group on the serine, threonine, or tyrosine residue of protein substrates and is often referred to as the "catalytic base," because it may help deprotonate the hydroxyl group on the protein substrate, although its precise role is controversial (Adams, 2001). Asp 166 is thought to help catalyze the phosphotransfer reaction by orienting the substrate for proper attack of the γ-phosphate group on ATP. Substrate coordination by the catalytic base (Asp 1132 in IRK) is nicely illustrated in the structure of IRK in complex with peptide substrate (PDB 1IR3) (Parang et al., 2001).

The activation loop is important for correctly positioning ATP in the kinase active site, but it also serves as a platform for binding protein substrates when it is in an open conformation that extends away from the site of phosphate transfer (see Fig. 2.1A). This active conformation is stabilized by phosphorylation of the activation loop in many kinases. In the absence of phosphorylation, the activation loop in these kinases can fold back into the active site and restrict the binding of ATP, protein substrate, or both, depending on the kinase.

There are several distinct structural features of kinases that can change upon transition from an inactive conformation to an active one. The activation loop has been observed in distinct inactive conformations in the crystal structures of various kinases (Nolen et al., 2004; Steichen et al., 2010). The active conformations of protein kinases are further characterized by the close packing of two arrays of side chains, referred to as the regulatory and catalytic spines (Fig. 2.1B; Kornev, Haste, Taylor, & Eyck, 2006, Kornev, Taylor, & Ten Eyck, 2008). These spines are distorted in various ways in the inactive conformations of kinases.

The DFG motif at the beginning of the activation loop can adopt at least two extreme conformations known as the "DFG-in" and the "DGF-out" conformations (Fig. 2.2). In the "DFG-in" orientation, the aspartic acid sidechain (Asp 184 in PKA) is pointed into the active site to coordinate magnesium and allows catalysis to proceed. This corresponds to the canonical active conformation. The crystal structure of imatinib (Gleevec™) bound to Abl revealed a conformation in which the DFG motif is flipped by 180° (Fig. 2.2A; Nagar et al., 2002; Schindler et al., 2000). In the "DFG-out" conformation, the aspartic acid side chain (Asp 381 in Abl) is pointed away from the active site, and thus this conformation represents an inactive state.

This "DFG-flip" corresponds to a change in the backbone torsion angles (ϕ and φ) from a more restricted region of the Ramachandran diagram in the active "DFG-in" conformation to a less restricted region in the inactive "DFG-out" state (Fig. 2.2B). Consistent with this, computer simulations indicate that the "DFG-in" conformation is strained locally around this motif—kinases may be spring loaded in this way to switch off unless held in an active conformation (Shan et al., 2009).

The αC helix can also make significant structural transitions between inactive and active states. To facilitate phosphorylation, the αC helix must be rotated inward so that it presents Glu 91 in PKA toward Lys 72, which coordinates ATP. To remove this critical interaction and to restrict activity, the αC helix can move away from the kinase core.

Thus far, most kinase inhibitors bind to a more active-like "DFG-in" conformation. These inhibitors are often referred to as Type I inhibitors, whereas compounds that bind to an inactive-like "DFG-out" conformation are described as Type II inhibitors (Dar & Shokat, 2011; Zhao et al., 2014). While these terms are often used to classify inhibitors, it must be emphasized that

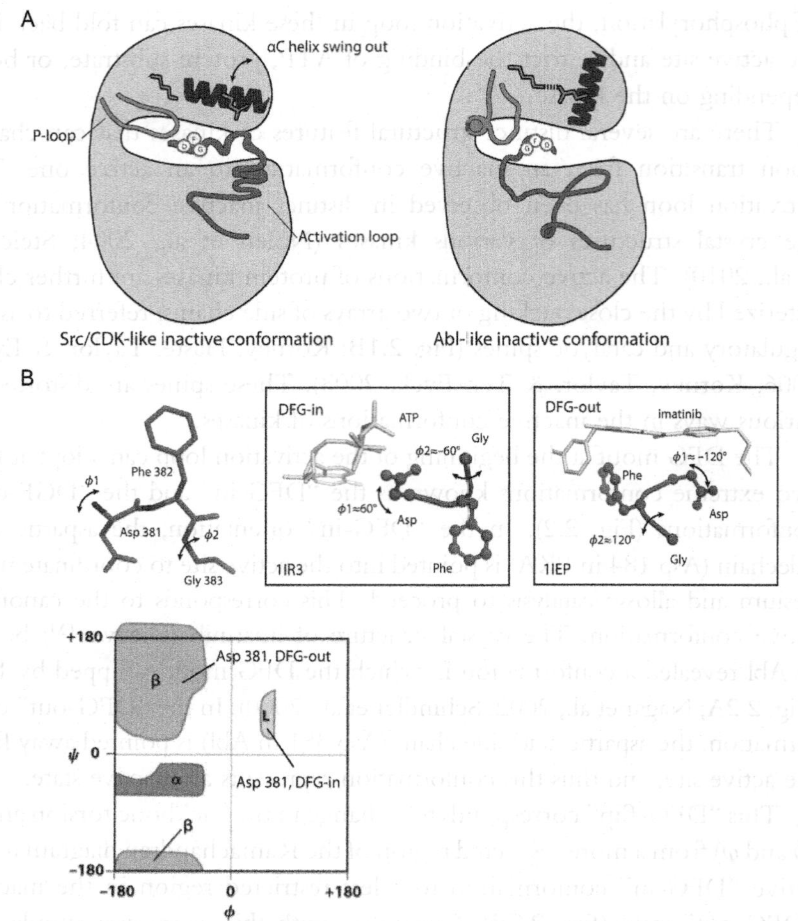

Figure 2.2 Structural rearrangements involved in kinase activation. (A) Two representative inactive conformations of tyrosine kinases. Left panel: Src/CDK-like inactive conformation (PDB: 1QCF); the activation loop stabilizes the αC helix in a distorted conformation. Right panel: Abl-like inactive conformation (PDB: 1IEP); the activation loop mimics substrate binding, and the DFG motif adopts a "DFG-out" conformation that is not compatible with ATP binding. (B) DFG-flip during Abl activation. Upper left: the torsion angle of Asp 381 and Gly 383 in the DFG motif. The Φ value changes about 180° from a "DFG-out" conformation (upper right) to a "DFG-in" conformation (upper middle). Lower panel: Ramachandran plot of peptide backbone torsion angles indicating the orientation of Asp 381 in the two conformations. *Adapted with permission from the Molecules of Life.* (See the color plate.)

inactive kinase conformations are quite variable (Chiu et al., 2013; van Linden et al., 2014). The DFG motif can be distorted away from the active conformation in more than one way in complexes with Type II inhibitors, leading to additional changes in the active site. It was proposed originally that Type II inhibitors would impart greater selectivity relative to Type I inhibitors, since kinases generally adopt very similar folds in the active conformation but with different inactive conformations (Liu & Gray, 2006; Schindler et al., 2000; Zhao et al., 2014). However, small molecules from both classes can target relatively few kinases by exploiting subtle variation within the active site. Furthermore, both types of inhibitors can also behave indiscriminately.

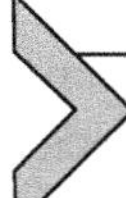

3. STAUROSPORINE: A PROMISCUOUS ATP-COMPETITIVE INHIBITOR

In spite of criticisms surrounding the "druggability" of protein kinases, several groups pushed programs forward to identify inhibitors against this class of enzymes (Cohen, 2002). A breakthrough came in 1986 when it was found that the natural product staurosporine could inhibit protein kinase C (PKC) potently (Tamaoki et al., 1986). This discovery ignited more widespread interest within the pharmaceutical industry regarding the potential of protein kinase inhibitors as therapeutic agents (Cohen, 2002).

Staurosporine was first isolated in 1977 from a Streptomyces strain in a study that focused on the identification of novel alkaloids (Omura et al., 1977). It was another 9 years until it was realized that this natural product could block the activity of protein kinases. This property of staurosporine was discovered when scientists in Japan began screening for inhibitors of PKC shortly after a report that tumor promoters, such as phorbol esters, could activate PKC (Castagna et al., 1982). Staurosporine exhibited a low nanomolar inhibitory constant against PKC *in vitro* and had a significant impact on cell viability (Tamaoki et al., 1986). While other compounds that inhibited protein kinases had been identified previously, they possessed weak potencies in biochemical assays, and their cellular activity could not necessarily be linked reliably to direct kinase inhibition (Davies, Reddy, Caivano, & Cohen, 2000).

Despite excitement over staurosporine inhibition of protein kinases, its binding mode remained elusive for over 10 years, hindering efforts to develop more selective analogs. The situation changed when the structures of staurosporine bound to two different kinases, PKA and CDK2, were reported (Lawrie et al., 1997; Prade et al., 1997; reviewed by Toledo & Lydon, 1997). Upon inspection of the structure of CDK2 kinase domain bound to staurosporine (PDB 1AQ1), it immediately becomes clear why this natural product exhibits such a high potency, with a low nanomolar K_i value against CDK (Omura, Sasaki, Iwai, & Takeshima, 1995). Staurosporine occupies the ATP-binding pocket and mimics several features of

ATP binding (Fig. 2.3). The planar indolocarbazole moiety of staurosporine forms interactions with hydrophobic side chains provided by both the N lobe and the C lobe of the kinase. In addition, the lactam nitrogen and the methylamine nitrogen form hydrogen bonds with the protein backbone. The methylamine nitrogen of staurosporine forms a hydrogen bond with the carboxyl oxygen of residue Asp 86, which is located near the C-terminal end of the hinge region.

Staurosporine interacts with residues in CDK and PKA that are nearly invariant in kinases and are important for ATP binding. In fact,

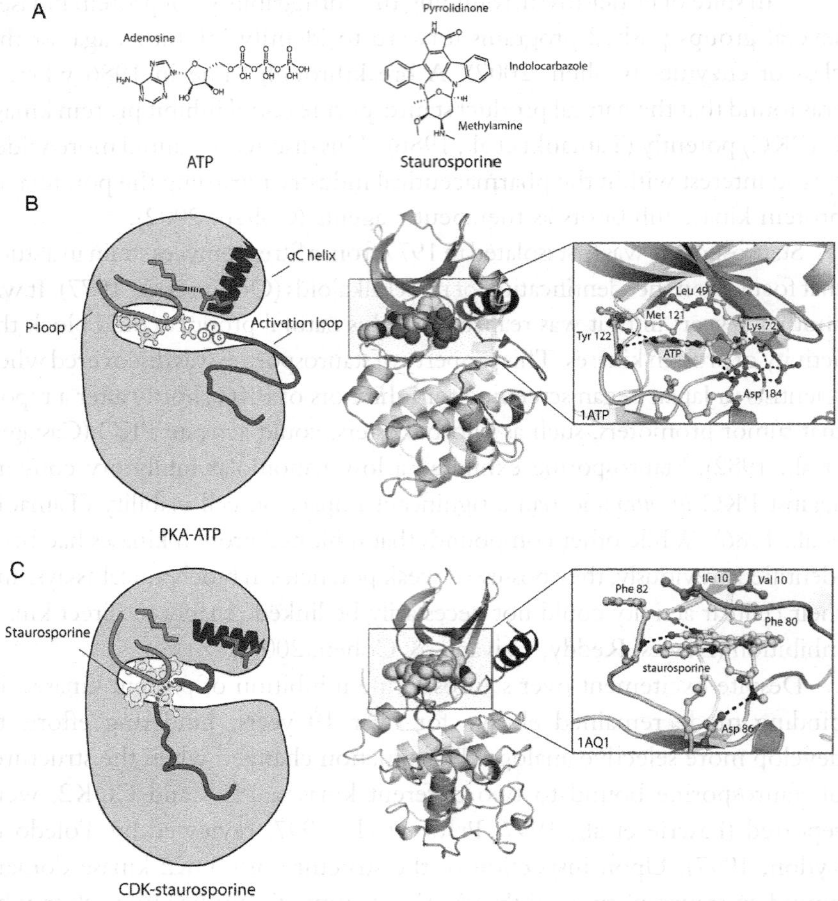

Figure 2.3 ATP and staurosporine. (A) A crystal structure of ATP-bound PKA (PDB: 1ATP). A detailed view of the ATP coordination between the N lobe and C lobe is shown on the right panel. (B) A crystal structure of staurosporine-bound CDK2 (PDB: 1AQ1). (See the color plate.)

staurosporine has been shown to potently inhibit >70% of human kinases with over 300 kinase structures containing staurosporine bound in the active site deposited in the PDB (Davis et al., 2011). Notably, staurosporine–kinase complexes can adopt either a "DFG-in" or a "DFG-out" conformation, although the "DFG-in" conformation is dominant (Chiu et al., 2013).

This lack of selectivity exhibited by staurosporine correlates with its high degree of toxicity in cells, making it unsuitable for clinical use. Staurosporine analogs, both synthetic as well as natural, have been identified that are more promising drug candidates (Cohen, 2002).

4. BCR-Abl INHIBITORS

As new strategies were being developed to selectively target protein kinases, the conformational plasticity of these enzymes was also starting to be revealed (Engh & Bossemeyer, 2002; Huse & Kuriyan, 2002). The crystal structures of inactive CDK and Src family kinases showed that these divergent kinases could adopt very similar inactive conformations (Jeffrey et al., 1995; Sicheri, Moarefi, & Kuriyan, 1997; Xu, Harrison, & Eck, 1997; reviewed by Jura et al., 2011). Crystal structures of the phosphorylated and unphosphorylsated forms of the kinase domains of fibroblast growth factor receptor (FGFR) and the insulin receptor revealed other inactive conformations (Hubbard, 1997; Mohammadi et al., 1997, Mohammadi, Schlessinger, & Hubbard, 1996). However, the potential of exploiting these conformations to inhibit kinases selectively was not fully realized until the development of imatinib (Gleevec™/Glivec™) (Druker et al., 1996; Schindler et al., 2000; Zimmermann, Buchdunger, Mett, Meyer, & Lydon, 1997; Zimmermann et al., 1996).

4.1. Imatinib binds to a "DFG-out" Abl conformation

The core phenylaminopyrimidine moiety of imatinib was originally identified in a screen for PKC inhibitors (Capdeville, Buchdunger, Zimmermann, & Matter, 2002). Medicinal chemistry around this scaffold led to derivatives that gained potency against tyrosine kinases, but lost activity against PKC (Zimmermann et al., 1996, 1997). This small molecule was further developed into an inhibitor that targeted the gene fusion product, BCR-Abl, a constitutively active tyrosine kinase associated with the progression of chronic myelogenous leukemia (CML). Clinical approval of this inhibitor came rapidly after the demonstration that imatinib preferentially induced apoptosis, or programmed cell death, in cells from patients with CML relative to control cells (Druker et al., 1996).

A structural rational for the selectivity of imatinib was revealed by our group in September 2000 (Nagar et al., 2003; Schindler et al., 2000). The crystal structure of an imatinib analog bound to Abl demonstrated that this scaffold recognized an inactive kinase conformation (Fig. 5B) (PDB 1IEP). In this conformation, the DFG motif is in a "DFG-out" conformation, which is not compatible with ATP coordination. The activation loop folds back on itself to adopt a substrate-mimicking conformation, seen previously in inactive IRK (PDB 1FPU) (Hubbard, Wei, Ellis, & Hendrickson, 1994). Furthermore, the P-loop adopts a distorted conformation that folds over imatinib. The pyridine–pyrimidine moiety of imatinib packs tightly in the active site against the side chains of residues Leu 248, Val 256, and Leu 370, which are also important for ATP binding. In addition, residue Phe 382 in the DFG motif, residue Tyr 253 in the P-loop, and residue Phe 317 in the hinge region between N lobe and C lobe each form π–π stacking interactions with the aromatic rings in imatinib.

Adjacent to the pyridine–pyrimidine moiety in imatinib is a tolyl group that packs against the "gatekeeper" residue, Thr 315. This residue is termed the gatekeeper because it is located between the ATP-binding site and an internal cavity that is seen in the inactive conformations of some kinases, referred to as the "specificity pocket" because it is one of the features that is the source of specificity in inhibitor binding. Many small-molecule kinase inhibitors bind in this cavity; however, the presence of a large side chain at the gatekeeper position can block the binding of such inhibitors. One example is PP1 (pyrazolopyrimidine 1), which specifically inhibits Src-family kinases but not serine–threonine kinases, such as PKA (Schindler et al., 1999). A crystal structure of the PP1 bound to the Src family kinase HCK shows that the gatekeeper residue Thr 338 and the residue Ala 403 upstream the DFG motif are the key determinants for PP1 selectivity. Mutation of these two residues to bulky residues, such as methionine and threonine as seen in PKA, will cause severe clashes between the inhibitor and the kinase (Fig. 2.4).

Imatinib makes good use of the specificity pocket by penetrating deeply into the core of the kinase. A hydrogen bond interaction also forms between the hydroxyl group of Thr 315 and the secondary amine linking the tolyl group to the pyrimidine moiety. The benzamide moiety packs against residue Leu 293 in this pocket, and the carbonyl group in the benzamide moiety forms a hydrogen bond with the amide group of residue Asp 381. In addition, an amine in the piperazine group forms a hydrogen bond with the carbonyl group of Ile 360.

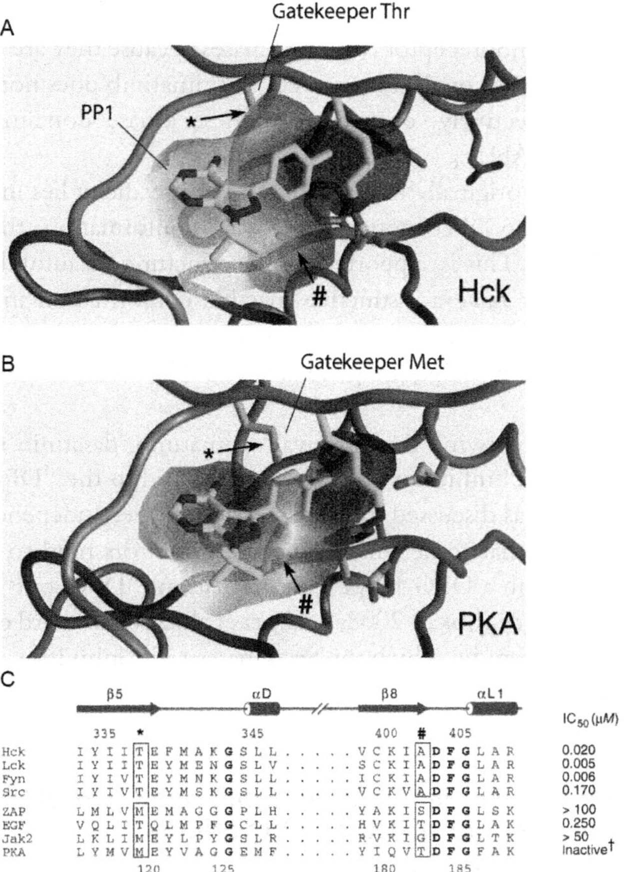

Figure 2.4 Selectivity determinants of ATP-analog PP1 binding. (A) PP1 bound to Hck. The Cα traces of the catalytic and activation segment are shown in green and magenta, respectively. The molecular surface of the binding site is superimposed over the traces in transparent gray. PP1 as well as selected residues of Hck are drawn as stick figures. (B) PP1 modeled into the ATP-binding site of PKA (PDB code 1ATP; Zheng et al., 1993). The model was generated by aligning the pyrazolopyrimidine moiety of PP1 and the adenine ring of ATP. (C) Sequence comparison of Src family tyrosine kinase members with a set of other protein kinases. The inhibition constants (IC50s) for PP1 were taken from Hanke et al. (1996) ([†] PP1 was reported to be essentially inactive against PKA). Thr-338 (*) and Ala 403 (#) are indicated in (A), as are the structurally equivalent residues in PKA, in (B). *Reprinted with permission from Schindler et al. (1999) Molecular Cell.* (See the color plate.)

While imatinib has a very narrow selectivity profile, it does potently inhibit several receptor tyrosine kinases, including c-Kit and platelet-derived growth factor receptor (PDGFR). The structural principles that underlie Abl recognition of imatinib also pertain to the inhibition of c-Kit by imatinib (Mol et al., 2004), which is discussed further in a later section on receptor tyrosine kinases. It is particularly striking that imatinib inhibits both receptor and nonreceptor tyrosine kinases because they are on different branches of the kinase evolutionary tree. Yet, imatinib does not inhibit Src family kinases effectively, even though these kinase domains are more closely related to Abl.

It was thought originally that the key to the specificity lies in the flipped conformation of the DFG motif ("DFG-out" conformation) that is recognized by imatinib. This is supported by the structures of autoinhibited Src-family kinases that have a distinctive inactive conformation in which the DFG motif is not flipped ("DFG-in" conformation) and that resemble the structure of inactive CDK, in general terms (Jura et al., 2011; Seeliger et al., 2007).

Immediately following the approval of imatinib, dasatinib was discovered as a potent Abl inhibitor and reported to bind to the "DFG-in" conformation of Abl, as discussed in a later section. Three independent studies revealed, further, that some small-molecule inhibitors bind to Src kinases with high affinity in a DFG-flipped conformation ("DFG-out" conformation) (Dar, Lopez, & Shokat, 2008; Seeliger et al., 2009; Simard et al., 2009). Thus, the mechanism by which the Src kinases resist inhibition by imatinib awaits further elucidation.

4.2. Nilotinib (Tasigna™): An imatinib analog effective against several imatinib-resistant Abl variants

The high efficacy and limited toxicity associated with imatinib led to its rapid advancement through clinical trials (Capdeville et al., 2002). In 2001, imatinib became the first drug that targeted a protein kinase to receive clinical approval in the United States. Despite high response rates, CML patients treated with imatinib often develop resistance to the drug. In the majority of cases, resistance arises from point mutations in the kinase domain of BCR-Abl that disrupt imatinib binding directly (Daub, Specht, & Ullrich, 2004; Krishnamurty & Maly, 2010).

An additional structure–activity relationship study around the core phenylaminopyrimidine scaffold of imatinib led to a more potent BCR-Abl inhibitor, nilotinib (Manley, Cowan-Jacob, & Mestan, 2005). Nilotinib contains an *N*-arylimidazole group in place of the piperazine and phenyl groups of imatinib. This drug not only potently inhibits wild-type BCR-Abl but also effectively inhibits the activity of many of the BCR-Abl resistance mutants that were identified in patients treated with imatinib (Weisberg et al., 2005). In 2007, the FDA approved nilotinib as a treatment for CML.

A crystal structure of nilotinib bound to Abl reveals additional interactions that this small molecule exploits in the active site that may explain its increased potency relative to imatinib (PDB 3CS9) (Fig. 2.5C). First, the presence of the triflouromethyl substituent on the phenyl ring allows the inhibitor to insert deeper into the specificity pocket relative to imatinib, tightening the hydrophobic interactions around the phenyl group (Fig. 2.5D). In addition, the imidazole moiety is in close contact with residue Phe 359 in the C lobe. The particular conformation of Phe 359 that enables this interaction is not compatible with the imatinib-bound Abl kinase structure, since clashes would occur between the Phe 359 side chain and the piperazine ring on imatinib. As a result, in the imatinib-bound Abl kinase domain, Phe 359 adopts a different conformation and is pointing away from the inhibitor. The basis for the efficacy of nilotinib against Abl variants that carry imatinib resistance mutations (E255V, M351T, and F486S) is, however, unclear (Weisberg et al., 2005).

4.3. Ponatinib (Iclusig™) overcomes an Abl gatekeeper resistance mutation

One of the most common mutations to appear following imatinib treatment is a gatekeeper mutation, T315I (Fig. 2.5E). While nilotinib is able to inhibit the vast majority of clinically observed resistance mutations, it is not effective against this gatekeeper mutation in BCR-Abl. Both nilotinib and imatinib use the gatekeeper residue (Thr 315) as an important binding element in the Abl kinase domain. Mutations of the threonine to more bulky hydrophobic residues are predicted to destabilize interactions between these drugs and the Abl kinase domain. A key hydrogen bond between the inhibitor and the threonine side chain is lost when the threonine is replaced by isoleucine. The bulky isoleucine residue also restricts access of nilotinib and imatinib to the specificity pocket (Fig. 2.5E).

Structure-based design in conjunction with modifications of the core imatinib scaffold led to the development of ponatinib (O'Hare et al., 2009; Zhou et al., 2011). Ponatinib shares several chemical features with nilotinib and imatinib, such as the presence of the piperazinyl group (a prominent moiety in imatinib) and the trifluoromethylphenyl group (a substituent in nilotinib). The key to the efficacy of ponatinib is the replacement of the amine linkage between the methylphenyl ring and a heterocycle seen in both imatinib and nilotinib with an extended alkyne linker.

In a crystal structure of ponatinib bound Abl-T315I kinase domain, the alkyne linker, which is hydrophobic, is packed tightly against Ile 315 (Fig. 2.5E) (PDB 3IK3) (Zhou et al., 2011). Neither imatinib nor nilotinib

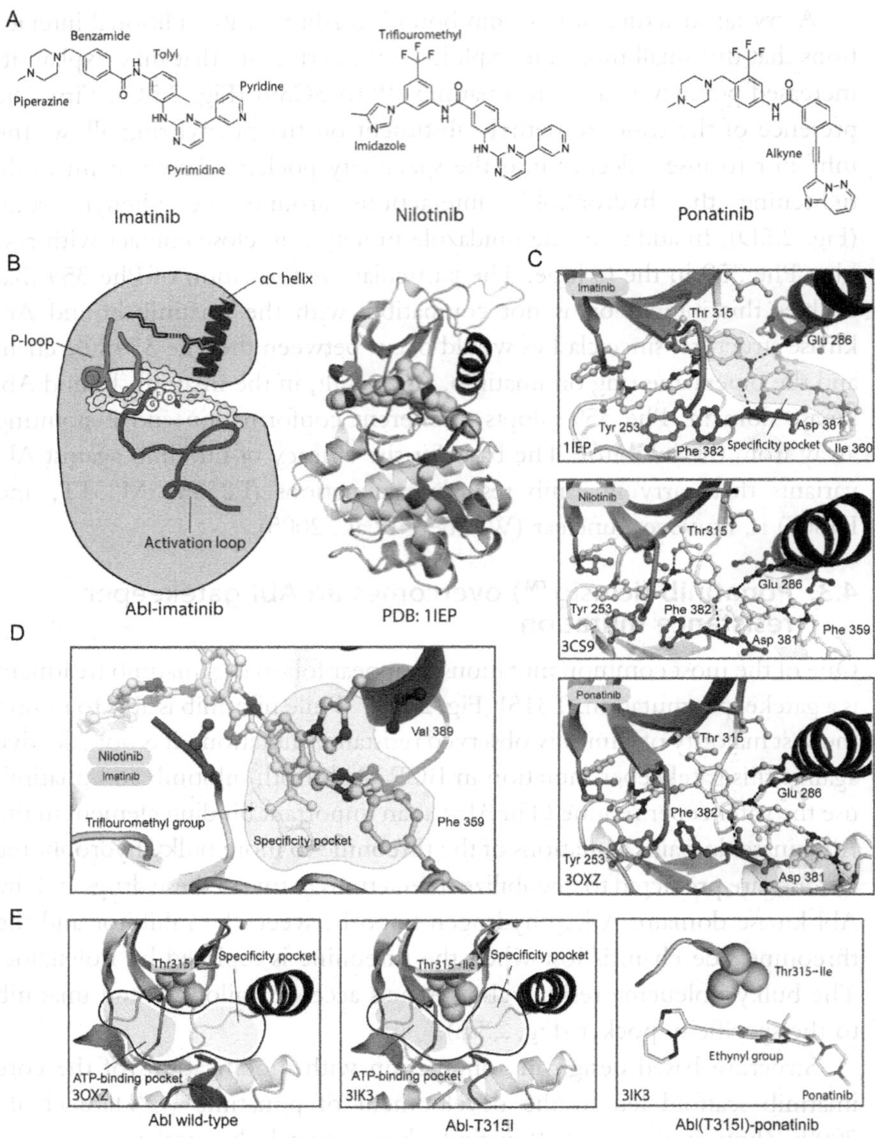

Figure 2.5 Inhibition of c-Abl kinase domain by imatinib and related analogs. (A) Chemical structures of imatinib, nilotinib, and ponatinib. (B) A crystal structure of the Abl kinase domain bound to an imatinib analog (PDB: 1IEP). (C) Detailed views of imatinib (upper panel), nilotinib (middle panel), and ponatinib (lower panel) coordination in Abl kinase domain. The specificity pocket of Abl kinase domain is shaded brown. (D) Structural overlay of nilotinib (yellow) and imatinib (grey) in the Abl specificity pocket (shaded brown). (E) The basis for inhibition of Abl-T315I mutant by ponatinib. Mutations of the gatekeeper threonine restrict access to the specificity pocket (shaded brown) from the ATP binding pocket (shaded purple) in the Abl-T315I crystal structure (middle panel) as compared to the wild-type Abl structure (left panel). Ponatinib is compatible with binding to both pockets of the Abl-T315I mutant. (See the color plate.)

can adopt this binding mode due to steric hindrance between the isoleucine side chain and the bulky pyrimidine group. In 2012, the FDA approved ponatinib to treat CML patients with resistance to imatinib.

4.4. Bosutinib (Bosulif™) inhibits BCR-Abl

Bosutinib is a potent Abl inhibitor and is built around a chemical scaffold that is different from that of imatinib (Boschelli et al., 2001; Golas et al., 2003). In the bosutinib-Abl cocrystal structure, the DFG motif adopts a "DFG-out" conformation that restricts important interactions with ATP (PDB 3UE4) (Levinson & Boxer, 2014). However, the activation loop is fully extended, and it is compatible with peptide substrate binding. This conformation is similar to that seen in the structures of several inhibitors bound to Abl (Nagar et al., 2002). In the structure of the bosutinib-bound Abl kinase domain, the quinoline moiety occupies the ATP-binding pocket. In addition, the nitrogen on the quinolone ring forms a hydrogen bond with the backbone amide of Met 318 in the hinge region. The dichloromethoxyaniline group packs against gatekeeper residue Thr 315 and forms π–π stacking interactions with the phenyl ring of Phe 382 in the DFG motif (Fig. 2.6C). The nitrile group on bosutinib forms a hydrogen bond with a structural water, which is coordinated by Asp 381 in the DFG motif.

Bosutinib is also seen bound to the "DFG-in" conformation of other kinases, including Src, Her3, and CaMKII (Chao et al., 2011; Levinson & Boxer, 2014; Littlefield, Moasser, & Jura, 2014; Remsing Rix et al., 2009). In these structures, the DFG motif is not directly involved in coordinating inhibitor binding as seen in the bosutinib-Abl structure. Interestingly, the nitrile–water interaction seen in the bosutinib–Abl complex is preserved in a structure of Src bound to bosutinib (PDB 4MXO) (Levinson & Boxer, 2014). Here, the nitrile group interacts with one of the two structural water molecules that are seen often in the crystal structures of active tyrosine kinases. This water-mediated hydrogen bond network has been shown to be important for bosutinib binding to the Src kinase domain, a distinct mechanism for bosutinib to achieve its selectivity.

4.5. Dasatinib (Sprycel™) binds to a "DFG-in" conformation of Abl

Dasatinib was the first of the so-called "second-generation" Abl inhibitor to be developed, and it is the only Abl kinase inhibitor currently in the clinic that interacts with a "DFG-in" conformation of Abl (Das et al., 2006;

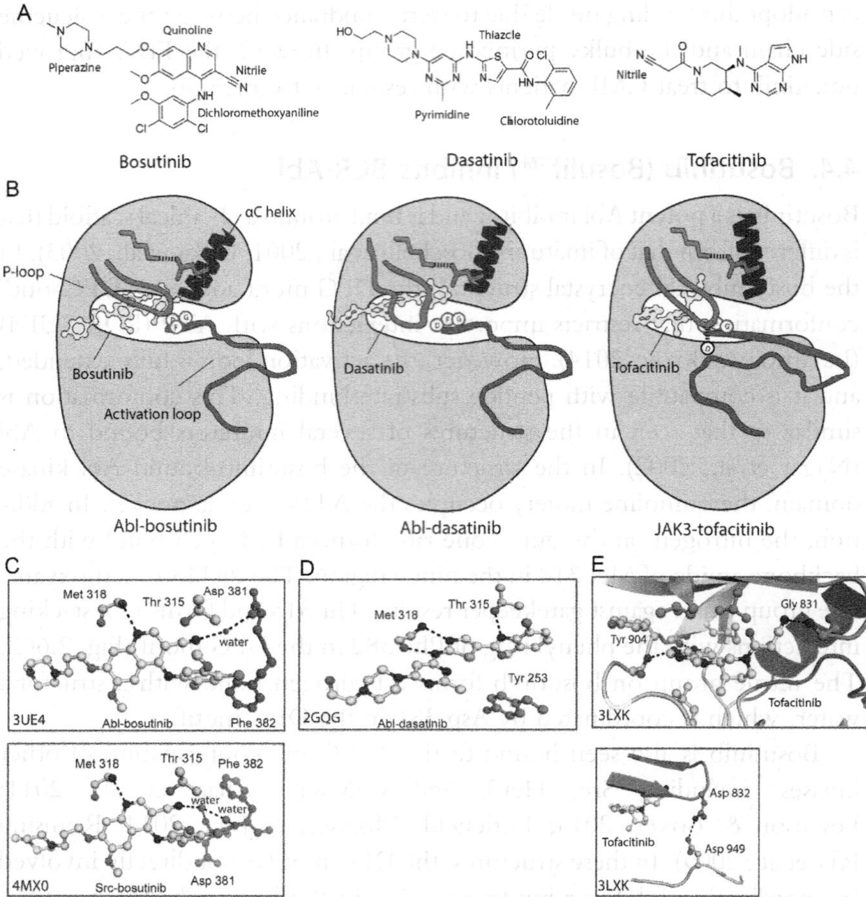

Figure 2.6 Clinical inhibitors that target an active conformation of Abl and JAK3. (A) Chemical structures of bosutinib, dasatinib, and tofacitinib. (B) The bosutinib-bound Abl kinase domain (PDB 3UE4), dasatinib-bound Abl kinase domain (PDB: 2GQG) and tofacitinib-bound JAK3 (PDB 3LXK). (C) Detailed view of the coordination of bosutinib in Abl kinase domain (upper panel) and in Src kinase domain (lower panel). The water molecules are shown as small red spheres. (D) Zoomed-in view of the coordination of dasatinib in the Abl kinase domain. (E) Zoomed-in view of the coordination of tofacitinib in the JAK3 kinase domain (upper panel). The P-loop residue Asn 832 in JAK3 forms hydrogen bonds to the catalytic base, Asp 949 (lower panel). (See the color plate.)

Schittenhelm et al., 2006; Weisberg, Manley, Cowan-Jacob, Hochhaus, & Griffin, 2007; Wityak et al., 2003). In a crystal structure of dasatinib bound to Abl, the DFG motif adopts a "DFG-in" conformation that is compatible with ATP binding; the activation loop is also extended for substrate binding (PDB 2GQG) (Tokarski et al., 2006). In addition, Lys 671, which is important for ATP coordination, is in the position it adopts in active kinases, and it forms a salt bridge with residue Glu 286 in the αC helix. The pyrimidinylaminothiazole moiety occupies the ATP-binding pocket (Fig. 2.6D). Tyr 253 in the P-loop also packs against the pyrimidinylaminothiazole group. The nitrogen in the thiazole ring and the amine group linking the pyrimidine moiety to the thiazole moiety each forms a hydrogen bond with the backbone amide of residue Met 318 and the carbonyl oxygen of Thr 319, respectively. In addition, the amide group in dasatinib forms a hydrogen bond with the hydroxyl group of the gatekeeper residue, Thr 315. The chlorotoluidine ring further extends deeper in the catalytic cleft and is located close to the gatekeeper residue, Thr 315. As a consequence, the gatekeeper mutation, T315I, also occurs in patients in response to dasatinib treatment.

5. TOFACITINIB (XELJANZ™) BINDS TO A "DFG-in" CONFORMATION OF JANUS KINASE

For kinases that have a large, bulky gatekeeper residue, it is particularly challenging to develop inhibitors that can interact with the kinase domain with high specificity, due to steric hindrance. Nevertheless, efficacious inhibitors have been developed for the Janus kinase (JAK)3, which has a methionine residue in the gatekeeper position. Tofacitinib is a JAK3 inhibitor that was approved for the treatment of rheumatoid arthritis in 2012.

Tofacitinib binds to a "DFG-in" conformation, with the activation loop extended away from the kinase, which is compatible with peptide substrate binding. The bulky side chain of Met 929 restricts access of the inhibitor to the deeper side of the catalytic cleft. Tofacitinib only occupies the ATP-binding pocket, with the nitrile group pointing toward the P-loop region in the N lobe (Fig. 2.6E).

For a kinase inhibitor that only targets the ATP-binding pocket, tofacitinib displays unusual high selectivity against JAK family kinases. This high degree of selectivity may be rationalized by the unique conformation of the tofacitinib-bound JAK3 kinase domain (Fig. 2.6B). Although the overall conformation of the kinase domain resembles that of active kinases, residue Asn 832 in the P-loop forms a hydrogen bond with the carboxyl group in Asp 949, the catalytic base. As a consequence, the N lobe of the kinase domain is twisted shut against the C lobe, and tofacitinib recognizes this closed conformation of the kinase domain. This distinct kinase conformation may not be adopted readily by other kinases, which might underlie the high selectivity of tofacitinib.

6. INHIBITION OF RECEPTOR TYROSINE KINASES

Receptor tyrosine kinases are attractive drug targets because of their central role in cellular signaling processes (Levitzki & Mishani, 2006). Shortly after the discovery of the first oncogene, the unregulated activation of receptor tyrosine kinases also became widely implicated in the progression of several disease states, including cancer. This led to many efforts to identify inhibitors specific to this kinase family.

Receptor tyrosine kinases have in common a single transmembrane helix that links an extracellular domain to a cytoplasmic tyrosine kinase domain. Different subfamilies of these receptors have quite distinct activation mechanisms (Lemmon & Schlessinger, 2010). One key difference concerns the function of the juxtamembrane segment, a 20–50 residue segment that links the transmembrane helix to the kinase domain (Hubbard, 2004). In the PDGFR and the vascular endothelial growth factor receptor (VEGFR) subfamilies, this segment stabilizes an inactive conformation of the kinase domain (Griffith et al., 2004; Mol et al., 2004; Wybenga-Groot et al., 2001). For this subfamily of receptors, phosphorylation of the juxtamembrane segment releases the autoinhibitory interactions. In contrast, the juxtamembrane segment in the epidermal growth factor receptor (EGFR) subfamily plays an activating role. The EGFR juxtamembrane segment latches two kinase domains together to form an activating asymmetric dimer (Brewer et al., 2009; Jura et al., 2009; Zhang, Gureasko, Shen, Cole, & Kuriyan, 2006). As discussed in later sections, the juxtamembrane segment not only influences kinase activity, but can also impact small molecule interactions with the kinase domain.

6.1. Imatinib binds to a "DFG-out" conformation of c-Kit

Imatinib, while initially approved for the treatment of cancers associated with BCR-Abl, also exhibits significant potency against two receptor tyrosine kinases, PDGFR and c-Kit. This reflects the origins of imatinib in a drug development program that targeted receptor tyrosine kinases as well as BCR-Abl (Lydon & Druker, 2004). In fact, one clinical indication for imatinib is for the treatment of gastrointestinal stromal tumors, where activating mutations in c-Kit are prominent (Hirota et al., 1998).

The structure of the c-Kit kinase domain bound to imatinib is very similar to that of the Abl kinase domain bound to imatinib (PDB 1T46) (Mol et al., 2004). One difference is that the P-loop segment of c-Kit has a

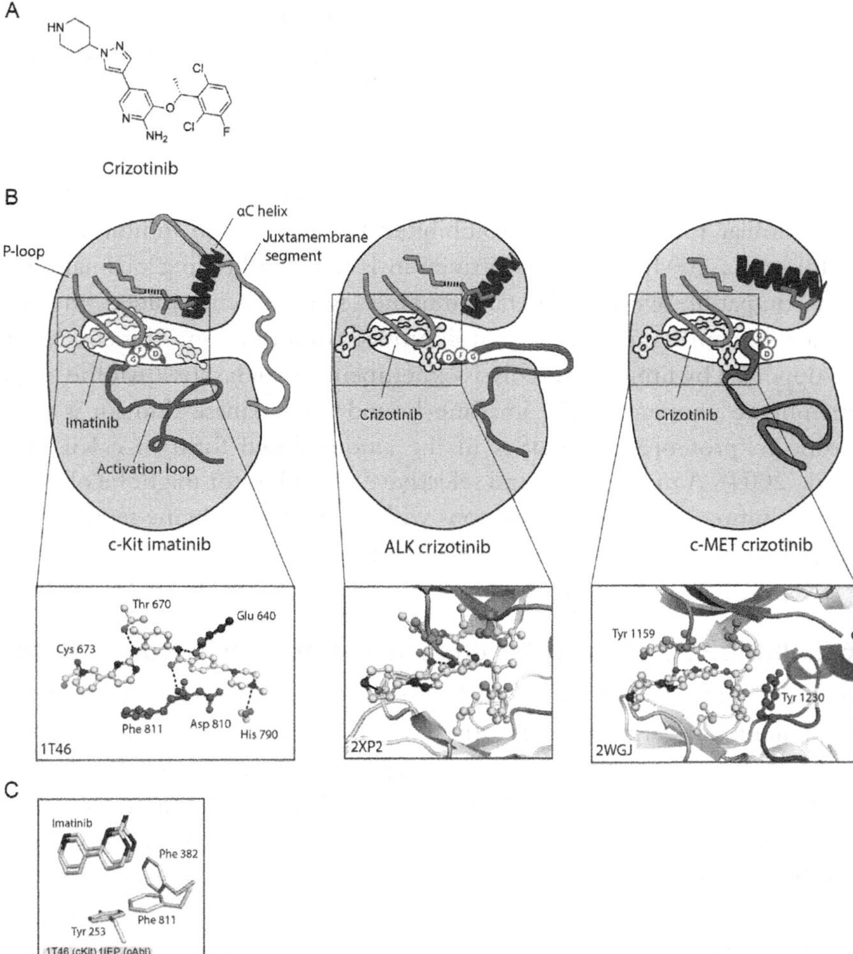

Figure 2.7 Inhibition of c-Kit, ALK, and c-MET by imatinib and crizotinib. (A) Chemical structure of crizotinib. (B) The c-Kit kinase domain bound to imatinib (PDB: 1T46), ALK kinase domain bound to crizotinib (PDB: 2XP2), and c-Met kinase domain bound to crizotinib (PDB 2WGJ). A detailed view of the interactions between the kinase and the inhibitors is shown in the lower panel. (C) Phe 811 in the DFG motif of c-Kit bound to imatinib adopts a different conformation compared to Phe 382 in the DFG motif of c-Abl bound to imatinib. (See the color plate.)

different conformation from that seen in the Abl–imatinib complex, where Tyr 253 in the P-loop of Abl interacts with imatinib (Fig. 2.7C). The corresponding residue in c-Kit is a phenylalanine that instead interacts with the αC helix. Phe 811 in the DFG motif in c-Kit also adopts a side chain conformation that is different from that of the corresponding phenylalanine residue in Abl (Phe 382). In particular, the phenyl ring of Phe 811 is perpendicular to the pyridine–pyrimidine moiety in imatinib in contrast to a parallel π–π stacking interaction seen for Phe 382 in Abl. Such perpendicular π–π stacking cannot be formed in Abl since the phenyl ring would clash with Tyr 253 in the P-loop that coordinates imatinib.

The conformation of the kinase domain of c-Kit bound to imatinib is very similar to that of the autoinhibited kinase with no inhibitor bound (Mol et al., 2004). One key distinction is that the binding of imatinib is not compatible with the juxtamembrane segment being folded onto the kinase domain as it does in the autoinhibited conformation. This conclusion is supported by limited proteolysis experiments, which show that the juxtamembrane segment in the imatinib-bound c-Kit kinase domain is more prone to proteolysis than it is in the autoinhibited form of c-Kit (Mol et al., 2004). A more potent and selective c-Kit inhibitor might be obtained by exploiting additional interactions with the juxtamembrane segment that help lock it down rather than displacing it.

6.2. Inhibitors of vascular endothelial growth factor receptor

Sorafenib (Nexavar™), sunitinib (Sutent™), pazopanib (Votrient™), regorafenib (Stivarga™), and axitinib (Inlyta™) are five clinically approved drugs that bind to "DFG-out" conformations of their targets, in some cases preferentially to recombinant kinase variants that include a juxtamembrane segment (McTigue et al., 2012; Solowiej et al., 2009). While these drugs inhibit several kinases, including many receptor tyrosine kinases, VEGFR has been implicated as the primary target, in the context of the approved drug indications for renal cell carcinoma and colorectal cancer. The ligand, VEGF, is overexpressed in the majority of renal cell carcinoma patients and is also believed to stimulate angiogenesis in colorectal cancer (Ellis & Hicklin, 2008). Thus, inhibitors of the VEGFR kinase domain offer a promising therapeutic strategy for treatment of these cancers.

The core biaryl-urea scaffold of sorafenib was originally identified in the mid-1990s from a high-throughput screen targeting the Ser/Thr kinase, RAF1 (Wilhelm et al., 2006). Oncogenic mutations in RAF isoforms are found in many human cancers, making the RAF kinase family an attractive drug target, as discussed later (Flaherty, 2007; Wan et al., 2004). Modification of this scaffold eventually led to a potent, nanomolar RAF inhibitor, sorafenib (Wilhelm et al., 2006). This small molecule also potently inhibited several receptor tyrosine kinases, including PDGFR, VEGFR, FGFR, Flt-3,

RET, and c-Kit. In 2005, sorafenib received FDA approval as a renal cell carcinoma therapy, with a closely related analog, regorafenib, receiving clinical approval 6 years later for colorectal cancer treatment (Wilhelm et al., 2011). While sorafenib's efficacy in renal cell carcinoma is associated with targeting multiple kinases, including RAF, a side effect of hypertension observed in clinical trials implicated VEGFR as a major effector (Flaherty, 2007).

In a crystal structure of sorafenib-bound VEGFR kinase domain, the kinase domain adopts a "DGF-out," activation loop-in conformation, resembling the inactive conformation of Abl (McTigue et al., 2012). Sorafenib forms extensive interactions with VEGFR in this conformation (Fig. 2.8B) (PDB 4ASD). Notably, the trifluoromethylphenyl ring is tightly packed into the specificity pocket. This binding mode of the trifluoromethylphenyl ring also resembles the one in the nilotinib-bound Abl kinase domain.

Sorafenib binding is not compatible with the conformation of the juxtamembrane segment seen in the crystal structure of the autoinhibited kinase domain. This rearrangement of the juxtamembrane segment required for sorafenib binding is reflected in biochemical assays that exhibit time-dependent inhibition of the VEGFR juxtamembrane-kinase domain variant (McTigue et al., 2012; Oguro et al., 2013; Solowiej et al., 2009). Up to a 10-minute preincubation time was necessary to observe a plateau in the inhibitory constant, but only for VEGFR constructs that included the juxtamembrane segment (McTigue et al., 2012). This suggests that the conformational rearrangements in the juxtamembrane domain required for drug binding can be extraordinarily slow.

In a crystal structure of VEGFR bound to axitinib, the conformation of the kinase domain is very similar to that adopted when sorafenib is bound to VEGFR (PDB 4AG8) (McTigue et al., 2012). The pyridine-vinyl-indazole moiety of axitinib occupies the ATP-binding pocket and interacts with Phe 1047 in the DFG motif. One amine in the indazole ring forms a hydrogen bond with the carbonyl oxygen group of Glu 917. The methyl benzamide moiety packs against the gatekeeper residue Val 916, and the amide group forms hydrogen bonds with the amide group of Asp 1046 and the carboxyl group of Glu 885.

One key difference between these structures is that the N-terminal portion of the juxtamembrane segment is folded back onto the kinase domain and partially occupies the specificity pocket when axitinib is bound (Fig. 2.8C). In contrast to sorafenib, which disrupts juxtamembrane interactions with the kinase domain, axitinib promotes a unique inhibitory conformation of VEGFR, with the juxtamembane segment folded against the

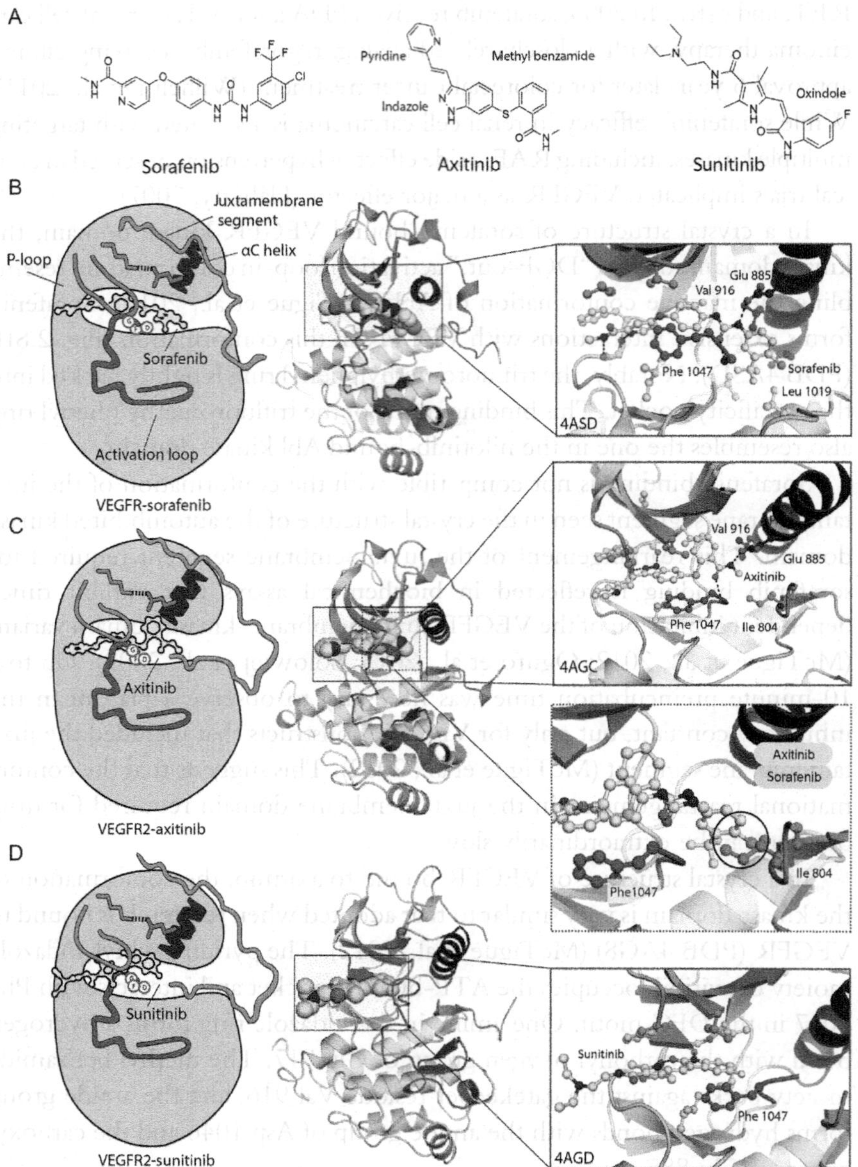

Figure 2.8 Inhibition of VEGFR by sorafenib, axitinib, and sunitinib. (A) Chemical structures of sorafenib, axitinib, and sunitinib. (B) A crystal structure of the VEGFR2 kinase domain bound to sorafenib. The juxtamembrane segment is colored orange. The zoomed-in view of the drug coordination is shown in the right panel. (C) A crystal structure of the VEGFR2 kinase domain bound to axitinib. A comparison of axitinib (yellow) to sorafenib (gray) in the VEGFR kinase domain is shown in the lower right panel. The black circle highlights where the juxtamembrane segment would clash with the phenyl group in sorafenib. (D) A crystal structure of the VEFGR kinase domain bound to sunitinib. (See the color plate.)

kinase domain (Fig. 2.8C). Ile 804 in the juxtamembrane segment of VEGFR is in close contact with the methyl group on axitinib. Interestingly, this interaction between the juxtamembrane segment, axitinib, and the kinase domain is consistent with the fact that axitinib is more potent against VEGFR variants containing the juxtamembrane segment than the kinase domain alone (McTigue et al., 2012; Solowiej et al., 2009). This likely contributes to axitinib being a more selective inhibitor for VEGFR than compounds identified previously.

Another VEGFR inhibitor that is in the clinic is sunitinib, which came out of a program that was aimed initially at the FGFR. A high-throughput screen against FGFR identified a scaffold, 3-substituted indolin-2-one (Sun et al., 1998, 1999, 2003). This small molecule potently inhibits receptor tyrosine kinases, and different substitution patterns on this scaffold result in distinct inhibitory profiles. In particular, sunitinib was found to have antiangiogenic and antitumor activities in clinical trials, due to inhibition of both the VEGFR and the PDGFR family of receptor tyrosine kinases (Laird et al., 2000).

In a crystal structure of sunitinib-bound VEGFR, the conformation of the kinase domain is almost identical to that observed when axinitib is bound (McTigue et al., 2012). Sunitinib occupies the smallest volume of the ATP-binding site among these three clinical drugs targeting VEGFR (Fig. 2.8D) (PDB 4AGD). It is also less potent and selective than axitinib and sorafenib. The oxindole moiety occupies the ATP-binding pocket, and the substituents make various interactions with residues in the kinase active site. While this inhibitor does bind to a "DFG-out" conformation of VEGFR, sunitinib can also bind to a "DFG-in" conformations of CDK2, ITK, and PAK6 (Gao et al., 2013; Kutach et al., 2010; Martin et al., 2012). As described earlier, the juxtamembrane segment stabilizes a "DFG-out" conformation of VEGFR, which likely constrains the interactions that sunitinib makes in the active site.

6.3. Crizotinib (Xalkori™) binds a "DFG-in" conformation of ALK and c-MET

Crizotinib is an inhibitor that makes minimal interactions with the ATP-binding pocket, and thus inhibits many kinases, including ALK and c-MET. In a crystal structure of crizotinib bound to the receptor tyrosine kinase ALK, the DFG motif is seen in a "DFG-in" conformation, and the activation loop is partially disordered. Crizotinib only occupies the ATP-binding pocket and does not extend into the specificity pocket.

Crizotinib also potently inhibits the receptor tyrosine kinase, c-MET. In the crizotinib bound c-MET structure, the position of crizotinib is largely comparable to the one in the ALK kinase structure described earlier. However, in the c-MET structure, the kinase domain adopts a "DFG-in" conformation and the activation loop lies against the kinase domain in an inactive conformation, which is distinct from the crizotinib bound to the ALK kinase domain.

6.4. Gefitinib (Iressa™) and erlotinib (Tarceva™) inhibit EGFR

EGFR and its family members are often overexpressed or contain activating mutations in patients with non-small cell lung cancer. Thus, EGFR family members were identified as targets early in the development of kinase inhibitors (Levitzki & Gazit, 1995; Levitzki & Mishani, 2006; Traxler & Furet, 1999). Several groups reported in close succession on a 4-anilinoquinazoline-based scaffold, which displayed increased specificity for EGFR family members relative to that exhibited by previously reported compounds (Barker et al., 2001; Moyer et al., 1997; Osherov & Levitzki, 1994; Rusnak et al., 2001; Ward et al., 1994; Fry et al., 1994). Four reversible small-molecule inhibitors based on this scaffold were clinically approved: gefitinib in 2003, erlotinib in 2005, lapatinib in 2007, and vandetanib in 2011. Lapatinib and vandetinib are discussed in more detail in later sections.

Gefitinib and erlotinib are both ATP-competitive inhibitors that not only exploit interactions important for ATP binding but also make extensive interactions within the selectivity pocket (Stamos, Sliwkowski, & Eigenbrot, 2002). In the structure of gefitinib-bound EGFR kinase domain, the quinazoline ring localizes to the ATP-binding pocket (Fig. 2.9B) (PDB 2ITY) (Yun et al., 2007). The N1 amine on the quinazoline ring forms hydrogen bond to the backbone amide of Met 769. The phenyl ring substituent extends deeper in the catalytic cleft, locating close to the gatekeeper residue Thr 790. This extension may provide an additional selectivity element for these EGFR inhibitors over the ones that only occupy the ATP-binding pocket. Additional substituents on the phenyl ring such as incorporation of a halogen (gefitinib) or an alkyne (erlotinib) group also enhance the binding of these inhibitors to the EGFR kinase domain.

6.5. Development of a more selective EGFR inhibitor: Lapatinib (Tykerb™)

Further medicinal chemistry surrounding this chemical scaffold led to even more potent and selective small-molecule inhibitors. In particular, lapatinib

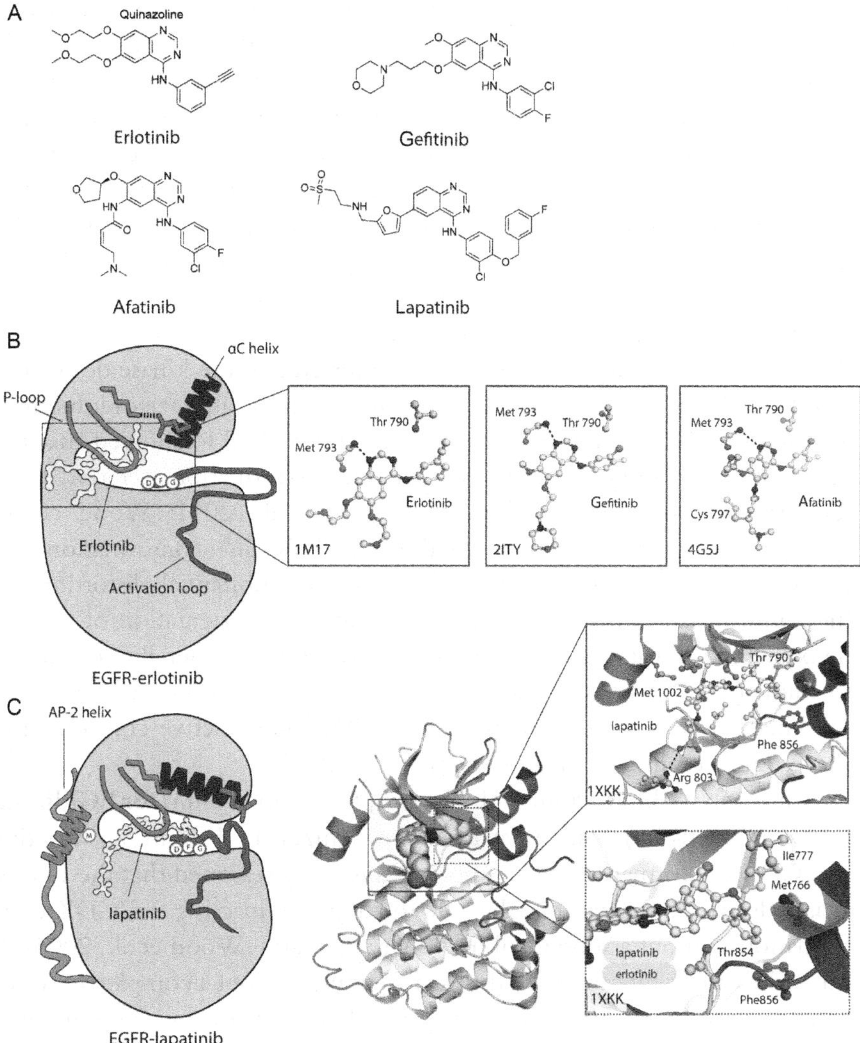

Figure 2.9 Inhibition of EGFR by lapatinib, erlotinib, gefitinib, and afatinib. (A) Chemical structures of erlotinib, gefitinib afatinib, and lapatinib. (B) A crystal structure of the EGFR kinase domain bound to erlotinib. The EGFR kinase domain is in an active conformation. Erlotinib, gefitinb, and afatinib all bind to this active conformation. (C) A crystal structure of the EGFR kinase domain bound to lapatinib. The EGFR kinase domain is in a Src/CDK-like inactive conformation. The C-terminal loop of the kinase domain and the AP-2 helix are colored orange. Met 1002 on the AP-2 helix is shown as an orange circle. A comparison of lapatinib (yellow) and erlotinib (gray) bound to the kinase domain of EGFR is displayed in the zoomed-in figure on the lower right panel. (See the color plate.)

was clinically approved in 2007 for the treatment of breast cancers that overexpress either EGFR or a related family member, Her2 (Frampton, 2009). The exquisite selectivity profile of lapatinib was, in part, due to more extensive interactions with the selectivity pocket, which has unique features in EGFR relative to the active site of other protein kinases (Karaman et al., 2008).

The extremely high degree of sequence similarity between EGFR family members means that there is considerable cross-reactivity of inhibitors that target this family. In the structure of erlotinib bound to EGFR, the kinase domain is in an active, "DFG-in" conformation, as also seen in the structure of gefitinib bound to EGFR (Stamos et al., 2002). The activation loop, however, is not phosphorylated; thus, it was inferred that the kinase domain of EGFR naturally favors the active conformation in the absence of inhibitory interactions. The structure of lapatinib bound to EGFR showed that the kinase domain adopts a distinctive inactive conformation, referred to as the Src/CDK inactive conformation (Jura et al., 2011; Wood et al., 2004). Although the DFG motif adopts a "DFG-in" conformation and points toward the ATP-binding site, there are substantial distortions in the conformation of the activation loop and in the orientation of the αC helix, which is rotated outward. While erlotinib is generally considered to bind to the active conformation, it has also been observed bound to the kinase domain of EGFR in this Src/CDK-like inactive conformation (Park, Liu, Lemmon, & Radhakrishnan, 2012).

Lapatinib is a high-affinity inhibitor ($K_i^{app} = 3$ and 13 nM for EGFR and Her2, respectively). Since lapatinib recognizes EGFR in an inactive Src/CDK conformation, the authors of this study reasoned that the EGFR kinase domain must naturally be able to adopt the inactive Src/CDK conformation without a substantial free-energy penalty (Wood et al., 2004). If this is the case, why can the EGFR kinase domain adopt a completely active conformation in the presence of erlotinib? The answer came from the realization that EGFR requires the formation of an asymmetric dimer of kinase domains to activate and that the crystal lattice of the erlotinib–EGFR complex drives the formation of the asymmetric dimer, which is otherwise not favored by the isolated kinase domain (Zhang et al., 2006). This idea was verified by the crystallization of a variant form of EGFR kinase domain in which the asymmetric dimer interface was mutated (Zhang et al., 2006). This variant EGFR kinase domain adopts an inactive Src/CDK conformation with no drug bound, and is virtually superimposable on the conformation of the lapatinib–EGFR complex. Thus, lapatinib does recognize an intrinsically stable, inactive conformation of EGFR.

A distinct feature of lapatinib is the presence of a 3-fluorobenzyloxy group attached to a 3-chloroaniline group off the quinazoline ring. In the cocrystal structure of lapatinib-bound EGFR kinase domain, this 3-fluorobenzyloxy moiety is inserted deep into a hydrophobic pocket adjacent to the ATP-binding pocket (Fig. 2.9C) (PDB 1XKK) (Wood et al., 2004). This hydrophobic pocket is formed by the gatekeeper residue Thr 790, Leu 777 on the β-sheet in the N lobe, Met 766 on the αC helix, Phe 856 from the DFG motif, and Leu 858 from the activation loop. In addition, Met 1002 on the AP-2 helix of EGFR coordinates lapatinib binding. This is in contrast to the crystal structures of other EGFR-inhibitor complex structures, in which the AP-2 helix is either disordered or not present.

While lapatinib does form extensive interactions with the active site of EGFR, one puzzling feature is its slow off-rate. For example, the half-life of a lapatinib–EGFR complex is estimated to be approximately 300 minutes, whereas the half-life for the gefitinib and erlotinib complexes with EGFR is less than 10 minutes (Wood et al., 2004). While lapatinib displays a slow off-rate for EGFR, it is interesting to note that sorafenib exhibits a slow on-rate against VEGFR. These slow-binding kinetics may suggest that dissociation of lapatinib from EGFR and association of sorafenib with VEGFR may require slow conformational changes. A simulation study of EGFR suggests that intrinsic disorder in the inactive conformation of the EGFR kinase domain is responsible for the slow-binding kinetics of lapatinib (Shan et al., 2012).

6.6. Afatinib (Gilotrif™): A selective kinase inhibitor that reacts covalently

Researchers soon began looking for additional selectivity filters to exploit to impart greater specificity and efficacy in disease states and possibly decrease toxicity associated with off-target effects. The presence of a cysteine residue (Cys 797) in the active site of EGFR inspired an approach of appending Michael acceptors to the anilinoquinazoline core to generate irreversible inhibitors that target Cys 797 on EGFR (Liu et al., 2013). This led eventually to the clinically approved drug, afatinib, which reacts irreversibly with Cys 797 in EGFR.

In the crystal structure of afatinib bound to the EGFR kinase domain, residue Cys 797 is linked covalently to the acrylamide moiety in afatinib (Fig. 2.9B) (PDB 4G5J) (Solca et al., 2012). Importantly, this covalent linkage does not disturb the canonical binding mode of the phenyl-quinazoline moiety of afatinib in the EGFR kinase domain. The corresponding scaffold in the parent compound, gefitinib, interacts with similar residues in the EGFR kinase domain. Notably, this cysteine residue in the active site of EGFR likely plays a physiological role in receptor activation. EGFR-mediated signaling events lead to production of hydrogen peroxide and subsequently, sulfenylation of Cys 797 (Paulsen et al., 2011). This posttranslational modification results in an increase in the tyrosine kinase activity of EGFR.

Several irreversible inhibitors that target protein kinases are now either in clinical trials or already clinically approved (Liu et al., 2013). The formation of a covalent bond with an active site cysteine residue that is not highly conserved provides another rational approach for the development of selective inhibitors (Cohen, Zhang, Shokat, & Taunton, 2005).

6.7. Vandetanib (Caprelsa™): A quinazoline analog that inhibits RET

Quinazolines were found to be inhibitors of not just EGFR, but many other receptor tyrosine kinases as well. For example, one reversible inhibitor with the quinazoline core scaffold potently inhibits both EGFR and VEGFR in addition to the rearranged during transfection (RET) receptor tyrosine kinase (Hennequin et al., 2002; Carlomagno et al., 2002). This small molecule inhibitor, termed vandetanib, was clinically approved in 2011 for treatment of medullary thyroid cancer, where activating RET mutations are often observed. Perhaps not surprisingly, the interactions between vandetanib and the RET kinase domain resemble those observed in the cocrystal structure of the related quinazoline analog, gefitinib, bound to the EGFR kinase domain (Knowles et al., 2006).

In a cocrystal structure of the RET kinase domain bound to vandetanib, the kinase domain adopts an active conformation, with the DFG motif in the "DFG-in" conformation, and the activation loop extended out fully (Fig. 2.10B) (PDB: 2IVU). The salt bridge between Glu 775 in the αC helix and Lys 758 in the N lobe is properly formed. The quinazoline ring of vandetanib localizes to the ATP binding pocket with the phenyl ring substituent packing against the gatekeeper residue Val 804 in the selectivity pocket. The quinazoline ring seems to utilize similar binding modes in the kinase active site, as demonstrated for EGFR and RET; however, distinct substituents on this ring allows the inhibitor to exploit differences among the active sites of receptor tyrosine kinases.

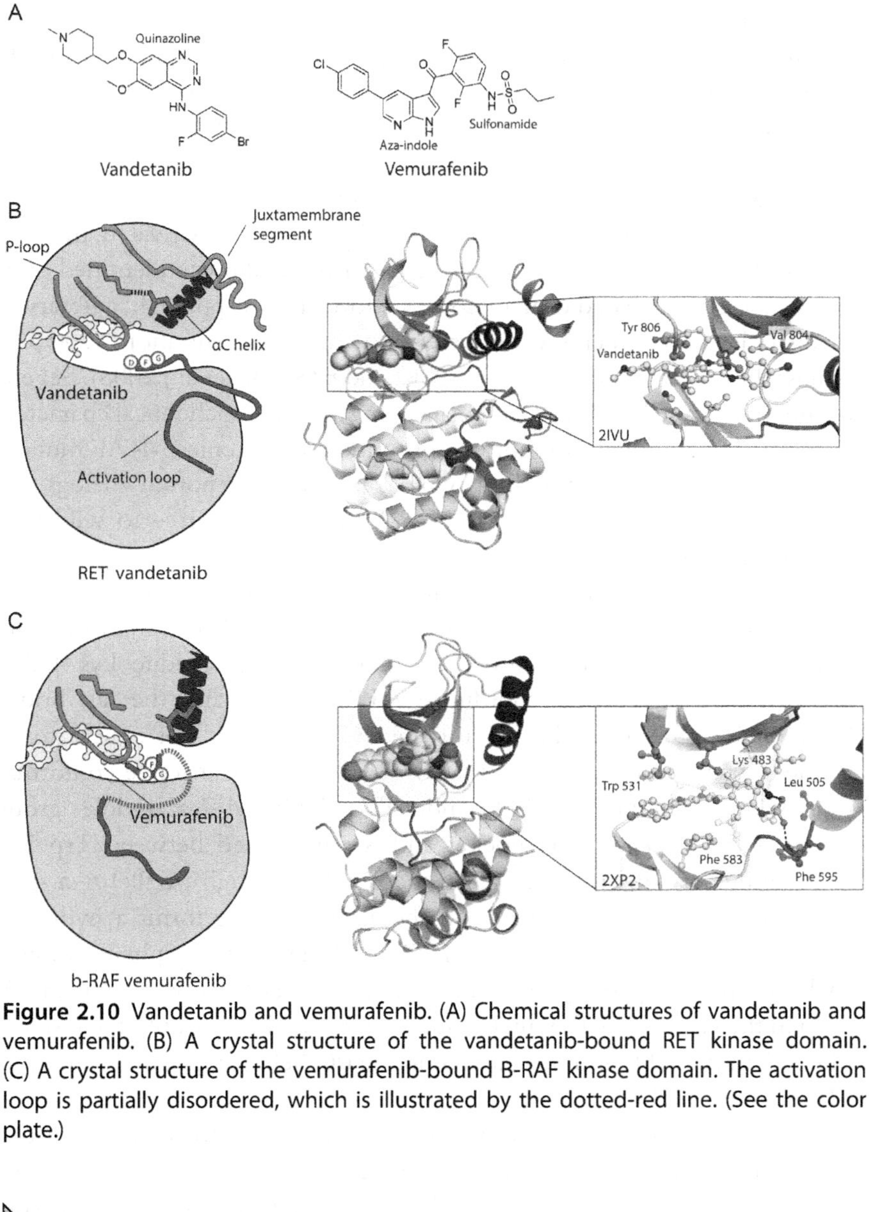

Figure 2.10 Vandetanib and vemurafenib. (A) Chemical structures of vandetanib and vemurafenib. (B) A crystal structure of the vandetanib-bound RET kinase domain. (C) A crystal structure of the vemurafenib-bound B-RAF kinase domain. The activation loop is partially disordered, which is illustrated by the dotted-red line. (See the color plate.)

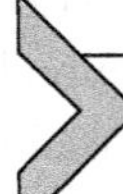

7. VEMURAFENIB (ZELBORAF™) BINDS TO A "DFG-in" CONFORMATION IN THE Ser/Thr KINASE RAF

RAF is a Ser/Thr kinase integral to the RAS/RAF/MEK/ERK signaling pathway that is activated in response to growth factors. Activation of this pathway stimulates cellular growth and proliferation and is often left unchecked in cancer cells. In fact, B-RAF is mutated to more active variants in over 40% of malignant melanomas (Shaw et al., 2014). Thus, inhibitors that target RAF provide viable strategies to combat melanoma.

An exciting development in the treatment of certain melanomas was the realization that small molecules could selectively target an activated, mutated form of B-RAF. The 7-azaindole core of vemurafenib was identified originally in a scaffold-based drug discovery effort. In this approach, a library of small-molecule fragments were screened at high concentrations (200 μ*M*) against three kinase models, Pim1, CSK, and p38 (Tsai et al., 2008). Several rounds of cocrystallization, medicinal chemistry, and biochemical characterization led to a potent inhibitor of a prominent oncogenic B-RAF mutant, V600E. Notably, vemurafenib as well as previously reported analogs also demonstrated some selectivity for B-RAF (V600E) relative to wild-type B-RAF (Bollag et al., 2010).

In a crystal structure of vemurafenib-bound mutant B-RAF (V600E) kinase domain, the kinase domain adopts a "DFG-in" conformation, but the activation loop is disordered (Fig. 2.10C). Residue Lys 483 is not oriented to properly coordinate ATP binding, and the salt bridge between Lys 483 and Glu 501 on the αC helix does not appear to be formed. These features suggest that B-RAF adopts an inactive conformation when vemurafenib is bound. The 5-phenyl-7-azaindole group occupies the ATP-binding pocket and is sandwiched between Trp 531 in the hinge region and Phe 583 in the C lobe through parallel π–π stacking interactions. The nitrogen on the azaindole ring forms a hydrogen bond with the carbonyl oxygen of Gln 530. The difluorophenyl moiety packs against the gatekeeper residue, Thr 529. The propyl group inserts deep in the catalytic cleft, which is surrounded by the hydrophobic residues Phe 595 in the DFG motif, Leu 505 on the αC helix, and Leu 527 and Phe 516 on the N lobe.

The mechanism by which these inhibitors block RAF signaling turned out to be unexpectedly complex, leading to an appreciation of the importance of homo- and heterodimerization in signaling by RAF kinases (Garnett, Rana, Paterson, Barford, & Marais, 2005; Hu et al., 2013). Several B-RAF inhibitors promote conformational changes in the kinase domain upon binding to the kinase active site. These structural changes enhance homo- and heterodimerization of RAF isoforms, leading to membrane localization and activation (Hatzivassiliou et al., 2010). The shift in the αC helix away from the active site observed in the cocrystal structure of mutant B-RAF (V600E) with vemurafenib affects mutant B-RAF (V600E) dimerization. This conformational change is not, however, sufficient to block the activity of the RAF dimer (Bollag et al., 2012).

8. INHIBITORS THAT OCCUPY POCKETS OTHER THAN THE ATP-BINDING SITE

Most of the clinically approved inhibitors bind to the cleft between the N lobe and C lobe of the kinase domain, which overlaps with the ATP-binding pocket. Several inhibitors that target additional binding pockets on the kinase domain also show promising results. Unlike the ATP-binding pocket, these alternative drug-binding pockets can be distinct in shape, volume, and residue composition. Conceptually, they represent a very exciting approach for developing next-generation, high-selectivity kinase inhibitors. Here, we summarize three representative cases for this mode of kinase inhibition.

8.1. Benzothiazines bind an allosteric site in focal adhesion kinase

Focal adhesion kinase (FAK) is a nonreceptor tyrosine kinase that is overexpressed in many cancers. A high-throughput screen against FAK, specifically searching for compounds that inhibited kinase activity under high concentrations of ATP, resulted in the identification of the benzothiazine compounds. Surprisingly, in a crystal structure of FAK3 bound to a benzothiazine derivative, the benzothiazine moiety occupies a pocket located in the C lobe of the kinase domain (PDB 4I4F) (Tomita et al., 2013). The phenyl moiety of this small molecule is located in the specificity pocket that is underneath the αC helix (Fig. 2.11B). The N terminus of the activation loop is distorted and inserted into the ATP-binding pocket. As a result, binding of the inhibitor may prevent the binding of both ATP and peptide substrates by trapping the activation loop in this distorted conformation.

8.2. PD318088 binds to MEK noncompetitively with ATP

PD318088 was identified as a highly selective, potent inhibitor of MEK1 and MEK2. Interestingly, this small molecule binds noncompetitively with

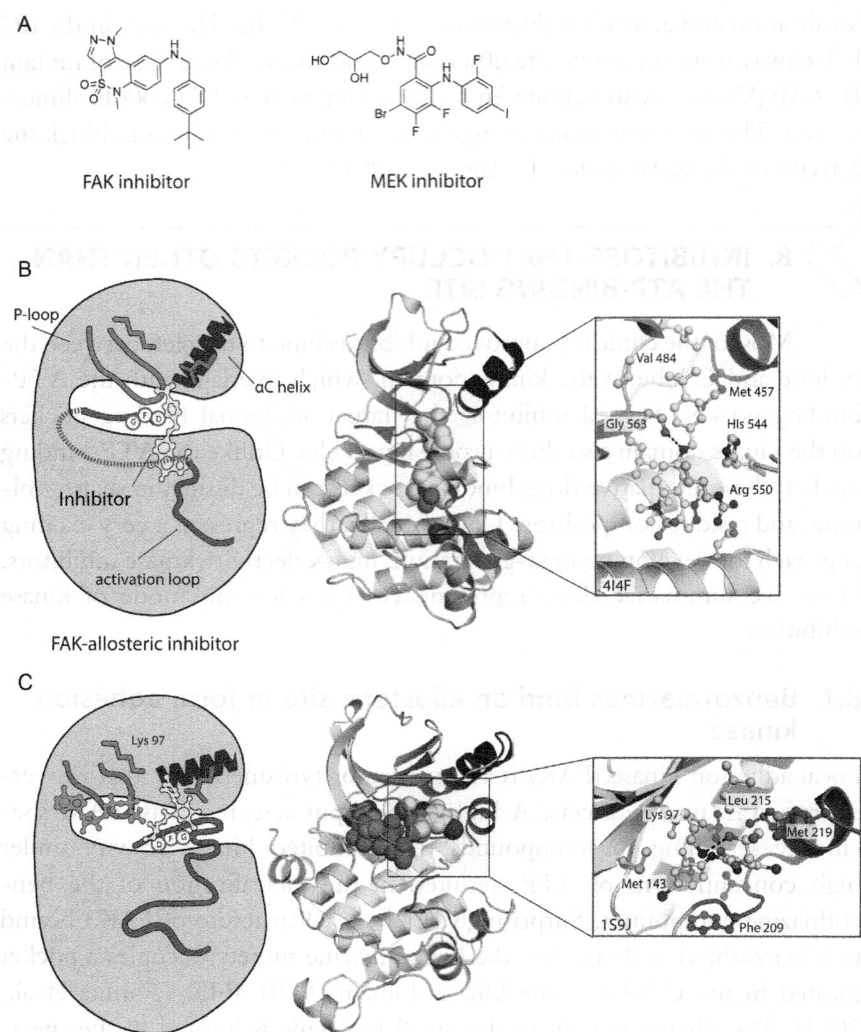

Figure 2.11 Inhibition of protein kinases by non-ATP-competitive compounds. (A) A crystal structure of the FAK kinase domain bound to an allosteric inhibitor. The activation loop is partially disordered, which is shown as a dotted-red line. (B) A crystal structure of the MEK1 kinase domain bound to an allosteric inhibitor and ATP. (See the color plate.)

ATP (Sebolt-Leopold et al., 1999). In a crystal structure of the MEK1 kinase domain, both ATP and PD318088 are bound simultaneously (PDB 1S9J) (Fig. 2.11C). The inhibitor lies in a pocket formed by the αC helix, the activation loop, and the N-lobe β-sheets (Ohren et al., 2004). This pocket is

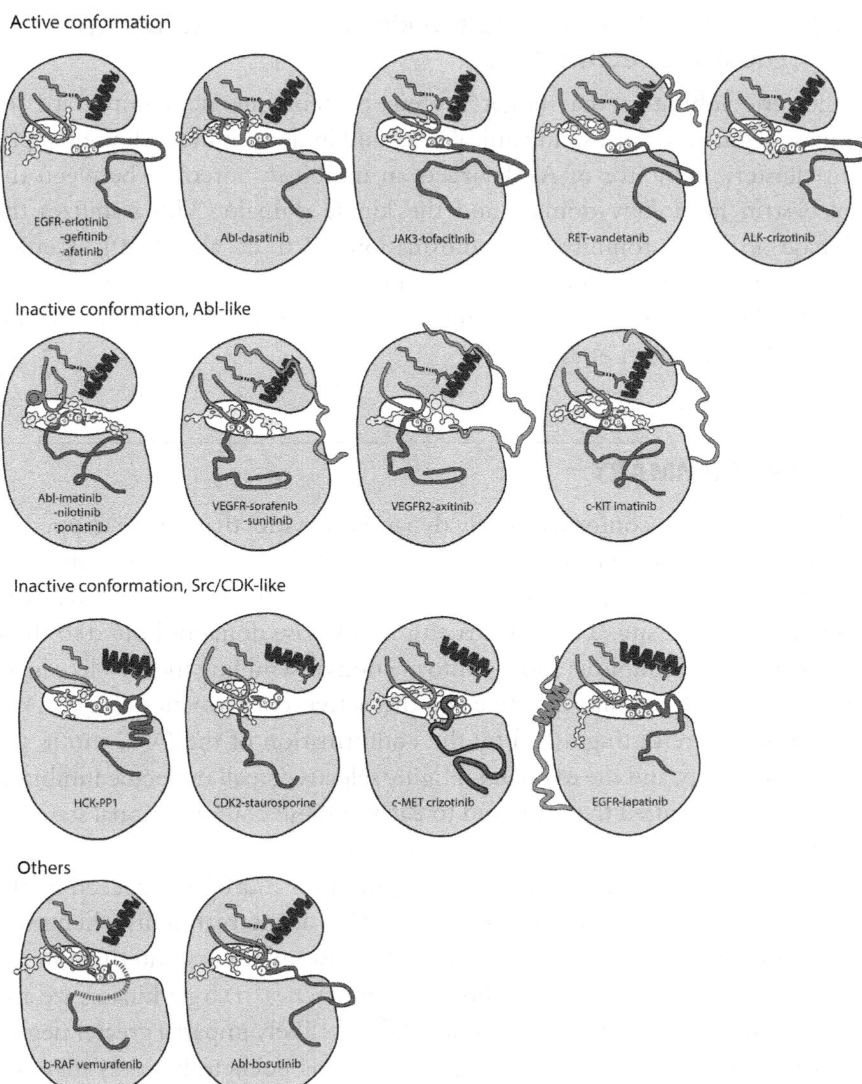

Figure 2.12 Summary of the kinase conformations that are targeted by clinically approved drugs. (See the color plate.)

next to, but not overlapped with the ATP-binding pocket. Although the DFG motif is in a "DFG-in" conformation, the activation loop interacts with the inhibitor to restrict peptide substrate binding. Many other clinically approved drugs, such as lapatinib, also use this alternative pocket in conjunction with the ATP-binding pocket to inhibit their targets.

8.3. Inhibitors that bind to the kinase domain to disrupt substrate recruitment

Allosteric inhibitors that target protein–protein interfaces important for kinase regulation are also in clinical trials and in development. For example, an allosteric inhibitor of Akt1 targets an inhibitory interface between the pleckstrin homology domain and the kinase domain. This stabilizes the kinase in an autoinhibited conformation (Wu et al., 2010). Certain anilino-naphthalene sulfonate compounds bind to the N lobe of the CDK2 kinase domain, and disrupt binding of cyclin A onto the N lobe of the CDK2 (Betzi et al., 2011).

9. SUMMARY

Kinases are conformationally dynamic enzymes that can be trapped by small molecules in many distinct states. Clinically approved kinase inhibitors block access of ATP, protein substrates, or both by binding competitively to the kinase active site. Thus far, structures of kinase domains bound to these small molecule inhibitors adopt conformations resembling the Abl-like inactive, the Src/CDK-like inactive, or the active conformation (Fig. 2.12). These states are distinguished by the conformation of the DFG motif, the activation loop, and the αC helix. Highly selective small molecule inhibitors have been identified that can bind to each of these conformational states by exploiting subtle differences among the active site of kinases.

While absolute selectivity is not required for a drug to succeed in the clinic, it does provide a rational path forward to adapt treatments when resistance mutations arise. As the number of kinase inhibitors entering clinical trials increases rapidly, many orthogonal approaches to target kinases are also becoming prevalent. These allosteric inhibitors likely impart a greater degree of selectivity by utilizing more divergent binding pockets, but they may also be more susceptible to resistance mutations. The clinical efficacy of these various modes of inhibition will inform future drug design efforts.

ACKNOWLEDGMENTS

We thank Tiago Barros and Alexander Warkentin for advice and critical feedback on this review. Q. W. is supported by an Irvington Institute postdoctoral fellowship from the Cancer Research Institute. J. A. Z. is supported by a postdoctoral fellowship from the National Institutes of Health National Cancer Institute (F32 CA177087-02).

REFERENCES

Adams, J. A. (2001). Kinetic and catalytic mechanisms of protein kinases. *Chemical Reviews*, *101*, 2271–2290.

Anastassiadis, T., Deacon, S. W., Devarajan, K., Ma, H., & Peterson, J. R. (2011). Comprehensive assay of kinase catalytic activity reveals features of kinase inhibitor selectivity. *Nature Biotechnology*, *29*, 1039–1045.

Barker, A. J., Gibson, K. H., Grundy, W., Godfrey, A. A., Barlow, J. J., Healy, M. P., et al. (2001). Studies leading to the identification of ZD1839 (IRESSA): An orally active, selective epidermal growth factor receptor tyrosine kinase inhibitor targeted to the treatment of cancer. *Bioorganic & Medicinal Chemistry Letters*, *11*, 1911–1914.

Barnett, S. F., Defeo-Jones, D., Fu, S., Hancock, P. J., Haskell, K. M., Jones, R. E., et al. (2005). Identification and characterization of pleckstrin-homology-domain-dependent and isoenzyme-specific Akt inhibitors. *The Biochemical Journal*, *385*, 399–408.

Betzi, S., Alam, R., Martin, M., Lubbers, D. J., Han, H., Jakkaraj, S. R., et al. (2011). Discovery of a potential allosteric ligand binding site in CDK2. *ACS Chemical Biology*, *6*, 492–501.

Blume-Jensen, P., & Hunter, T. (2001). Oncogenic kinase signalling. *Nature*, *411*, 355–365.

Bollag, G., Hirth, P., Tsai, J., Zhang, J., Ibrahim, P. N., Cho, H., et al. (2010). Clinical efficacy of a RAF inhibitor needs broad target blockade in BRAF-mutant melanoma. *Nature*, *467*, 596–599.

Bollag, G., Tsai, J., Zhang, J., Zhang, C., Ibrahim, P., Nolop, K., et al. (2012). Vemurafenib: The first drug approved for BRAF-mutant cancer. *Nature Reviews. Drug Discovery*, *11*, 873–886.

Boschelli, D. H., Ye, F., Wang, Y. D., Dutia, M., Johnson, S. L., Wu, B., et al. (2001). Optimization of 4-phenylamino-3-quinolinecarbonitriles as potent inhibitors of Src kinase activity. *Journal of Medicinal Chemistry*, *44*, 3965–3977.

Brewer, M. R., Choi, S. H., Alvarado, D., Moravcevic, K., Pozzi, A., Lemmon, M. A., et al. (2009). The juxtamembrane region of the EGF receptor functions as an activation domain. *Molecular Cell*, *34*, 641–651.

Capdeville, R., Buchdunger, E., Zimmermann, J., & Matter, A. (2002). Glivec (STI571, imatinib), a rationally developed, targeted anticancer drug. *Nature Reviews. Drug Discovery*, *1*, 493–502.

Carlomagno, F., Vitagliano, D., Guida, T., Ciardiello, F., Giampaolo, T., Vecchio, G., et al. (2002). ZD6474, an orally available inhibitor of KDR tyrosine kinase activity, efficiently blocks oncogenic RET kinases. *Cancer Research*, *62*, 7284–7290.

Castagna, M., Takai, Y., Kaibuchi, K., Sano, K., Kikkawa, U., & Nishizuka, Y. (1982). Direct activation of calcium-activated, phospholipid-dependent protein kinase by tumor-promoting phorbol esters. *The Journal of Biological Chemistry*, *257*, 7847–7851.

Chao, L. H., Stratton, M. M., Lee, I.-H., Rosenberg, O. S., Levitz, J., Mandell, D. J., et al. (2011). A mechanism for tunable autoinhibitionin the structure of a human Ca. *Cell*, *146*, 732–745.

Chiu, Y.-Y., Lin, C.-T., Huang, J.-W., Hsu, K.-C., Tseng, J.-H., You, S.-R., et al. (2013). KIDFamMap: A database of kinase-inhibitor-disease family maps for kinase inhibitor selectivity and binding mechanisms. *Nucleic Acids Research*, *41*, D430–D440.

Churchill, E. N., Qvit, N., & Mochly-Rosen, D. (2009). Rationally designed peptide regulators of protein kinase C. *Trends in Endocrinology and Metabolism*, *20*, 25–33.

Cohen, P. (2002). Protein kinases—The major drug targets of the twenty-first century? *Nature Reviews. Drug Discovery*, *1*, 309–315.

Cohen, P., & Alessi, D. R. (2013). Kinase drug discovery—What's next in the field? *ACS Chemical Biology*, *8*, 96–104.

Cohen, M. S., Zhang, C., Shokat, K. M., & Taunton, J. (2005). Structural bioinformatics-based design of selective, irreversible kinase inhibitors. *Science, 308*, 1318–1321.

Collett, M. S., & Erikson, R. L. (1978). Protein kinase activity associated with the avian sarcoma virus src gene product. *Proceedings of the National Academy of Sciences of the United States of America, 75*, 2021–2024.

Dar, A. C., Lopez, M. S., & Shokat, K. M. (2008). Brief communication. *Chemistry & Biology, 15*, 1015–1022.

Dar, A. C., & Shokat, K. M. (2011). The evolution of protein kinase inhibitors from antagonists to agonists of cellular signaling. *Annual Review of Biochemistry, 80*, 769–795.

Das, J., Chen, P., Norris, D., Padmanabha, R., Lin, J., Moquin, R. V., et al. (2006). 2-Aminothiazole as a novel kinase inhibitor template. Structure-activity relationship studies toward the discovery of N-(2-chloro-6-methylphenyl)-2-[[6-[4-(2-hydroxyethyl)-1-piperazinyl)]-2-methyl-4-pyrimidinyl]amino)]-1,3-thiazole-5-carboxamide (dasatinib, BMS-354825) as a potent pan-Src kinase inhibitor. *Journal of Medicinal Chemistry, 49*, 6819–6832.

Daub, H., Specht, K., & Ullrich, A. (2004). Strategies to overcome resistance to targeted protein kinase inhibitors. *Nature Reviews. Drug Discovery, 3*, 1001–1010.

Davies, S. P., Reddy, H., Caivano, M., & Cohen, P. (2000). Specificity and mechanism of action of some commonly used protein kinase inhibitors. *The Biochemical Journal, 351*, 95–105.

Davis, M. I., Hunt, J. P., Herrgard, S., Ciceri, Pietro, Wodicka, L. M., Pallares, G., et al. (2011). Comprehensive analysis of kinase inhibitor selectivity. *Nature Biotechnology, 29*, 1046–1051.

Druker, B. J., Tamura, S., Buchdunger, E., Ohno, S., Segal, G. M., Fanning, S., et al. (1996). Effects of a selective inhibitor of the Abl tyrosine kinase on the growth of Bcr|[ndash]|Abl positive cells. *Nature Medicine, 2*, 561–566.

Eckhart, W., Hutchinson, M. A., & Hunter, T. (1979). An activity phosphorylating tyrosine in polyoma T antigen immunoprecipitates. *Cell, 18*, 925–933.

Ellis, L. M., & Hicklin, D. J. (2008). VEGF-targeted therapy: Mechanisms of anti-tumour activity. *Nature Reviews. Cancer, 8*, 579–591.

Engel, M., Hindie, V., Lopez-Garcia, L. A., Stroba, A., Schaeffer, F., Adrian, I., et al. (2006). Allosteric activation of the protein kinase PDK1 with low molecular weight compounds. *The EMBO Journal, 25*, 5469–5480.

Engh, R. A., & Bossemeyer, D. (2002). Structural aspects of protein kinase control—Role of conformational flexibility. *Pharmacology and Therapeutics, 93*, 99–111.

Fabian, M. A., Biggs, W. H., Treiber, D. K., Atteridge, C. E., Azimioara, M. D., Benedetti, M. G., et al. (2005). A small molecule-kinase interaction map for clinical kinase inhibitors. *Nature Biotechnology, 23*, 329–336.

Flaherty, K. T. (2007). Sorafenib in renal cell carcinoma. *Clinical Cancer Research, 13*, 747s–752s.

Frampton, J. E. (2009). Lapatinib: A review of its use in the treatment of HER2-overexpressing, trastuzumab-refractory, advanced or metastatic breast cancer. *Drugs, 69*, 2125–2148.

Fry, D. W., Kraker, A. J., McMichael, A., Ambroso, L. A., Nelson, J. M., Leopold, W. R., et al. (1994). A specific inhibitor of the epidermal growth factor receptor tyrosine kinase. *Science, 265*, 1093–1095.

Gao, J., Ha, B. H., Lou, H. J., Morse, E. M., Zhang, R., Calderwood, D. A., et al. (2013). Substrate and inhibitor specificity of the type II p21-activated kinase, pak6. *PLoS One, 8*, e77818.

Garnett, M. J., Rana, S., Paterson, H., Barford, D., & Marais, R. (2005). Wild-type and mutant B-RAF activate C-RAF through distinct mechanisms involving heterodimerization. *Molecular Cell, 20*, 963–969.

Golas, J. M., Arndt, K., Etienne, C., Lucas, J., Nardin, D., Gibbons, J., et al. (2003). SKI-606, a 4-anilino-3-quinolinecarbonitrile dual inhibitor of Src and Abl kinases, is a potent antiproliferative agent against chronic myelogenous leukemia cells in culture and causes regression of K562 xenografts in nude mice. *Cancer Research*, *63*, 375–381.
Gower, C. M., Chang, M. E. K., & Maly, D. J. (2014). Bivalent inhibitors of protein kinases. *Critical Reviews in Biochemistry and Molecular Biology*, *49*, 102–115.
Griffith, J., Black, J., Faerman, C., Swenson, L., Wynn, M., Lu, F., et al. (2004). The structural basis for autoinhibition of FLT3 by the juxtamembrane domain. *Molecular Cell*, *13*, 169–178.
Hanke, J. H., Gardner, J. P., Dow, R. L., Changelian, P. S., Brissette, W. H., Weringer, E. J., et al. (1996). Discovery of a novel, potent, and Src family-selective tyrosine kinase inhibitor. Study of Lck- and FynT-dependent T cell activation. *The Journal of Biological Chemistry*, *271*, 695–701.
Hatzivassiliou, G., Song, K., Yen, I., Brandhuber, B. J., Anderson, D. J., Alvarado, R., et al. (2010). RAF inhibitors prime wild-type RAF to activate the MAPK pathway and enhance growth. *Nature*, *464*, 431–435.
Hennequin, L. F., Stokes, E. S. E., Thomas, A. P., Johnstone, C., Ple, P. A., Ogilvie, D. J., et al. (2002). Novel 4-anilinoquinazolines with C-7 basic side chains: Design and structure activity relationship of a series of potent, orally active, VEGF receptor tyrosine kinase inhibitors. *Journal of Medicinal Chemistry*, *45*, 1300–1312.
Hirota, S., Isozaki, K., Moriyama, Y., Hashimoto, K., Nishida, T., Ishiguro, S., et al. (1998). Gain-of-function mutations of c-Kit in human gastrointestinal stromal tumors. *Science*, *279*, 577–580.
Hu, J., Stites, E. C., Yu, H., Germino, E. A., Meharena, H. S., Stork, P. J. S., et al. (2013). Allosteric activation of functionally asymmetric RAF kinase dimers. *Cell*, *154*, 1036–1046.
Hubbard, S. R. (1997). Crystal structure of the activated insulin receptor tyrosine kinase in complex with peptide substrate and ATP analog. *The EMBO Journal*, *16*, 5572–5581.
Hubbard, S. R. (2004). Juxtamembrane autoinhibition in receptor tyrosine kinases. *Nature Reviews. Molecular Cell Biology*, *5*, 464–471.
Hubbard, S. R., Mohammadi, M., & Schlessinger, J. (1998). Autoregulatory mechanisms in protein-tyrosine kinases. *The Journal of Biological Chemistry*, *273*, 11987–11990.
Hubbard, S. R., Wei, L., Ellis, L., & Hendrickson, W. A. (1994). Crystal structure of the tyrosine kinase domain of the human insulin receptor. *Nature*, *372*, 746–754.
Hunter, T. (1995). Protein kinases and phosphatases: The yin and yang of protein phosphorylation and signaling. *Cell*, *80*, 225–236.
Huse, M., & Kuriyan, J. (2002). The conformational plasticity of protein kinases. *Cell*, *109*, 275–282.
Jeffrey, P. D., Russo, A. A., Polyak, K., Gibbs, E., Hurwitz, J., Massagué, J., et al. (1995). Mechanism of CDK activation revealed by the structure of a cyclinA-CDK2 complex. *Nature*, *376*, 313–320.
Johnson, L. N. (2009a). The regulation of protein phosphorylation. *Biochemical Society Transactions*, *37*, 627–641.
Johnson, L. N. (2009b). Protein kinase inhibitors: Contributions from structure to clinical compounds. *Quarterly Reviews of Biophysics*, *42*, 1.
Jura, N., Endres, N. F., Engel, K., Deindl, S., Das, R., Lamers, M. H., et al. (2009). Mechanism for activation of the EGF receptor catalytic domain by the juxtamembrane segment. *Cell*, *137*, 1293–1307.
Jura, N., Zhang, X., Endres, N. F., Seeliger, M. A., Schindler, T., & Kuriyan, J. (2011). Catalytic control in the EGF receptor and its connection to general kinase regulatory mechanisms. *Molecular Cell*, *42*, 9–22.

Karaman, M. W., Herrgard, S., Treiber, D. K., Gallant, P., Atteridge, C. E., Campbell, B. T., et al. (2008). A quantitative analysis of kinase inhibitor selectivity. *Nature Biotechnology*, *26*, 127–132.

Knight, Z. A., Lin, H., & Shokat, K. M. (2010). *Nature Reviews Cancer*, *10*(2), 130–137.

Knight, Z. A., & Shokat, K. M. (2005). Features of selective kinase inhibitors. *Chemistry & Biology*, *12*, 621–637.

Knighton, D. R., Zheng, J. H., Ten Eyck, L. F., Ashford, V. A., Xuong, N. H., Taylor, S. S., et al. (1991). Crystal structure of the catalytic subunit of cyclic adenosine monophosphate-dependent protein kinase. *Science*, *253*, 407–414.

Knowles, P. P., Murray-Rust, J., Kjaer, S., Scott, R. P., Hanrahan, S., Santoro, M., et al. (2006). Structure and chemical inhibition of the RET tyrosine kinase domain. *The Journal of Biological Chemistry*, *281*(44), 33577–33587.

Kornev, A. P., Haste, N. M., Taylor, S. S., & Eyck, L. F. T. (2006). Surface comparison of active and inactive protein kinases identifies a conserved activation mechanism. *Proceedings of the National Academy of Sciences of the United States of America*, *103*, 17783–17788.

Kornev, A. P., Taylor, S. S., & Ten Eyck, L. F. (2008). A helix scaffold for the assembly of active protein kinases. *Proceedings of the National Academy of Sciences*, *105*, 14377–14382.

Krishnamurty, R., & Maly, D. J. (2010). Biochemical mechanisms of resistance to small-molecule protein kinase inhibitors. *ACS Chemical Biology*, *5*, 121–138.

Kuriyan, J., & Eisenberg, D. (2007). The origin of protein interactions and allostery in colocalization. *Nature*, *450*, 983–990.

Kutach, A. K., Villaseñor, A. G., Lam, D., Belunis, C., Janson, C., Lok, S., et al. (2010). Crystal structures of IL-2-inducible T cell kinase complexed with inhibitors: insights into rational drug design and activity regulation. *Chemical Biology & Drug Design*, *76*, 154–163.

Laird, A. D., Vajkoczy, P., Shawver, L. K., Thurnher, A., Liang, C., Mohammadi, M., et al. (2000). SU6668 is a potent antiangiogenic and antitumor agent that induces regression of established tumors. *Cancer Research*, *60*, 4152–4160.

Laudet, B., Barette, C., Dulery, V., Renaudet, O., Dumy, P., Metz, A., et al. (2007). Structure-based design of small peptide inhibitors of protein kinase CK2 subunit interaction. *The Biochemical Journal*, *408*, 363–373.

Lawrie, A. M., Noble, M. E., Tunnah, P., Brown, N. R., Johnson, L. N., & Endicott, J. A. (1997). Protein kinase inhibition by staurosporine revealed in details of the molecular interaction with CDK2. *Nature Structural Biology*, *4*, 796–801.

Lemmon, M. A., & Schlessinger, J. (2010). Cell signaling by receptor tyrosine kinases. *Cell*, *141*, 1117–1134.

Levinson, N. M., & Boxer, S. G. (2014). A conserved water-mediated hydrogen bond network defines bosutinib's kinase selectivity. *Nature Chemical Biology*, *10*, 127–132.

Levinson, A. D., Oppermann, H., Levintow, L., Varmus, H. E., & Bishop, J. M. (1978). Evidence that the transforming gene of avian sarcoma virus encodes a protein kinase associated with a phosphoprotein. *Cell*, *15*, 561–572.

Levitzki, A. (2003). Protein kinase inhibitors as a therapeutic modality. *Accounts of Chemical Research*, *36*, 462–469.

Levitzki, A., & Gazit, A. (1995). Tyrosine kinase inhibition: An approach to drug development. *Science*, *267*, 1782–1788.

Levitzki, A., & Mishani, E. (2006). Tyrphostins and other tyrosine kinase inhibitors. *Annual Review of Biochemistry*, *75*, 93–109.

Littlefield, P., Moasser, M. M., & Jura, N. (2014). An ATP-competitive inhibitor modulates the allosteric function of the HER3 pseudokinase. *Chemistry & Biology*, *21*, 453–458.

Liu, Y., & Gray, N. S. (2006). Rational design of inhibitors that bind to inactive kinase conformations. *Nature Chemical Biology*, *2*, 358–364.

Liu, Q., Sabnis, Y., Zhao, Z., Zhang, T., Buhrlage, S. J., Jones, L. H., et al. (2013). Developing irreversible inhibitors of the protein kinase cysteinome. *Chemistry & Biology, 20,* 146–159.

Lydon, N. B., & Druker, B. J. (2004). Lessons learned from the development of imatinib. *Leukemia Research, 28*(Suppl. 1), S29–S38.

Manley, P. W., Cowan-Jacob, S. W., & Mestan, J. (2005). Advances in the structural biology, design and clinical development of Bcr-Abl kinase inhibitors for the treatment of chronic myeloid leukaemia. *Biochimica Et biophysica Acta, 1754,* 3–13.

Manning, G., Whyte, D. B., Martinez, R., Hunter, T., & Sudarsanam, S. (2002). The protein kinase complement of the human genome. *Science, 298,* 1912–1934.

Martin, M. P., Alam, R., Betzi, S., Ingles, D. J., Zhu, J. Y., & Schönbrunn, E. (2012). A novel approach to the discovery of small-molecule ligands of CDK2. *ChemBioChem, 13,* 2128–2136.

McTigue, M., Murray, B. W., Chen, J. H., Deng, Y.-L., Solowiej, J., & Kania, R. S. (2012). Molecular conformations, interactions, and properties associated with drug efficiency and clinical performance among VEGFR TK inhibitors. *Proceedings of the National Academy of Sciences, 109,* 18281–18289.

Mohammadi, M., McMahon, G., Sun, L., Tang, C., Hirth, P., Yeh, B. K., et al. (1997). Structures of the tyrosine kinase domain of fibroblast growth factor receptor in complex with inhibitors. *Science, 276,* 955–960.

Mohammadi, M., Schlessinger, J., & Hubbard, S. R. (1996). Structure of the FGF receptor tyrosine kinase domain reveals a novel autoinhibitory mechanism. *Cell, 86,* 577–587.

Mol, C. D., Dougan, D. R., Schneider, T. R., Skene, R. J., Kraus, M. L., Scheibe, D. N., et al. (2004). Structural basis for the autoinhibition and STI-571 inhibition of c-Kit tyrosine kinase. *The Journal of Biological Chemistry, 279,* 31655–31663.

Moyer, J. D., Barbacci, E. G., Iwata, K. K., Arnold, L., Boman, B., Cunningham, A., et al. (1997). Induction of apoptosis and cell cycle arrest by CP-358,774, an inhibitor of epidermal growth factor receptor tyrosine kinase. *Cancer Research, 57,* 4838–4848.

Nagar, B., Bornmann, W. G., Pellicena, P., Schindler, T., Veach, D. R., Miller, W. T., et al. (2002). Crystal structures of the kinase domain of c-Abl in complex with the small molecule inhibitors PD173955 and imatinib (STI-571). *Cancer Research, 62,* 4236–4243.

Nagar, B., Hantschel, O., Young, M. A., Scheffzek, K., Veach, D. R., Bornmann, W., et al. (2003). Structural basis for the autoinhibition of c-Abl tyrosine kinase. *Cell, 112,* 859–871.

Nolen, B., Taylor, S., & Ghosh, G. (2004). Regulation of protein kinases; controlling activity through activation segment conformation. *Molecular Cell, 15,* 661–675.

Oguro, Y., Cary, D. R., Miyamoto, N., Tawada, M., Iwata, H., Miki, H., et al. (2013). Design, synthesis, and evaluation of novel VEGFR2 kinase inhibitors: discovery of [1,2,4]triazolo[1,5-a]pyridine derivatives with slow dissociation kinetics. *Bioorganic & Medicinal Chemistry, 21,* 4714–4729.

O'Hare, T., Shakespeare, W. C., Zhu, X., Eide, C. A., Rivera, V. M., Wang, F., et al. (2009). Ap24534, a Pan-Bcr-Abl inhibitor for chronic myeloid leukemia, potently inhibits the t315i mutant and overcomes mutation-based resistance. *Cancer Cell, 16,* 401–412.

Ohren, J. F., Chen, H., Pavlovsky, A., Whitehead, C., Zhang, E., Kuffa, P., et al. (2004). Structures of human MAP kinase kinase 1 (MEK1) and MEK2 describe novel noncompetitive kinase inhibition. *Nature Structural & Molecular Biology, 11,* 1192–1197.

Omura, S., Iwai, Y., Hirano, A., Nakagawa, A., Awaya, J., Tsuchya, H., et al. (1977). A new alkaloid AM-2282 OF Streptomyces origin. Taxonomy, fermentation, isolation and preliminary characterization. *The Journal of Antibiotics, 30,* 275–282.

Omura, S., Sasaki, Y., Iwai, Y., & Takeshima, H. (1995). Staurosporine, a potentially important gift from a microorganism. *The Journal of Antibiotics, 48,* 535–548.

Osherov, N., & Levitzki, A. (1994). Epidermal-growth-factor-dependent activation of the src-family kinases. *European Journal of Biochemistry*, *225*, 1047–1053.

Parang, K., Till, J. H., Ablooglu, A. J., Kohanski, R. A., Hubbard, S. R., & Cole, P. A. (2001). Mechanism-based design of a protein kinase inhibitor. *Nature Structural Biology*, *8*, 37–41.

Park, J. H., Liu, Y., Lemmon, M. A., & Radhakrishnan, R. (2012). Erlotinib binds both inactive and active conformations of the EGFR tyrosine kinase domain. *The Biochemical Journal*, *448*, 417–423.

Paulsen, C. E., Truong, T. H., Garcia, F. J., Homann, A., Gupta, V., Leonard, S. E., et al. (2011). Peroxide-dependent sulfenylation of the eGFr catalytic site enhances kinase activity. *Nature Chemical Biology*, *8*, 57–64.

Pawson, T., & Hunter, T. (1994). Signal transduction and growth control in normal and cancer cells. *Current Opinion in Genetics & Development*, *4*, 1–4.

Pawson, T., & Kofler, M. (2009). Kinome signaling through regulated protein–protein interactions in normal and cancer cells. *Current Opinion in Cell Biology*, *21*, 147–153.

Pawson, T., & Scott, J. D. (1997). Signaling through scaffold, anchoring, and adaptor proteins. *Science*, *278*, 2075–2080.

Pellicena, P., & Kuriyan, J. (2006). Protein–protein interactions in the allosteric regulation of protein kinases. *Current Opinion in Structural Biology*, *16*, 702–709.

Prade, L., Engh, R. A., Girod, A., Kinzel, V., Huber, R., & Bossemeyer, D. (1997). Staurosporine-induced conformational changes of cAMP-dependent protein kinase catalytic subunit explain inhibitory potential. *Structure*, *5*, 1627–1637.

Remsing Rix, L. L., Rix, U., Colinge, J., Hantschel, O., Bennett, K. L., Stranzl, T., et al. (2009). Global target profile of the kinase inhibitor bosutinib in primary chronic myeloid leukemia cells. *Leukemia*, *23*, 477–485.

Rini, B. I. (2008). Temsirolimus, an inhibitor of mammalian target of rapamycin. *Clinical Cancer Research*, *14*, 1286–1290.

Rusnak, D. W., Affleck, K., Cockerill, S. G., Stubberfield, C., Harris, R., Page, M., et al. (2001). The characterization of novel, dual ErbB-2/EGFR, tyrosine kinase inhibitors: Potential therapy for cancer. *Cancer Research*, *61*, 7196–7203.

Sawyers, C. L. (2003). Opportunities and challenges in the development of kinase inhibitor therapy for cancer. *Genes & Development*, *17*, 2998–3010.

Schindler, T., Bornmann, W., Pellicena, P., Miller, W. T., Clarkson, B., & Kuriyan, J. (2000). Structural mechanism for STI-571 inhibition of abelson tyrosine kinase. *Science*, *289*, 1938–1942.

Schindler, T., Sicheri, F., Pico, A., Gazit, A., Levitzki, A., & Kuriyan, J. (1999). Crystal structure of Hck in complex with a Src family-selective tyrosine kinase inhibitor. *Molecular Cell*, *3*, 639–648.

Schittenhelm, M. M., Shiraga, S., Schroeder, A., Corbin, A. S., Griffith, D., Lee, F. Y., et al. (2006). Dasatinib (BMS-354825), a dual SRC/ABL kinase inhibitor, inhibits the kinase activity of wild-type, juxtamembrane, and activation loop mutant KIT isoforms associated with human malignancies. *Cancer Research*, *66*, 473–481.

Schwartz, P. A., & Murray, B. W. (2011). Protein kinase biochemistry and drug discovery. *Bioorganic Chemistry*, *39*, 192–210.

Sebolt-Leopold, J. S., Dudley, D. T., Herrera, R., Van Becelaere, K., Wiland, A., Gowan, R. C., et al. (1999). Blockade of the MAP kinase pathway suppresses growth of colon tumors in vivo. *Nature Medicine*, *5*, 810–816.

Seeliger, M. A., Nagar, B., Frank, F., Cao, X., Henderson, M. N., & Kuriyan, J. (2007). c-Src binds to the cancer drug imatinib with an inactive Abl/c-Kit conformation and a distributed thermodynamic penalty. *Structure*, *15*, 299–311.

Seeliger, M. A., Ranjitkar, P., Kasap, C., Shan, Y., Shaw, D. E., Shah, N. P., et al. (2009). Equally potent inhibition of c-Src and Abl by compounds that recognize inactive kinase conformations. *Cancer Research*, *69*, 2384–2392.

Shan, Y., Eastwood, M. P., Zhang, X., Kim, E. T., Arkhipov, A., Dror, R. O., et al. (2012). Oncogenic mutations counteract intrinsic disorder in the EGFR kinase and promote receptor dimerization. *Cell*, *149*, 860–870.

Shan, Y., Seeliger, M. A., Eastwood, M. P., Frank, F., Xu, H., Jensen, M. Ø., et al. (2009). A conserved protonation-dependent switch controls drug binding in the Abl kinase. *Proceedings of the National Academy of Sciences*, *106*, 139–144.

Shaw, A. S., Kornev, A. P., Hu, J., Ahuja, L. G., & Taylor, S. S. (2014). Kinases and pseudokinases: Lessons from RAF. *Molecular and Cellular Biology*, *34*, 1538–1546.

Sicheri, F., Moarefi, I., & Kuriyan, J. (1997). Crystal structure of the Src family tyrosine kinase Hck. *Nature*, *385*, 602–609.

Simard, J. R., Klüter, S., Grütter, C., Getlik, M., Rabiller, M., Rode, H. B., et al. (2009). A new screening assay for allosteric inhibitors of cSrc. *Nature Chemical Biology*, *5*, 394–396.

Solca, F., Dahl, G., Zoephel, A., Bader, G., Sanderson, M., Klein, C., et al. (2012). Target binding properties and cellular activity of afatinib (BIBW 2992), an irreversible ErbB family blocker. *The Journal of Pharmacology and Experimental Therapeutics*, *343*, 342–350.

Solowiej, J., Bergqvist, S., McTigue, M. A., Marrone, T., Quenzer, T., Cobbs, M., et al. (2009). Characterizing the effects of the juxtamembrane domain on vascular endothelial growth factor receptor-2 enzymatic activity, autophosphorylation, and inhibition by axitinib. *Biochemistry*, *48*, 7019–7031.

Stamos, J., Sliwkowski, M. X., & Eigenbrot, C. (2002). Structure of the epidermal growth factor receptor kinase domain alone and in complex with a 4-anilinoquinazoline inhibitor. *The Journal of Biological Chemistry*, *277*, 46265–46272.

Steichen, J. M., Iyer, G. H., Li, S., Saldanha, S. A., Deal, M. S., Woods, V. L., et al. (2010). Global consequences of activation loop phosphorylation on protein kinase A. *Journal of Biological Chemistry*, *285*, 3825–3832.

Sun, L., Liang, C., Shirazian, S., Zhou, Y., Miller, T., Cui, J., et al. (2003). Discovery of 5-[5-fluoro-2-oxo-1,2- dihydroindol-(3Z)-ylidenemethyl]-2,4- dimethyl-1H-pyrrole-3-carboxylic acid (2-diethylaminoethyl)amide, a novel tyrosine kinase inhibitor targeting vascular endothelial and platelet-derived growth factor receptor tyrosine kinase. *Journal of Medicinal Chemistry*, *46*, 1116–1119.

Sun, L., Tran, N., Liang, C., Tang, F., Rice, A., Schreck, R., et al. (1999). Design, synthesis, and evaluations of substituted 3-[(3- or 4-carboxyethylpyrrol-2-yl)methylidenyl]indolin-2-ones as inhibitors of VEGF, FGF, and PDGF receptor tyrosine kinases. *Journal of Medicinal Chemistry*, *42*, 5120–5130.

Sun, L., Tran, N., Tang, F., App, H., Hirth, P., McMahon, G., et al. (1998). Synthesis and biological evaluations of 3-substituted indolin-2-ones: A novel class of tyrosine kinase inhibitors that exhibit selectivity toward particular receptor tyrosine kinases. *Journal of Medicinal Chemistry*, *41*, 2588–2603.

Tamaoki, T., Nomoto, H., Takahashi, I., Kato, Y., Morimoto, M., & Tomita, F. (1986). Staurosporine, a potent inhibitor of phospholipid/Ca++dependent protein kinase. *Biochemical and Biophysical Research Communications*, *135*, 397–402.

Tokarski, J. S., Newitt, J. A., Chang, C. Y. J., Cheng, J. D., Wittekind, M., Kiefer, S. E., et al. (2006). The structure of Dasatinib (BMS-354825) bound to activated ABL kinase domain elucidates its inhibitory activity against imatinib-resistant ABL mutants. *Cancer Research*, *66*, 5790–5797.

Toledo, L. M., & Lydon, N. B. (1997). Structures of staurosporine bound to CDK2 and cAPK—New tools for structure-based design of protein kinase inhibitors. *Structure*, *5*, 1551–1556.

Tomita, N., Hayashi, Y., Suzuki, S., Oomori, Y., Aramaki, Y., Matsushita, Y., et al. (2013). Structure-based discovery of cellular-active allosteric inhibitors of FAK. *Bioorganic & Medicinal Chemistry Letters*, *23*, 1779–1785.

Traxler, P., & Furet, P. (1999). Strategies toward the design of novel and selective protein tyrosine kinase inhibitors. *Pharmacology and Therapeutics*, *82*, 195–206.

Tsai, J., Lee, J. T., Wang, W., Zhang, J., Cho, H., Mamo, S., et al. (2008). Discovery of a selective inhibitor of oncogenic B-Raf kinase with potent antimelanoma activity. *Proceedings of the National Academy of Sciences*, *105*, 3041–3046.

van Linden, O. P. J., Kooistra, A. J., Leurs, R., de Esch, I. J. P., & de Graaf, C. (2014). KLIFS: A knowledge-based structural database to navigate kinase-ligand interaction space. *Journal of Medicinal Chemistry*, *57*, 249–277.

Vogelstein, B., & Kinzler, K. W. (2004). Cancer genes and the pathways they control. *Nature Medicine*, *10*, 789–799.

Wan, P. T. C., Garnett, M. J., Roe, S. M., Lee, S., Niculescu-Duvaz, D., Good, V. M., et al. (2004). Mechanism of activation of the RAF-ERK signaling pathway by oncogenic mutations of B-RAF. *Cell*, *116*, 855–867.

Wang, H., Kadlecek, T. A., Au-Yeung, B. B., Goodfellow, H. E. S., Hsu, L. Y., Freedman, T. S., et al. (2010). ZAP-70: An essential kinase in T-cell signaling. *Cold Spring Harbor Perspectives in Biology*, *2*, a002279.

Ward, W. H., Cook, P. N., Slater, A. M., Davies, D. H., Holdgate, G. A., & Green, L. R. (1994). Epidermal growth factor receptor tyrosine kinase. Investigation of catalytic mechanism, structure-based searching and discovery of a potent inhibitor. *Biochemical Pharmacology*, *48*, 659–666.

Weisberg, E., Manley, P. W., Breitenstein, W., Brüggen, J., Cowan-Jacob, S. W., Ray, A., et al. (2005). Characterization of AMN107, a selective inhibitor of native and mutant Bcr-Abl. *Cancer Cell*, *7*, 129–141.

Weisberg, E., Manley, P. W., Cowan-Jacob, S. W., Hochhaus, A., & Griffin, J. D. (2007). Second generation inhibitors of BCR-ABL for the treatment of imatinib-resistant chronic myeloid leukaemia. *Nature Reviews. Cancer*, *7*, 345–356.

Wilhelm, S., Carter, C., Lynch, M., Lowinger, T., Dumas, J., Smith, R. A., et al. (2006). Discovery and development of sorafenib: A multikinase inhibitor for treating cancer. *Nature Reviews. Drug Discovery*, *5*, 835–844.

Wilhelm, S. M., Dumas, J., Adnane, L., Lynch, M., Carter, C. A., Schütz, G., et al. (2011). Regorafenib (BAY 73-4506): A new oral multikinase inhibitor of angiogenic, stromal and oncogenic receptor tyrosine kinases with potent preclinical antitumor activity. *International Journal of Cancer*, *129*, 245–255.

Wityak, J., Das, J., Moquin, R. V., Shen, Z., Lin, J., Chen, P., et al. (2003). Discovery and initial SAR of 2-amino-5-carboxamidothiazoles as inhibitors of the Src-family kinase p56(Lck). *Bioorganic & Medicinal Chemistry Letters*, *13*, 4007–4010.

Wood, E. R., Truesdale, A. T., McDonald, O. B., Yuan, D., Hassell, A., Dickerson, S. H., et al. (2004). A unique structure for epidermal growth factor receptor bound to GW572016 (Lapatinib): Relationships among protein conformation, inhibitor off-rate, and receptor activity in tumor cells. *Cancer Research*, *64*, 6652–6659.

Wu, W.-I., Voegtli, W. C., Sturgis, H. L., Dizon, F. P., Vigers, G. P. A., & Brandhuber, B. J. (2010). Crystal structure of human AKT1 with an allosteric inhibitor reveals a new mode of kinase inhibition. *PLoS One*, *5*, e12913.

Wybenga-Groot, L. E., Baskin, B., Ong, S. H., Tong, J., Pawson, T., & Sicheri, F. (2001). Structural basis for autoinhibition of the Ephb2 receptor tyrosine kinase by the unphosphorylated juxtamembrane region. *Cell*, *106*, 745–757.

Xu, W., Harrison, S. C., & Eck, M. J. (1997). Three-dimensional structure of the tyrosine kinase c-Src. *Nature*, *385*, 595–602.

Yamaguchi, H., & Hendrickson, W. A. (1996). Structural basis for activation of human lymphocyte kinase Lck upon tyrosine phosphorylation. *Nature*, *384*, 484–489.

Yun, C. H., Boggon, T. J., Li, Y., Woo, M. S., Greulich, H., Eck, M. J., et al. (2007). Structures of lung cancer-derived EGFR mutants and inhibitor complexes: mechanism of activation and insights into differential inhibitor sensitivity. *Cancer Cell, 11*(3), 217–227.

Zhang, X., Gureasko, J., Shen, K., Cole, P. A., & Kuriyan, J. (2006). An allosteric mechanism for activation of the kinase domain of epidermal growth factor receptor. *Cell, 125*, 1137–1149.

Zhang, J., Yang, P. L., & Gray, N. S. (2009). Targeting cancer with small molecule kinase inhibitors. *Nature Reviews Cancer, 9*, 28–39.

Zhao, Z., Wu, H., Wang, L., Liu, Y., Knapp, S., Liu, Q., et al. (2014). Exploration of type II binding mode: A privileged approach for kinase inhibitor focused drug discovery? *ACS Chemical Biology, 6*, 1230–1241.

Zheng, J., Trafny, E. A., Knighton, D. R., Xuong, N. H., Taylor, S. S., Ten Eyck, L. F., et al. (1993). 2.2 A refined crystal structure of the catalytic subunit of cAMP-dependent protein kinase complexed with MnATP and a peptide inhibitor. *Acta Crystallographica. Section D, Biological Crystallography, 49*, 362–365.

Zhou, T., Commodore, L., Huang, W.-S., Wang, Y., Thomas, M., Keats, J., et al. (2011). Structural mechanism of the Pan-BCR-ABL inhibitor ponatinib (AP24534): Lessons for overcoming kinase inhibitor resistance. *Chemical Biology & Drug Design, 77*, 1–11.

Zimmermann, J., Buchdunger, E., Mett, H., Meyer, T., & Lydon, N. B. (1997). Potent and selective inhibitors of the Abl-kinase: Phenylamino-pyrimidine (PAP) derivatives. *Bioorganic & Medicinal Chemistry Letters, 7*, 187–192.

Zimmermann, J., Buchdunger, E., Mett, H., Meyer, T., Lydon, N. B., & Traxler, P. (1996). Phenylamino-pyrimidine (PAP)-derivatives: A new class of potent and highly selective PDGF-receptor autophosphorylation inhibitors. *Bioorganic & Medicinal Chemistry Letters, 6*, 1221–1226.

CHAPTER THREE

Fragment-Based Approaches to the Discovery of Kinase Inhibitors

Paul N. Mortenson[1], Valerio Berdini, Marc O'Reilly
Astex Pharmaceuticals, Cambridge, United Kingdom
[1]Corresponding author: e-mail address: paul.mortenson@astx.com

Contents

Abstract

Protein kinases are one of the most important families of drug targets, and aberrant kinase activity has been linked to a large number of disease areas. Although eminently targetable using small molecules, kinases present a number of challenges as drug targets, not least obtaining selectivity across such a large and relatively closely related target family. Fragment-based drug discovery involves screening simple, low-molecular weight compounds to generate initial hits against a target. These hits are then optimized to more potent compounds via medicinal chemistry, usually facilitated by

Methods in Enzymology, Volume 548
ISSN 0076-6879
http://dx.doi.org/10.1016/B978-0-12-397918-6.00003-3

structural biology. Here, we will present a number of recent examples of fragment-based approaches to the discovery of kinase inhibitors, detailing the construction of fragment-screening libraries, the identification and validation of fragment hits, and their optimization into potent and selective lead compounds. The advantages of fragment-based methodologies will be discussed, along with some of the challenges associated with using this route. Finally, we will present a number of key lessons derived both from our own experience running fragment screens against kinases and from a large number of published studies.

ABBREVIATIONS

ATP adenosine triphosphate
CDK cyclin-dependent kinase
FBDD fragment-based drug discovery
FDA US Food and Drug Administration
HCS high-concentration screening
HTS high-throughput screening
IC_{50} half-maximal inhibitory concentration
K_d dissociation constant
LE ligand efficiency
LLE lipophilic ligand efficiency
LLE_{AT} size-scaled variant of lipophilic ligand efficiency
MEK mitogen-activated protein kinase kinase
MW molecular weight
PDB Brookhaven protein data bank
SPR surface plasmon resonance
STD saturation transfer difference

1. INTRODUCTION

Protein kinases catalyze the transfer of the terminal phosphate group of adenosine triphosphate (ATP) onto an amino acid residue within a polypeptide chain. In eukaryotes, protein kinase-mediated phosphorylation is critical for intracellular signal transduction, and aberrant regulation of protein kinase pathways has been linked to a diverse set of disease states including cancer, inflammation and metabolic, autoimmune, and neurological conditions (Cohen, 2002).

Protein kinases share a common bilobal fold (Taylor, Knighton, Zheng, Ten Eyck, & Sowadski, 1992) comprising an N-terminal, predominantly beta-sheet, domain, and a helical C-terminal domain (see Fig. 3.1). The

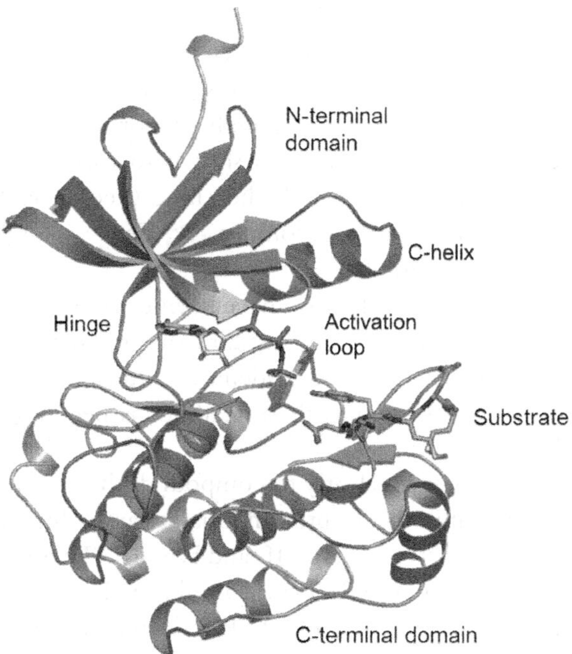

Figure 3.1 Architecture of a typical protein kinase domain. The protein backbone is shown in cartoon form in gray, and the peptide substrate is shown in orange. The non-hydrolyzable ATP mimetic AMP-PNP is shown in yellow. From PDB structure 1ir3. (See the color plate).

kinase catalytic site resides in the cleft between these two domains, which are linked by a "hinge polypeptide". Sequencing of the human genome has revealed that there are 518 genes encoding protein kinases (Manning, Whyte, Martinez, Hunter, & Sudarsanam, 2002).

1.1. Challenges of kinases as drug targets

The fact that kinases utilize ATP as a cofactor leads to a number of key challenges when considering them as drug targets.

1. Competition with ATP. Any inhibitor that targets the kinase active site (and thus directly competes with ATP) must overcome millimolar levels of intracellular ATP.
2. Selectivity. As stated earlier, there are over 500 human protein kinases, which bind ATP in a very similar manner. This presents obvious challenges when attempting to design a compound that selectively inhibits just one kinase.

3. Physicochemical properties. The main part of the ATP-binding site is relatively flat and generally lined on two sides by hydrophobic residues. This means that unless care is taken, it is possible to end up with inhibitors that are lipophilic and have a high proportion of aromatic atoms, both characteristics that have been shown to carry a poor prognosis for progression into and through the clinic (Leeson & Springthorpe, 2007; Lipinski, Lombardo, Dominy, & Feeney, 1997; Lovering, Bikker, & Humblet, 2009).
4. Intellectual property (IP). Kinases have been extensively targeted by the pharmaceutical industry over the last 20 years, and many templates have been discovered that mimic the adenine group in ATP. As a consequence, the IP space around these scaffolds is becoming increasingly congested.

These challenges apply most clearly to compounds that attempt to compete directly with ATP. Inhibitors of protein kinases can be broadly partitioned into three major classes: type I, type II, and type III. These modes of inhibition are illustrated in Fig. 3.2. Type I inhibitors bind within the ATP-binding cleft, are competitive with ATP, and interact directly with the kinase-hinge motif. Type II binders often contain a type I moiety but bind to an inactive conformation of the kinase, typically with the DFG loop in the "out" configuration. This conformation allows access to a hydrophobic pocket beyond the gatekeeper residue, but is only accessible in a subset of kinases. Type III ligands bind in a hydrophobic pocket proximal to the ATP-binding site, are uncompetitive with ATP, and inhibit via an allosteric mechanism. There are also examples of other modes of inhibition, and these are typically specific to individual kinases, or closely related kinase families. There may be advantages associated with non-type I modes of inhibition, and these are discussed further later.

2. FRAGMENT-BASED DRUG DISCOVERY

Fragment-based drug discovery (FBDD) involves the screening of relatively small chemical compounds—typically with molecular weight (MW) less than 300 Da (Congreve, Carr, Murray, & Jhoti, 2003) and often much smaller—in order to discover leads against a protein target. These leads are then optimized using medicinal chemistry, often with significant structural input, either from X-ray crystallography or from NMR spectroscopy. The genesis of modern FBDD is usually attributed to two publications from

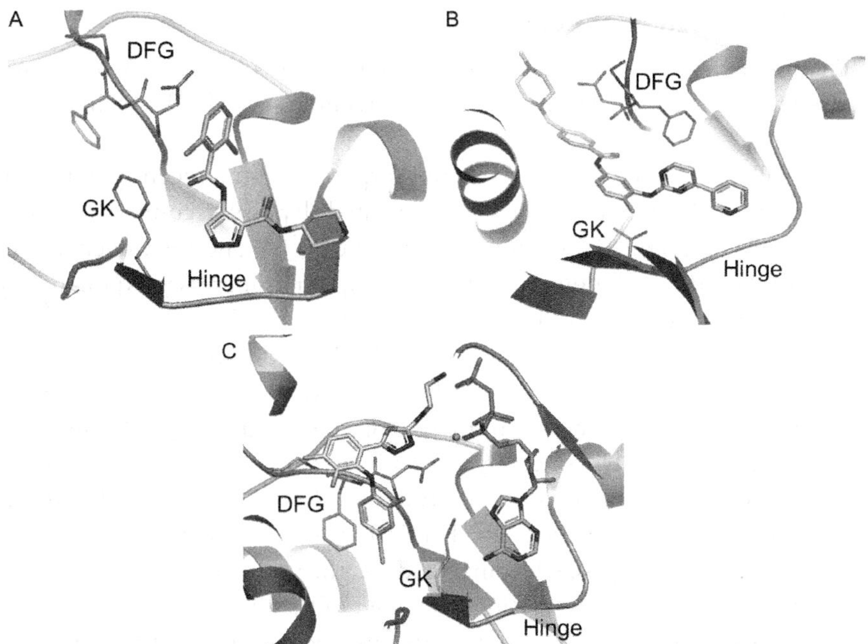

Figure 3.2 X-ray structures of exemplary type I, II, and III inhibitors. In each case, the ligand is shown in yellow, the protein is shown in cartoon form in gray, and the gatekeeper (GK) and DFG-loop residues are highlighted in green. (A) AT7519, a type I inhibitor, bound to CDK2 (in-house structure). (B) Imatinib, a type II inhibitor, bound to c-Abl (PDB structure 1iep). (C) A type III inhibitor bound to MEK1 (PDB structure 3eqb). ATP is also present in the structure and is shown in orange. (See the color plate).

researchers at Abbott describing the "SAR by NMR" method (Hajduk et al., 1997; Shuker, Hajduk, Meadows, & Fesik, 1996).

A number of excellent reviews of general practice in FBDD have been published (Chessari & Woodhead, 2009; Congreve, Chessari, Tisi, & Woodhead, 2008; Coyne, Scott, & Abell, 2010; de Kloe, Bailey, Leurs, & de Esch, 2009; Hajduk & Greer, 2007; Murray, Verdonk, & Rees, 2012; Schulz & Hubbard, 2009; Scott, Coyne, Hudson, & Abell, 2012), along with several books (Davies & Hyvönen, 2012; Jahnke & Erlanson, 2006; Kuo, 2011; Zartler & Shapiro, 2008) and the reader is referred to these for many examples of the successful application of FBDD, as well as details of the many different screening methodologies used. Reviews of FBDD as applied to kinases have also been published (Erlanson, 2009; Gill, 2004).

2.1. Advantages of FBDD

The fragment-based approach carries a number of advantages when compared to, for example, high-throughput screening (HTS). The main advantages are:

1. Smaller screening libraries. Chemical space is vast; a recent publication (Blum & Reymond, 2009) has enumerated all possible compounds (subject to certain rules) containing up to 13 heavy (i.e., nonhydrogen) atoms and shown that the number of such compounds increases exponentially with the number of heavy atoms. So while for 12 heavy atoms (MW $\sim$170), there are approximately 10^8 compounds possible, for 24 heavy atoms (MW $\sim$340) there would be around 10^{19}, and for 36 heavy atoms (MW $\sim$500) there would be around 10^{30}. These numbers are obviously extremely approximate; however, it is clear that a library of 1000 fragments with around 12 heavy atoms each has much more complete coverage of chemical space than a library of 10^6 or 10^7 lead- or drug-sized compounds. Other researchers have reached similar conclusions by starting from a set of fragment hits, and considering how many lead-like compounds could be constructed that contained the fragments as a substructure (Teotico et al., 2009).
2. Higher hit rates. In addition to being small, the other defining characteristic of fragments is their simplicity. Hann, Leach, and Harper (2001) elegantly showed over a decade ago that theoretically such simple compounds should be more likely to bind to an arbitrary target. Experimentally it has proved challenging to validate this argument (Leach & Hann, 2011), possibly due to the confounding effect of lipophilicity. However, we and others (Chen & Hubbard, 2009) have found that hit rates for fragment screens typically range from 1% to 10%, significantly higher than would be expected for HTS.
3. Improved physicochemical properties. Inappropriate values for molecular properties such as MW and log P have been shown to carry significant risks in terms of progression into and through the clinic (Leeson & Springthorpe, 2007; Lipinski et al., 1997; Vieth et al., 2004; Wenlock, Austin, Barton, Davis, & Leeson, 2003). Fragments by definition have low MW and will generally have low log P compared to a typical HTS hit. Starting from a fragment hit offers an opportunity to control these properties during optimization, and there is evidence that when fragments are optimized carefully, the resulting leads compare favorably to leads generated from HTS (Leeson & St-Gallay, 2011; Murray et al., 2012).

4. Opportunities for chemical novelty. These can arise from two sources. Firstly, fragment hits are by definition small, and they are frequently simple molecules with many possible avenues for elaboration. This flexibility will often provide ample scope to steer away from areas of congested IP. Secondly, the sampling advantage of fragments can lead to the identification of templates that HTS has missed. An example might be where a fragment binds in a mode, which will require chemically unusual substitution patterns for optimization. In this situation, the larger compounds in an HTS library are unlikely to be appropriately decorated in order to bind.

2.2. Challenges of FBDD

Unsurprisingly, there are challenges associated with FBDD:

1. Detection of binding. Smaller compounds will in general bind more weakly than larger compounds. A typical HTS hit may bind with a dissociation constant (K_d) in the range of 1–10 μ*M*, and thus be easily detectable by conventional assay techniques. A fragment hit will more commonly have a K_d in the range of 100 μ*M* to 10 m*M*, which typically requires more specialized detection methods. However, the plethora of academic and small industrial groups currently practicing FBDD would imply that this challenge, while significant, is not an insurmountable one.
2. Comparing hits of different sizes. Fragment-screening libraries may contain compounds ranging in size from fewer than 10 to around 20 heavy atoms. In addition, it is not uncommon for a fragment-based screen to be run alongside a more conventional HTS campaign, involving compounds that are larger still. Practitioners of FBDD, therefore, often need to compare the relative merits of a small fragment hit (likely to have a potency in the high micromolar or millimolar range), and a larger hit that is significantly more potent. The concept of ligand efficiency (LE), defined as binding free energy per heavy atom (Hopkins, Groom, & Alex, 2004; Kuntz, 1999), is invaluable here as it effectively scales the potency of a compound according to its size, putting fragments, and larger compounds on a more level playing field. A number of other efficiency indices have been defined that either use different scaling techniques (Bembenek, Tounge, & Reynolds, 2009) or incorporate additional properties such as log P and polar surface area (PSA) (Abad-Zapatero & Metz, 2005; Keserü & Makara, 2009). Internally, we have found a scaled version of lipophilic ligand efficiency (LLE)

(Leeson & Springthorpe, 2007) called LLE_{AT} (Mortenson & Murray, 2011) to be a useful addition to LE in this context.

3. Optimization of weak initial hits. In order to be clinically efficacious, small molecules usually need to bind to their protein target with a $K_d < 100$ n*M*, and often significantly more tightly than this. Thus, from an initial fragment hit we may need to improve the potency by many orders of magnitude. Combined with the fact that fragments are often relatively simple compounds with many opportunities for elaboration, it may be hard to plan a route for optimization using traditional medicinal chemistry. One solution to this problem is to obtain information about how the initial hit is binding the protein target, and indeed many FBDD programs are associated with significant structural biology efforts. This information (if used carefully) enables the development of a focussed chemistry plan that can quickly improve potency, and some examples of rapid structure-guided optimization will be presented later.

2.3. Discovering kinase inhibitors with FBDD

FBDD offers potential solutions to at least two of the challenges involved in developing kinase inhibitors—developing inhibitors with favorable physicochemical properties and navigating a congested IP space. Firstly, as described earlier, fragment-based approaches (combined with diligent medicinal chemistry) can lead to compounds with improved physicochemical properties compared to HTS. Secondly, due to the sampling advantage of fragment libraries, fragment screens can identify novel inhibitory chemotypes.

Additionally, one of the significant challenges of FBDD—the optimization of weak initial hits—is mitigated for kinases by the large quantity of structural information available. As will be detailed later, this wealth of data can make it possible to make extremely rapid progress from fragment to lead. Kinases have proven to be a fertile source of targets for FBDD, as evidenced by the number of fragment-derived kinase inhibitors in the clinic.

2.4. FBDD-derived kinase inhibitors in the clinic

A significant FBDD milestone was passed in 2011, when the US Food and Drug Administration (FDA) approved vemurafenib for the treatment of malignant melanoma (US Food & Drug Administration, 2012). The target for vemurafenib is the oncogenic V600E mutant of B-Raf kinase (Bollag et al., 2010). It is the first marketed drug to have its origins in fragment-based screening.

The progression from initial fragment hit to vemurafenib is illustrated in Fig. 3.3. It started with a biochemical screen of 20,000 compounds with MW between 150 and 350 against multiple structurally characterized kinases (Tsai et al., 2008). Around 200 compounds were shown to inhibit the kinase activity of Pim-1, p38, and CSK by at least 30% at 200 μ*M* and were taken into crystallographic studies. These latter studies revealed the fragment 7-azaindole binding to the ATP site, and this led to the synthesis of a more elaborated analogue, showing 100 μ*M* activity against Pim-1. Structural overlays with a number of kinases indicated that the scaffold had the potential to inhibit multiple kinase targets, and a number of further analogues were synthesized, one of which (PLX4720) inhibits B-Raf V600E with a half-maximal inhibitory concentration (IC_{50}) of 13 n*M*. PLX4720 also shows good selectivity against wild-type B-Raf (IC_{50} = 160 n*M*) and a number of other kinases (IC_{50} > 1 μ*M*). Further, optimization led to the progression of the close analogue PLX4032 into the clinic and now onto the market.

As well as this notable success, a number of other kinase inhibitors having their origins in fragment-based approaches have entered clinical trials, including AT7519 (Wyatt et al., 2008), AT9283 (Howard et al., 2009), AT13148 (Yap et al., 2012), AZD5363 (Davies et al., 2012), PLX3397

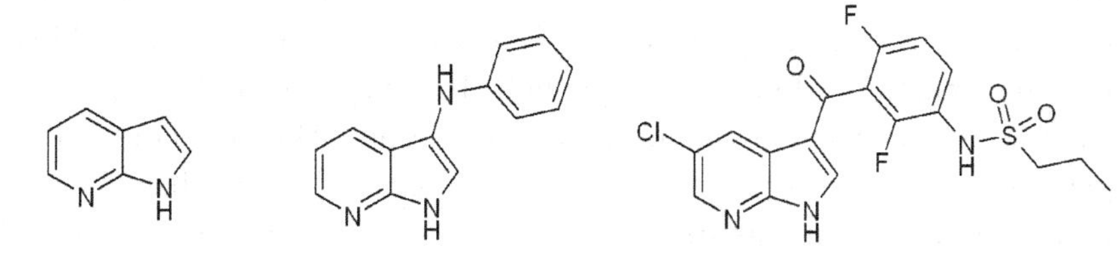

1
Pim-1 IC_{50} > 200 μ*M*

2
Pim-1 IC_{50} ~ 100 μ*M*

PLX4720
B-Raf V600E IC_{50} = 0.013 μ*M*

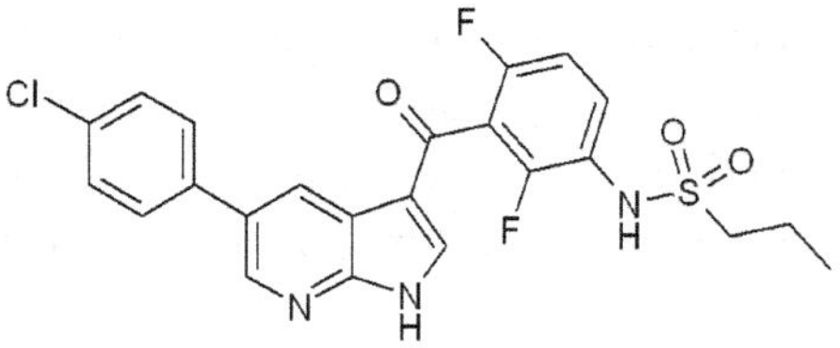

PLX4032/vemurafenib
B-Raf V600E IC_{50} = 0.031 μ*M*

Figure 3.3 Fragment to lead progression for vemurafenib.

(Denardo et al., 2011), PLX5622 (Yu et al., 2012), SGX523 (Buchanan et al., 2009), and SNS-314 (VanderPorten et al., 2009). It seems likely that there are many more examples whose mode of discovery has yet to be disclosed.

3. IDENTIFYING FRAGMENT HITS

3.1. Library construction

It is beyond the scope of this review to discuss fragment library construction in great depth, but it would be remiss of us not to emphasize the importance of this step. A variety of strategies exist for constructing a fragment-screening library, and the correct choice will depend on the screening methods that will be employed, and the size of the compounds in the library. Practitioners of high-concentration screening (HCS), for example, will typically be screening relatively "large" fragments (due to sensitivity limitations), and so (from arguments related to coverage of chemical space) will often have relatively large libraries of 10,000 compounds or more. Those employing more sensitive (but more labor intensive) screening methodologies such as X-ray screening will typically screen much smaller numbers (perhaps only a few hundred) of smaller fragments.

Two recent publications describe approaches to fragment library construction—one from Pfizer describing a general purpose fragment library (Lau et al., 2011) and one from GlaxoSmithKline describing a kinase-focussed fragment library (Bamborough, Brown, Christopher, Chung, & Mellor, 2011). Another publication from Abbott (Akritopoulou-Zanze & Hajduk, 2009) describes the custom synthesis of more than 50 novel hinge-binding fragments to augment their library. These papers describe in great detail the effort that those organizations have put into constructing and maintaining their fragment libraries, along with the complexity of the process. While their methods differ from our own, we share the belief that it is essential to construct a fragment-screening library with great care. Our library is currently on its sixth generation and is constantly under review, informed by over 1000 fragment crystal structures. We have also invested significant effort in synthesizing novel fragments for our library.

3.2. Hit identification

There is a wealth of publications describing fragment-based approaches to discover kinase inhibitors. A wide variety of screening methodologies have been used, including X-ray screening (Betzi et al., 2011; Cho et al., 2012; Howard et al., 2009; Schulz et al., 2011; Wyatt et al., 2008), HCS (Hughes et al., 2011; Iwata et al., 2012; Li et al., 2012; Matthews et al., 2009; Medina et al., 2010), surface plasmon resonance (SPR) (Soth et al., 2011; Xiang et al., 2011), NMR spectroscopy (Jahnke et al., 2010; Johnson et al., 2011; Lee et al., 2012; Ray et al., 2011), virtual screening (Giordanetto, Kull, & Dellsén, 2011; Moffett et al., 2011), and even *de novo* design (Charrier et al., 2011).

The breadth of approaches used is nicely illustrated by four recent publications describing fragment-based campaigns against phosphoinositide-dependent kinase 1 (PDK1), a member of the AGC family of serine/threonine kinases.

Researchers at Pfizer (Johnson et al., 2011) screened a library of approximately 10,000 low-MW fragments by NMR using the saturation transfer difference (STD) method and identified a number of hits, including isoquinolone, which bound with an IC_{50} of 870 μM (LE = 0.39). Efforts to obtain a crystal structure of this compound were unsuccessful, but docking predicted two possible binding modes. Analogues were designed to determine which of these binding modes was correct, and one of these proved sufficiently potent and soluble to obtain an X-ray structure, confirming one of the predicted modes. Further, structure-based optimization led to a compound with $IC_{50} = 1.8$ μM and LE = 0.42 kcal mol^{-1} per heavy atom (all LE values will be given in these units).

Workers at S*BIO (Lee et al., 2012) also employed NMR screening, using the waterLOGSY method, and identified a thienopyrimidine starting point with $IC_{50} = 55$ μM and LE = 0.39. Again, docking studies were used to guide optimization, which led to a compound that was 10-fold more potent, albeit with some loss of LE ($IC_{50} = 5.2$ μM, LE = 0.33).

Medina et al. (2010) used the same methods but in the opposite order, running an initial high-concentration bioassay, and confirming the hits via NMR. This study is discussed later in the section on hit validation. Finally researchers at Sunesis applied their "Tethering with extenders" approach to this target (Erlanson et al., 2011), and this will be discussed in the section later on type II inhibitors.

3.3. Hit validation

Similarly to HTS, the critical first step after a fragment screen is to identify (or "validate") the true hits and conversely to remove the false positives. The preferred method of validation in most cases is to obtain an X-ray structure of the fragment bound to the target. This also provides invaluable information for the fragment-optimization process. In cases where this is impractical or even impossible, binding can be confirmed using an orthogonal screening technique. Often a combination of these techniques is used, with two (or more) screening techniques used to select a set of compounds for which X-ray structure determination will be attempted.

This approach has been effectively employed in a recent publication from Medina et al. (2010). They screened around 1000 fragments against PDK1 in a high-concentration (400 μ*M*) bioassay and confirmed the hits via NMR STD experiments. They also attempted to obtain X-ray structures of a number of the fragments with PDK1 and were successful in three cases. The most ligand efficient fragments identified by HCS did not give crystal structures and were also shown by STD to not interact directly with PDK1, highlighting the usefulness of orthogonal screening methods.

4. FROM FRAGMENTS TO LEADS

4.1. Selection of hits for optimization

Having obtained and validated primary screening hits, the next task is to select which of these hits one should attempt to optimize. A number of factors need to be considered at this point: LE, lipophilicity (which can be incorporated into an efficiency index such as LLE_{AT}), growth vectors (positions from which substitution might lead to improved potency), chemical tractability, and potential for novelty being just some of the considerations. Knowledge of fragment-binding modes is extremely helpful when performing this analysis. This is most obviously true when looking for growth vectors, but such knowledge can also be useful for assessing chemical tractability and novelty, as different possible substitution patterns may have different degrees of synthetic accessibility and precedence.

Our preference is initially to progress multiple fragment series in parallel, as it can be hard to predict which will be most successful. An example of this is presented in the discovery of our cyclin-dependent kinase (CDK) 2 inhibitor, AT7519 (Wyatt et al., 2008). A library of around 500 fragments was soaked into apo crystals of CDK2, in cocktails of four. From this effort, more than 30 hits were identified, all binding to the ATP site and interacting with the hinge.

Four examples are shown in Fig. 3.4, which were pursued to varying degrees. Fragment 4a, while showing an interesting binding mode and reasonable LE, did not have obvious growth vectors and was considered less

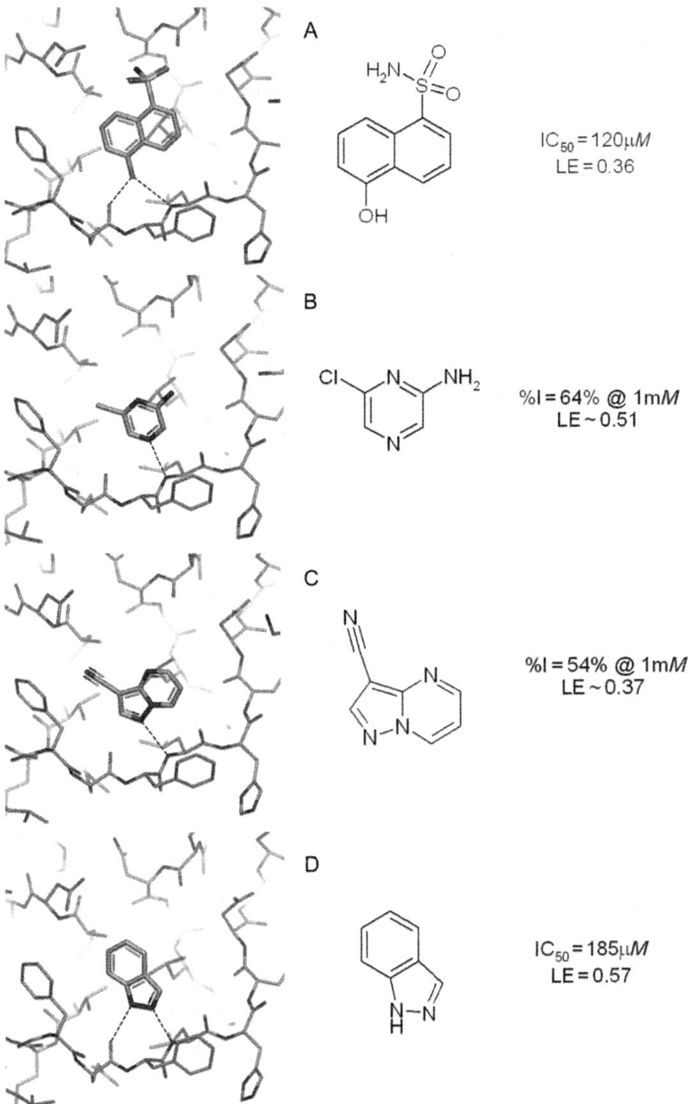

Figure 3.4 Four fragment hits found in our screening campaign against CDK2, discussed in further detail in the text. The X-ray structures are presented in a consistent frame of reference, with the hinge at the bottom and the gatekeeper (Phe 80) on the left. Hydrogen bonds to the hinge are illustrated by dotted lines. (See the color plate).

chemically tractable than some other hits, so did not progress into hits-to-leads chemistry. Fragment 4b was optimized to micromolar potency, but structural information and SAR suggested that it would be hard to improve potency significantly beyond this, so the series was not pursued further. Fragment 4c was progressed into an interesting lead series, with 30 n*M* potency and good cellular activity. Finally, fragment 4d was eventually optimized to AT7519, which is currently being investigated in phase 2 clinical studies. Throughout the process, structural information was critical not just for optimizing the fragments, but also deciding which series to pursue.

4.2. Structure-guided optimization

Atomic resolution X-ray crystal structures have now been determined for more than 200 different human protein kinases (data obtained from UniProt) (The UniProt Consortium, 2012; The Universal Protein Resource, 2012) and there are over 3000 protein kinase structures deposited within the PDB (PDB; Berman, 2000; The Protein Data Bank, 2012). This wealth of structural data has shed light on the subtle and varied mechanisms by which protein kinase activity is regulated and on how protein kinases achieve their substrate specificities.

High-quality structural information can greatly assist in the early stages of fragment optimization, which are normally focussed on improving potency. A recent example comes from researchers at Pfizer (Hughes et al., 2011). These researchers used an HCS approach to identify inhibitors of the lipid kinase phosphatidylinositol-3-kinase γ (PI3K γ). A relatively large fragment library (5960 compounds) was screened, identifying 312 hits, of which 150 were confirmed (i.e., gave a dose–response curve). These 150 compounds were further cross-screened by isothermal titration calorimetry and X-ray. A total of five X-ray structures of fragment complexes were obtained, and an amino pyrazolopyrimidine hit was selected for progression based on its LE, potential for selectivity, and synthetic tractability. The overlay of the X-ray structure of the initial hit with an X-ray structure of a literature compound suggested an obvious modification and quickly led to a compound with an IC_{50} of 35 n*M*, submicromolar activity in cells, and a promising selectivity profile against a panel of 43 other kinases.

4.3. Achieving selective inhibition

Structural information can be even more critical when selectivity against a closely related target is required. Achieving selective inhibition of just one target is a particular problem in the field of kinase inhibitors. Using a fragment-based approach at first seems a counterintuitive way to tackle this problem—following the complexity argument outlined earlier (Hann et al., 2001), one would expect fragments to be less selective than larger compounds. A recent publication explores this issue in some detail

(Bamborough et al., 2011). These researchers assembled a kinase-focussed fragment library of around 1000 compounds and screened it by HCS against 30 different kinase targets. This showed that commonly occurring hinge-binding fragments such as 7-azaindole, 2-anilinopyridine, and even adenine displayed surprising variability in the extent to which they inhibited different kinases. They also compared the selectivity profiles of the fragments with those of lead compounds containing the fragments as substructures. In general, as expected, the larger lead-like compounds were more selective than the fragments. Interestingly, little correlation was seen between the selectivity of the leads, and that of their component fragments; unselective fragments were often found in selective leads, and unselective leads often contained selective hinge-binding fragments.

There are two caveats to this analysis—firstly, the two sets of compounds were tested against different kinase panels, and secondly, the leads were not produced by optimization of the fragments. However, these findings mirror our own experience, where we have frequently taken an unselective fragment hit and (via careful structure-guided optimization) produced a highly selective lead compound, even against closely related targets. An example of this has been published recently (Cho et al., 2012).

Perhaps the poor correlation between fragment selectivity and lead selectivity should not be surprising, given that binding affinities can be highly sensitive to small structural changes. An interesting example comes from Roche (Soth et al., 2011) and relates to p38α kinase. These researchers used SPR to identify an indazole fragment that bound relatively weakly ($K_d = 2$ mM from SPR). A modeled structure of the compound combined with X-ray structures of previous hit compounds enabled rapid development into a very efficient and selective lead compound ($IC_{50} = 100$ nM, binds >50% at 10 μM only to p38α and β out of a panel of 363 kinases). The exquisite selectivity of the lead compound can be traced (among other features) to the presence of a chiral center with a methyl group pointing toward the back of the ATP pocket. X-ray structures revealed that this methyl group sits snugly in a small hydrophobic pocket in the vicinity of Ala157, which is an uncommon residue in that position. Both the enantiomer and the des-methyl compound are more than 10-fold less potent against p38α.

5. ALTERNATIVE INHIBITION STRATEGIES

The examples described so far have been type I kinase inhibitors. As described earlier, there is much interest in finding alternative ways to inhibit kinases, exemplified by type II, type III, and allosteric inhibitors. These modes of inhibition (where available) may have advantages over type I inhibition, including improved selectivity (via accessing less well-conserved regions of the protein), reduced potential for IP issues (for the same reason), and slow binding kinetics (where protein conformational change is required for binding). Fragment-based approaches (particularly those with a structural component) are ideally positioned to identify these more unusual types of inhibitor, as illustrated later by a number of recent publications.

5.1. Type II inhibition

Researchers at Sunesis (Erlanson et al., 2011) have employed their approach known as "Tethering with extenders" (Erlanson et al., 2000, 2003) to discover type II inhibitors of PDK1. Using this technology, they screened approximately 3000 compounds against both a phosphorylated and an unphosphorylated forms of the protein. A pyridone-containing fragment was identified which was a much stronger hit against the latter form of the protein. A noncovalent analogue was synthesized and gave an IC_{50} of 200 nM in a PDK1–AKT cascade assay. Further optimization led to a compound with low-nanomolar potency and submicromolar activity in a cellular assay. An X-ray structure of this optimized compound confirmed the predicted DFG-out-binding mode.

Other approaches have also been taken to specifically design type II inhibitors, including an HCS-based approach (Iwata et al., 2012) and an *in silico* strategy (Moffett et al., 2011). In both of these cases, fragments binding exclusively in the back pocket were identified and then grown toward the ATP site. Given the level of interest in identifying type II inhibitors, it seems likely that further fragment-based examples will be published over the next few years.

5.2. Type III inhibition

A number of inhibitors of mitogen-activated protein kinase kinase (MEK) have been discovered that utilize a novel allosteric-binding site proximal to the ATP site (Barrett et al., 2008; Heald et al., 2012). These compounds have been shown to form a ternary complex with MEK and ATP (Ohren et al., 2004) and have been termed type III inhibitors. The inhibitors in these publications were discovered by HTS and rational SBDD. Currently, we are not aware of any published type III inhibitors that have been discovered via a fragment-based approach, although we see no inherent reason why FBDD could not discover such compounds. To our knowledge, it has not yet been shown that this mode of inhibition is available in kinases other than MEK, although some have speculated that it may be possible (Tecle et al., 2009).

5.3. Other modes of inhibition

Development of allosteric kinase inhibitors remains a significant challenge. Fragment-based screens are well suited to identify "novel" ligand-binding sites on a kinase, but significant additional work may be required to understand the biological relevance and utility of these sites. In particular, multiple iterations of chemistry may be required before sufficient ligand potencies are achieved to determine the functional relevance of such sites. Sites may also be solvent exposed, small, and exhibit extremely specific SAR, making optimization challenging.

Some of these issues arose in a recent study by Jahnke et al. (2010), looking for allosteric inhibitors of Bcr–Abl binding to the myristate pocket. This pocket is involved in the regulation of c-Abl activity—when a myristoyl group covalently bound at the N-terminus of the protein binds in this pocket, it causes the SH2 and SH3 domains to dock against the kinase domain, forming an assembled inactive state. The oncogenic fusion protein Bcr–Abl lacks the myristoylation site and is therefore constitutively active.

A library of 500 fragments was screened by NMR against Abl kinase complexed with imatinib to block the ATP site. A number of hits were found, the most potent of which had a K_d of 6 μM (LE = 0.59). Surprisingly, even the most potent hits were not active in cell-based proliferation and autophosphorylation assays, despite other similarly potent myristate binders showing good activities in these same assays. This is explained by considering the conformation of the C-terminal helix I of the kinase domain. In order to allow the SH2 and SH3 domains to dock, helix I needs to bend toward the myristate pocket, which is what happens when myristate itself binds. However, some of the bound fragments would clash with the helix in this bent conformation (shown in Fig. 3.5), and so they push into its more usual linear configuration, which prevents formation of the inactive state.

NMR observations associated with residue Val525 were used to determine whether helix I was in the straight or bent configuration, and this method was able to distinguish between compounds that simply bound to the myristate pocket, and those that had activity in the functional assays

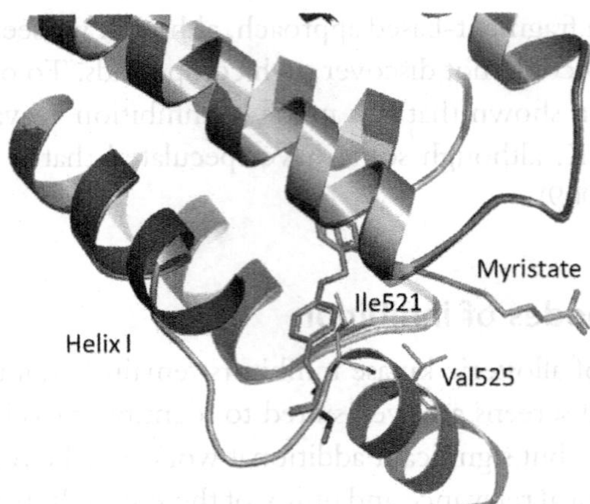

Figure 3.5 Overlay of (gray) X-ray structure of c-Abl (PDB code 1OPK) with myristate bound and helix I bent with (purple) X-ray structure of c-Abl with a fragment (pink) bound in the myristate pocket. In this latter structure, the second half of helix I is not resolved, but the bent conformation is not accessible due to the clashes between the fragment and Ile521. (See the color plate).

as well. Intriguingly, some of the fragments that bound to the myristate site without inducing the bent conformation of helix I acted as agonists of c-Abl activity in a biochemical assay.

Allosteric inhibitors have also recently been identified for CDK2 using X-ray screening (Betzi et al., 2011) and p38 using a variety of screening methods (Pollack et al., 2011).

6. SUMMARY

We hope we have shown that fragment-based screening is a versatile, effective, and proven methodology for finding kinase inhibitors. We have described several recent examples of the discovery of type I and II inhibitors, as well as allosteric inhibitors, using an array of different screening technologies.

Finally, we present a number of key lessons, derived from our own experience as well as numerous published studies:

1. Think carefully about how you wish to inhibit your target. Screens for type I, II, III, or allosteric inhibitors may need to be set up differently and even require different forms of the protein. For multidomain kinases, the other domains may contain allosteric sites, or even represent targets in their own right.

2. Invest time and effort in constructing the screening library. Many considerations go into making up a high-quality fragment-screening library, such as chemical and pharmacophoric diversity, physicochemical properties, IP, potential growth vectors, and synthetic tractability. Ensure that your library is well matched to the screening methods you plan to use in terms of the size, properties, and number of compounds.
3. Validate primary hits thoroughly. Early identification of false positives is critical to the efficient running of a fragment-screening campaign. Nearly all the examples presented here employed orthogonal screening methods to minimize this problem. Internally, we do not consider a fragment to be a validated hit until we have an X-ray structure of it bound to the target or a close family member.
4. Select the right hits for progression. There are many factors to consider when comparing hits from a fragment screen—LE, LLE_{AT}, growth vectors, novelty, and tractability being attributes that we pay particular attention to. It can be hard to predict which fragments will turn into the best leads, and so we prefer to progress as many hits as is practicable, at least initially.
5. Obtain structural information as early as possible. In almost every example, we have described, experimental structures of either the fragment hits or elaborated versions were obtained, and used to drive optimization. In some cases, information was also brought in from structures of other compounds binding to the same target, or even to different but related targets. In many of the examples, it is clear that this data enabled very rapid progression of the fragment hits. Structural information is even more critical where selectivity against closely related proteins is desired.
6. Optimize hits with care. A well-run fragment screen using a carefully designed fragment library will provide a number of promising starting points displaying efficient binding to the target. Diligent medicinal chemistry, with careful monitoring of binding efficiency and properties, will be required to optimize these to attractive lead molecules. Unless there are compelling structural overlays with other compounds, we try to add only a few atoms at a time and evaluate the contribution made by each one. Group efficiency (Verdonk & Rees, 2008) can be a helpful concept here.

Over the last 15 years, FBDD has transformed from an academic curiosity to a mainstream approach with practitioners across industry and academia. Kinases have played a significant part in this story, and it is unsurprising that the first FBDD-derived drug on the market is a kinase inhibitor. This success, along with the number of other fragment-derived kinase inhibitors currently in the clinic, and a wealth of publications all highlight how well suited FBDD is to discovering kinase inhibitors. We anticipate the approval of more fragment-derived drugs targeting kinases over the coming years.

ACKNOWLEDGMENTS

The authors would like to thank Drs. Chris Murray, David Rees, and Michelle Jones for helpful discussions and critical reading of the chapter.

REFERENCES

Abad-Zapatero, C., & Metz, J. T. (2005). Ligand efficiency indices as guideposts for drug discovery. *Drug Discovery Today, 10*, 464–469.

Akritopoulou-Zanze, I., & Hajduk, P. J. (2009). Kinase-targeted libraries: The design and synthesis of novel, potent, and selective kinase inhibitors. *Drug Discovery Today, 14*, 291–297.

Bamborough, P., Brown, M. J., Christopher, J. A., Chung, C., & Mellor, G. W. (2011). Selectivity of kinase inhibitor fragments. *Journal of Medicinal Chemistry, 54*, 5131–5143.

Barrett, S. D., Bridges, A. J., Dudley, D. T., Saltiel, A. R., Fergus, J. H., Flamme, C. M., et al. (2008). The discovery of the benzhydroxamate MEK inhibitors CI-1040 and PD 0325901. *Bioorganic & Medicinal Chemistry Letters, 18*, 6501–6504.

Bembenek, S. D., Tounge, B. A., & Reynolds, C. H. (2009). Ligand efficiency and fragment-based drug discovery. *Drug Discovery Today, 14*, 278–283.

Berman, H. M. (2000). The protein data bank. *Nucleic Acids Research, 28*, 235–242.

Betzi, S., Alam, R., Martin, M., Lubbers, D. J., Han, H., Jakkaraj, S. R., et al. (2011). Discovery of a potential allosteric ligand binding site in CDK2. *ACS Chemical Biology, 6*, 492–501.

Blum, L. C., & Reymond, J.-L. (2009). 970 million druglike small molecules for virtual screening in the chemical universe database GDB-13. *Journal of the American Chemical Society, 131*, 8732–8733.

Bollag, G., Hirth, P., Tsai, J., Zhang, J., Ibrahim, P. N., Cho, H., et al. (2010). Clinical efficacy of a RAF inhibitor needs broad target blockade in BRAF-mutant melanoma. *Nature, 467*, 596–599.

Buchanan, S. G., Hendle, J., Lee, P. S., Smith, C. R., Bounaud, P.-Y., Jessen, K. A., et al. (2009). SGX523 is an exquisitely selective, ATP-competitive inhibitor of the MET receptor tyrosine kinase with antitumor activity in vivo. *Molecular Cancer Therapeutics, 8*, 3181–3190.

Charrier, J.-D., Miller, A., Kay, D. P., Brenchley, G., Twin, H. C., Collier, P. N., et al. (2011). Discovery and structure-activity relationship of 3-aminopyrid-2-ones as potent and selective interleukin-2 inducible T-cell kinase (Itk) inhibitors. *Journal of Medicinal Chemistry, 54*, 2341–2350.

Chen, I.-J., & Hubbard, R. E. (2009). Lessons for fragment library design: Analysis of output from multiple screening campaigns. *Journal of Computer-Aided Molecular Design, 23*, 603–620.

Chessari, G., & Woodhead, A. J. (2009). From fragment to clinical candidate—A historical perspective. *Drug Discovery Today, 14*, 668–675.

Cho, Y. S., Angove, H., Brain, C., Chen, C. H.-T., Cheng, H., Cheng, R., et al. (2012). Fragment-based discovery of 7-azabenzimidazoles as potent, highly selective, and orally active CDK4/6 inhibitors. *ACS Medicinal Chemistry Letters, 3*, 445–449.

Cohen, P. (2002). Protein kinases—The major drug targets of the twenty-first century? *Nature Reviews. Drug Discovery*, *1*, 309–315.

Congreve, M., Carr, R., Murray, C., & Jhoti, H. (2003). A "Rule of Three" for fragment-based lead discovery? *Drug Discovery Today*, *8*, 876–877.

Congreve, M., Chessari, G., Tisi, D., & Woodhead, A. J. (2008). Recent developments in fragment-based drug discovery. *Journal of Medicinal Chemistry*, *51*, 3661–3680.

Coyne, A. G., Scott, D. E., & Abell, C. (2010). Drugging challenging targets using fragment-based approaches. *Current Opinion in Chemical Biology*, *14*, 299–307.

Davies, B. R., Greenwood, H., Dudley, P., Crafter, C., Yu, D.-H., Zhang, J., et al. (2012). Preclinical pharmacology of AZD5363, an inhibitor of AKT: Pharmacodynamics, antitumor activity, and correlation of monotherapy activity with genetic background. *Molecular Cancer Therapeutics*, *11*, 873–887.

Davies, T. G., & Hyvönen, M. (Eds.), (2012). *Fragment-based drug discovery and X-ray crystallography*. Berlin/Heidelberg: Springer-Verlag.

de Kloe, G. E., Bailey, D., Leurs, R., & de Esch, I. J. P. (2009). Transforming fragments into candidates: Small becomes big in medicinal chemistry. *Drug Discovery Today*, *14*, 630–646.

Denardo, D. G., Brennan, D. J., Rexhepaj, E., Ruffell, B., Shiao, S. L., Madden, S. F., et al. (2011). Leukocyte complexity predicts breast cancer survival and functionally regulates response to chemotherapy. *Cancer Discovery*, *1*, 54–67.

Erlanson, D. A. (2009). Fragment-based drug discovery of kinase inhibitors. In R. Li, & J. A. Stafford (Eds.), *Kinase inhibitor drugs* (pp. 461–483). Hoboken, NJ, USA: John Wiley & Sons Inc.

Erlanson, D. A., Arndt, J. W., Cancilla, M. T., Cao, K., Elling, R. A., English, N., et al. (2011). Discovery of a potent and highly selective PDK1 inhibitor via fragment-based drug discovery. *Bioorganic & Medicinal Chemistry Letters*, *21*, 3078–3083.

Erlanson, D. A., Braisted, A. C., Raphael, D. R., Randal, M., Stroud, R. M., Gordon, E. M., et al. (2000). Site-directed ligand discovery. *Proceedings of the National Academy of Sciences of the United States of America*, *97*, 9367–9372.

Erlanson, D. A., Lam, J. W., Wiesmann, C., Luong, T. N., Simmons, R. L., DeLano, W. L., et al. (2003). In situ assembly of enzyme inhibitors using extended tethering. *Nature Biotechnology*, *21*, 308–314.

Gill, A. (2004). New lead generation strategies for protein kinase inhibitors—Fragment based screening approaches. *Mini-Reviews in Medicinal Chemistry*, *4*, 301–311.

Giordanetto, F., Kull, B., & Dellsén, A. (2011). Discovery of novel class 1 phosphatidylinositide 3-kinases (PI3K) fragment inhibitors through structure-based virtual screening. *Bioorganic & Medicinal Chemistry Letters*, *21*, 829–835.

Hajduk, P. J., & Greer, J. (2007). A decade of fragment-based drug design: Strategic advances and lessons learned. *Nature Reviews. Drug Discovery*, *6*, 211–219.

Hajduk, P. J., Sheppard, G., Nettesheim, D. G., Olejniczak, E. T., Shuker, S. B., Meadows, R. P., et al. (1997). Discovery of potent nonpeptide inhibitors of stromelysin using SAR by NMR. *Journal of the American Chemical Society*, *119*, 5818–5827.

Hann, M. M., Leach, A. R., & Harper, G. (2001). Molecular complexity and its impact on the probability of finding leads for drug discovery. *Journal of Chemical Information and Modeling*, *41*, 856–864.

Heald, R. A., Jackson, P., Savy, P., Jones, M., Gancia, E., Burton, B., et al. (2012). Discovery of novel allosteric mitogen-activated protein kinase kinase (MEK) 1,2 inhibitors possessing bidentate Ser212 interactions. *Journal of Medicinal Chemistry*, *55*, 4594–4604.

Hopkins, A. L., Groom, C. R., & Alex, A. (2004). Ligand efficiency: A useful metric for lead selection. *Drug Discovery Today*, *9*, 430–431.

Howard, S., Berdini, V., Boulstridge, J. A., Carr, M. G., Cross, D. M., Curry, J., et al. (2009). Fragment-based discovery of the pyrazol-4-yl urea (AT9283), a multitargeted kinase inhibitor with potent aurora kinase activity. *Journal of Medicinal Chemistry*, *52*, 379–388.

Hughes, S. J., Millan, D. S., Kilty, I. C., Lewthwaite, R. A., Mathias, J. P., O'Reilly, M. A., et al. (2011). Fragment based discovery of a novel and selective PI3 kinase inhibitor. *Bioorganic & Medicinal Chemistry Letters, 21*, 6586–6590.

Iwata, H., Oki, H., Okada, K., Takagi, T., Tawada, M., Miyazaki, Y., et al. (2012). A back-to-front fragment-based drug design search strategy targeting the DFG-out pocket of protein tyrosine kinases. *ACS Medicinal Chemistry Letters, 3*, 342–346.

Jahnke, W., & Erlanson, D. A. (Eds.), (2006). *Fragment-based approaches in drug discovery*. Meinheim, Germany: Wiley-VCH.

Jahnke, W., Grotzfeld, R. M., Pellé, X., Strauss, A., Fendrich, G., Cowan-Jacob, S. W., et al. (2010). Binding or bending: Distinction of allosteric Abl kinase agonists from antagonists by an NMR-based conformational assay. *Journal of the American Chemical Society, 132*, 7043–7048.

Johnson, M. C., Hu, Q., Lingardo, L., Ferre, R. A., Greasley, S., Yan, J., et al. (2011). Novel isoquinolone PDK1 inhibitors discovered through fragment-based lead discovery. *Journal of Computer-Aided Molecular Design, 25*, 689–698.

Keserü, G. M., & Makara, G. M. (2009). The influence of lead discovery strategies on the properties of drug candidates. *Nature Reviews. Drug Discovery, 8*, 203–212.

Kuntz, I. D. (1999). The maximal affinity of ligands. *Proceedings of the National Academy of Sciences of the United States of America, 96*, 9997–10002.

Kuo, L. C. (Ed.), (2011). *Methods in enzymology, Vol. 493. Fragment-based drug design: Tools, practical approaches, and examples*. Amsterdam: Elsevier.

Lau, W. F., Withka, J. M., Hepworth, D., Magee, T. V., Du, Y. J., Bakken, G. A., et al. (2011). Design of a multi-purpose fragment screening library using molecular complexity and orthogonal diversity metrics. *Journal of Computer-Aided Molecular Design, 25*, 621–636.

Leach, A. R., & Hann, M. M. (2011). Molecular complexity and fragment-based drug discovery: Ten years on. *Current Opinion in Chemical Biology, 15*, 489–496.

Lee, A. C.-H., Ramanujulu, P. M., Poulsen, A., Williams, M., Blanchard, S., Ma, D. M., et al. (2012). Thieno[3,2-d]pyrimidin-4(3H)-one derivatives as PDK1 inhibitors discovered by fragment-based screening. *Bioorganic & Medicinal Chemistry Letters, 22*, 4023–4027.

Leeson, P. D., & Springthorpe, B. (2007). The influence of drug-like concepts on decision-making in medicinal chemistry. *Nature Reviews. Drug Discovery, 6*, 881–890.

Leeson, P. D., & St-Gallay, S. A. (2011). The influence of the "organizational factor" on compound quality in drug discovery. *Nature Reviews. Drug Discovery, 10*, 749–765.

Li, R., Martin, M. P., Liu, Y., Wang, B., Patel, R. A., Zhu, J.-Y., et al. (2012). Fragment-based and structure-guided discovery and optimization of Rho kinase inhibitors. *Journal of Medicinal Chemistry, 55*, 2474–2478.

Lipinski, C. A., Lombardo, F., Dominy, B. W., & Feeney, P. J. (1997). Experimental and computational approaches to estimate solubility and permeability in drug discovery and development settings. *Advanced Drug Delivery Reviews, 23*, 3–25.

Lovering, F., Bikker, J., & Humblet, C. (2009). Escape from flatland: Increasing saturation as an approach to improving clinical success. *Journal of Medicinal Chemistry, 52*, 6752–6756.

Manning, G., Whyte, D. B., Martinez, R., Hunter, T., & Sudarsanam, S. (2002). The protein kinase complement of the human genome. *Science (New York, N.Y.), 298*, 1912–1934.

Matthews, T. P., Klair, S., Burns, S., Boxall, K., Cherry, M., Fisher, M., et al. (2009). Identification of inhibitors of checkpoint kinase 1 through template screening. *Journal of Medicinal Chemistry, 52*, 4810–4819.

Medina, J. R., Blackledge, C. W., Heerding, D. A., Campobasso, N., Ward, P., Briand, J., et al. (2010). Aminoindazole PDK1 inhibitors: A case study in fragment-based drug discovery. *ACS Medicinal Chemistry Letters, 1*, 439–442.

Moffett, K., Konteatis, Z., Nguyen, D., Shetty, R., Ludington, J., Fujimoto, T., et al. (2011). Discovery of a novel class of non-ATP site DFG-out state p38 inhibitors utilizing computationally assisted virtual fragment-based drug design (vFBDD). *Bioorganic & Medicinal Chemistry Letters*, *21*, 7155–7165.

Mortenson, P. N., & Murray, C. W. (2011). Assessing the lipophilicity of fragments and early hits. *Journal of Computer-Aided Molecular Design*, *25*, 663–667.

Murray, C. W., Verdonk, M. L., & Rees, D. C. (2012). Experiences in fragment-based drug discovery. *Trends in Pharmacological Sciences*, *33*, 224–232.

Ohren, J. F., Chen, H., Pavlovsky, A., Whitehead, C., Zhang, E., Kuffa, P., et al. (2004). Structures of human MAP kinase kinase 1 (MEK1) and MEK2 describe novel non-competitive kinase inhibition. *Nature Structural & Molecular Biology*, *11*, 1192–1197.

Pollack, S. J., Beyer, K. S., Lock, C., Müller, I., Sheppard, D., Lipkin, M., et al. (2011). A comparative study of fragment screening methods on the p38α kinase: New methods, new insights. *Journal of Computer-Aided Molecular Design*, *25*, 677–687.

Ray, P., Wright, J., Adam, J., Bennett, J., Boucharens, S., Black, D., et al. (2011). Fragment-based discovery of 6-substituted isoquinolin-1-amine based ROCK-I inhibitors. *Bioorganic & Medicinal Chemistry Letters*, *21*, 97–101.

Schulz, M. N., Fanghänel, J., Schäfer, M., Badock, V., Briem, H., Boemer, U., et al. (2011). A crystallographic fragment screen identifies cinnamic acid derivatives as starting points for potent Pim-1 inhibitors. *Acta Crystallographica Section D: Biological Crystallography*, *67*, 156–166.

Schulz, M. N., & Hubbard, R. E. (2009). Recent progress in fragment-based lead discovery. *Current Opinion in Pharmacology*, *9*, 615–621.

Scott, D. E., Coyne, A. G., Hudson, S. A., & Abell, C. (2012). Fragment-based approaches in drug discovery and chemical biology. *Biochemistry*, *51*, 4990–5003.

Shuker, S. B., Hajduk, P. J., Meadows, R. P., & Fesik, S. W. (1996). Discovering high-affinity ligands for proteins: SAR by NMR. *Science*, *274*, 1531–1534.

Soth, M., Abbot, S., Abubakari, A., Arora, N., Arzeno, H., Billedeau, R., et al. (2011). 3-Amino-pyrazolo[3,4-d]pyrimidines as p38α kinase inhibitors: Design and development to a highly selective lead. *Bioorganic & Medicinal Chemistry Letters*, *21*, 3452–3456.

Taylor, S. S., Knighton, D. R., Zheng, J., Ten Eyck, L. F., & Sowadski, J. M. (1992). Structural framework for the protein kinase family. *Annual Review of Cell Biology*, *8*, 429–462.

Tecle, H., Shao, J., Li, Y., Kothe, M., Kazmirski, S., Penzotti, J., et al. (2009). Beyond the MEK-pocket: Can current MEK kinase inhibitors be utilized to synthesize novel type III NCKIs? Does the MEK-pocket exist in kinases other than MEK? *Bioorganic & Medicinal Chemistry Letters*, *19*, 226–229.

Teotico, D. G., Babaoglu, K., Rocklin, G. J., Ferreira, R. S., Giannetti, A. M., & Shoichet, B. K. (2009). Docking for fragment inhibitors of AmpC beta-lactamase. *Proceedings of the National Academy of Sciences of the United States of America*, *106*, 7455–7460.

The Protein Data Bank. (2012). www.pdb.org Accessed 04.07.12 [WWW Document].

The UniProt Consortium. (2012). Reorganizing the protein space at the Universal Protein Resource (UniProt). *Nucleic Acids Research*, *40*, D71–D75.

The Universal Protein Resource (UniProt). (2012) www.uniprot.org Accessed 04.07.12 [WWW Document].

Tsai, J., Lee, J. T., Wang, W., Zhang, J., Cho, H., Mamo, S., et al. (2008). Discovery of a selective inhibitor of oncogenic B-Raf kinase with potent antimelanoma activity. *Proceedings of the National Academy of Sciences of the United States of America*, *105*, 3041–3046.

US Food and Drug Administration. (2012) *FDA labeling information—Zelboraf*. FDA website: http://www.accessdata.fda.gov/drugsatfda_docs/label/2011/202429s000lbl.pdf Accessed 03.07.12 [WWW Document].

VanderPorten, E. C., Taverna, P., Hogan, J. N., Ballinger, M. D., Flanagan, W. M., & Fucini, R. V. (2009). The aurora kinase inhibitor SNS-314 shows broad therapeutic potential with chemotherapeutics and synergy with microtubule-targeted agents in a colon carcinoma model. *Molecular Cancer Therapeutics*, *8*, 930–939.

Verdonk, M. L., & Rees, D. C. (2008). Group efficiency: A guideline for hits-to-leads chemistry. *ChemMedChem*, *3*, 1179–1180.

Vieth, M., Siegel, M. G., Higgs, R. E., Watson, I. A., Robertson, D. H., Savin, K. A., et al. (2004). Characteristic physical properties and structural fragments of marketed oral drugs. *Journal of Medicinal Chemistry*, *47*, 224–232.

Wenlock, M. C., Austin, R. P., Barton, P., Davis, A. M., & Leeson, P. D. (2003). A comparison of physiochemical property profiles of development and marketed oral drugs. *Journal of Medicinal Chemistry*, *46*, 1250–1256.

Wyatt, P. G., Woodhead, A. J., Berdini, V., Boulstridge, J. A., Carr, M. G., Cross, D. M., et al. (2008). Identification of N-(4-piperidinyl)-4-(2,6-dichlorobenzoylamino)-1H-pyrazole-3-carboxamide (AT7519), a novel cyclin dependent kinase inhibitor using fragment-based X-ray crystallography and structure based drug design. *Journal of Medicinal Chemistry*, *51*, 4986–4999.

Xiang, Y., Hirth, B., Asmussen, G., Biemann, H.-P., Bishop, K. A., Good, A., et al. (2011). The discovery of novel benzofuran-2-carboxylic acids as potent Pim-1 inhibitors. *Bioorganic & Medicinal Chemistry Letters*, *21*, 3050–3056.

Yap, T. A., Walton, M. I., Grimshaw, K. M., Te Poele, R. H., Eve, P. D., Valenti, M. R., et al. (2012). AT13148 is a novel, oral multi-AGC kinase inhibitor with potent pharmacodynamic and antitumor activity. *Clinical Cancer Research*, *18*, 3912–3923.

Yu, W., Chen, J., Xiong, Y., Pixley, F. J., Yeung, Y.-G., & Stanley, E. R. (2012). Macrophage proliferation is regulated through CSF-1 receptor tyrosines 544, 559, and 807. *The Journal of Biological Chemistry*, *287*, 13694–13704.

Zartler, E., & Shapiro, M. (Eds.), (2008). *Fragment-based drug discovery: A practical approach.* Hoboken, NJ, USA: Wiley.

CHAPTER FOUR

Targeting Protein Kinases with Selective and Semipromiscuous Covalent Inhibitors

Rand M. Miller*, **Jack Taunton**[†,1]
*Chemistry and Chemical Biology Graduate Program, University of California, San Francisco, California, USA
[†]Howard Hughes Medical Institute and Department of Cellular and Molecular Pharmacology, University of California, San Francisco, California, USA
[1]Corresponding author: e-mail address: jack.taunton@ucsf.edu

Contents

Abstract

Protein kinase inhibitors are an important class of therapeutics. In addition, selective kinase inhibitors can often reveal unexpected biological insights, augmenting genetic approaches and playing a decisive role in preclinical target validation studies. Nevertheless, developing protein kinase inhibitors with sufficient selectivity and pharmacodynamic potency presents significant challenges. Targeting noncatalytic cysteines with covalent inhibitors is a powerful approach to address both challenges simultaneously. Here, we describe our efforts to design irreversible and reversible electrophilic inhibitors with varying degrees of kinase selectivity. Highly selective covalent inhibitors have been used to elucidate the roles of p90 ribosomal protein S6 kinases in animal models of atherosclerosis and diabetes. By contrast, semipromiscuous covalent inhibitors have revealed new therapeutic targets in disease-causing parasites and have shown utility as chemoproteomic probes for interrogating kinase occupancy in living cells.

Methods in Enzymology, Volume 548
ISSN 0076-6879
http://dx.doi.org/10.1016/B978-0-12-397918-6.00004-5

1. INTRODUCTION

Protein phosphorylation by kinases regulates nearly every aspect of cellular physiology (Manning, Whyte, Martinez, Hunter, & Sudarsanam, 2002). Because protein kinases are misregulated in many diseases, they have been hotly pursued as therapeutic targets for drug discovery. Out of 518 human protein kinases, only a small fraction have been targeted with selective inhibitors. By virtue of their high structural homology, particularly in the ATP binding site targeted by most small-molecule inhibitors, selective inhibition of distinct protein kinases remains a major challenge.

Covalent targeting of poorly conserved, noncatalytic cysteine residues with electrophilic kinase inhibitors has emerged as a powerful strategy for increasing potency and selectivity (Barf & Kaptein, 2012). Structural bioinformatics analysis of the human kinome has revealed ~200 kinase domains that have a solvent-exposed cysteine within striking distance of the ATP binding site (Leproult, Barluenga, Moras, Wurtz, & Winssinger, 2011; Liu et al., 2013). In spite of the potential for improved selectivity, covalent inhibitors are typically avoided by the pharmaceutical industry out of fear of idiosyncratic toxicity resulting from drug–protein adducts (Evans, Watt, Nicoll-Griffith, & Baillie, 2004; Uetrecht, 2008). Nevertheless, covalent drugs are making a comeback, especially in the context of molecularly targeted therapies for cancer (Singh, Petter, Baillie, & Whitty, 2011).

An early example of a cysteine-targeted kinase inhibitor is 2′-thioadenosine, a proof-of-concept compound designed to form a disulfide bond with a poorly conserved cysteine (Cys797) in the EGFR kinase domain (Singh et al., 1997). Subsequently, derivatization of an EGFR-selective quinazoline scaffold with an acrylamide electrophile resulted in an irreversible inhibitor with enhanced antitumor activity (Fry et al., 1998). The success of this approach is evident in the recent FDA approval of two irreversible cysteine-targeted kinase inhibitors. Afatinib, a quinazoline-based EGFR inhibitor (Solca et al., 2012), is approved for advanced non-small cell lung cancer, and ibrutinib, a Bruton's tyrosine kinase (BTK) inhibitor based on a pyrazolopyrimidine scaffold (Pan et al., 2007), is approved for mantle cell lymphoma and chronic lymphocytic leukemia. Both compounds use an acrylamide electrophile to irreversibly target a cysteine shared by EGFR and BTK.

In this chapter, we describe our chemistry-focused approach to the design of electrophilic kinase inhibitors. Case studies illustrate strategies for targeting noncatalytic cysteines as well as the catalytic lysine in protein kinases. We discuss our recent discovery of reversible covalent kinase inhibitors and discuss a novel method for targeting noncatalytic cysteines with reversible electrophilic fragments. In addition to designing selective inhibitors, we have also developed covalent probes that are "semipromiscuous," targeting a relatively common cysteine or the conserved catalytic lysine. We have used these probes to identify new therapeutic kinase targets and quantify kinase target engagement in living cells by other inhibitors, including ponatinib, a leukemia drug recently associated with life-threatening side effects.

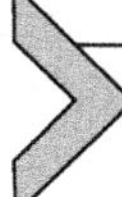

2. DESIGN OF IRREVERSIBLE CYSTEINE-TARGETED KINASE INHIBITORS

Identification of a suitable kinase target and a noncovalent recognition scaffold are primary considerations when embarking on a covalent inhibitor design project. If the goal is to obtain highly selective inhibitors, the kinase should have a solvent-exposed cysteine near a "druggable" pocket, typically within ~10 Å of the ATP binding site. Ideally, this cysteine should be found in only a handful of other human kinases. The design process usually starts with the identification of a noncovalent kinase-recognition scaffold that (1) inhibits the desired target with at least micromolar potency (e.g., $IC_{50} < 100\ \mu M$) and (2) can be modified to place an electrophilic moiety in close proximity to the targeted cysteine. Although a cocrystal structure of the scaffold bound to the kinase target is ideal for guiding the placement of electrophilic substituents, it is not essential; in a pinch, docking to a homology model may suffice. With the appropriate choice of electrophile, it is possible to transform a recognition scaffold with weak affinity and poor selectivity into a selective covalent inhibitor with potency in the low nanomolar to picomolar range.

Because IC_{50} values for irreversible inhibitors usually depend on the incubation time and ATP concentration, we hold these parameters constant to compare potencies across a series of compounds (e.g., 200 μM ATP, 30 min incubation). If desired, a real-time assay format (e.g., ADP Quest, DiscoveRX) can be used to measure the kinetics of kinase inhibition (K_i and k_{inact}). To increase the stringency of the biochemical kinase assay and our ability to differentiate between highly thiol-reactive inhibitors, we often include 10 mM glutathione (GSH) as a physiologically relevant competing nucleophile.

2.1. Irreversible covalent inhibitors of RSK1/2/4

The p90 ribosomal protein S6 kinases comprise four closely related paralogs (RSK1–4) activated downstream of the Ras-MAPK pathway (Hauge & Frödin, 2006; Romeo, Zhang, & Roux, 2012). RSKs contain two kinase domains, an AGC-family N-terminal kinase domain (NTD) and a CAMK-family C-terminal kinase domain (CTD) connected by a short linker with several regulatory phosphorylation sites. Following phosphorylation of Thr577 (human RSK2 numbering) by ERK, the activated CTD autophosphorylates Ser386 within the linker segment (Dalby, Morrice, Caudwell, Avruch, & Cohen, 1998, Fisher & Blenis, 1996). This serves as a docking site for PDK1, which phosphorylates the NTD activation loop, leading to full activation of the NTD and downstream signaling (Frödin, Jensen, Merienne, & Gammeltoft, 2000). RSKs have been shown to phosphorylate dozens of proteins involved in diverse cellular processes, including the sodium/hydrogen exchanger NHE1 (Cuello, Snabaitis, Cohen, Taunton, & Avkiran, 2006; Takahashi et al., 1999), the translation initiation factor eIF4B (Shahbazian et al., 2006), the tumor suppressor kinase LKB1 (Doehn et al., 2009; Sapkota et al., 2007), and the transcription factor c-Fos (David et al., 2005). Many if not most phosphorylation sites attributed to RSK have also been linked to other kinases (e.g., kinases downstream of PI3K and p38 MAPK), depending on the cellular context. RSK hyperactivity has been implicated in tumor cell invasion (Doehn et al., 2009; Kang et al., 2010; Smolen et al., 2010), as well as endothelial dysfunction and atherosclerosis (Le et al., 2013).

We used a sequence alignment of the human kinome (Buzko & Shokat, 2002) to identify poorly conserved, noncatalytic cysteines in the ATP binding site. Structure-guided analysis of this alignment revealed 11 kinases (Fig. 4.1A), including the CTDs of RSK1–4, with a cysteine projecting down from the "ceiling" of the ATP pocket, near the C-terminal end of the glycine-rich loop. Motivating our studies at the time, few useful inhibitors were known for any of these kinases. We were also motivated by the distance and geometry constraints imposed by this cysteine, which lies deeper in the pocket than the EGFR cysteine. These constraints forced us to consider alternatives to the acrylamide electrophiles used to target EGFR. Of the 11 kinases with this particular cysteine, we noticed that only RSK1/2/4 (but not RSK3) have a threonine in the "gatekeeper" position (Fig. 4.1A), which in principle could open up an extended hydrophobic pocket that is less accessible in kinases with larger gatekeepers (Liu et al.,

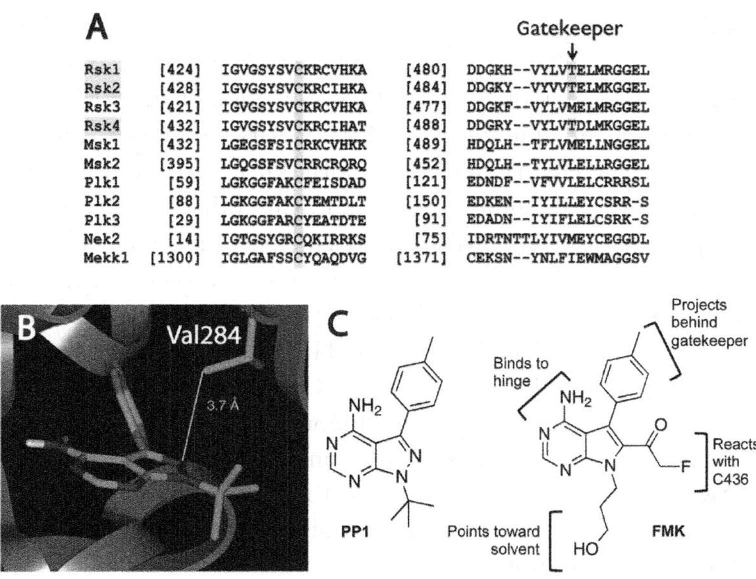

Figure 4.1 Structural bioinformatics-based design of cysteine-targeted RSK inhibitors. (A) Sequence alignment reveals two selectivity filters unique to RSK1/2/4: a noncatalytic Cys and a Thr gatekeeper. (B) Multikinase inhibitor PP1 bound to the Src-family kinase HCK (PDB code: 1QCF), with N2 of PP1 proximal to Val284, corresponding to Cys436 in RSK2. (C) Chemical structure of PP1 and the irreversible RSK inhibitor, **FMK**. (See the color plate.)

1999). Although we lacked knowledge of any RSK CTD inhibitors or crystal structures, our goal was to design an inhibitor that simultaneously exploited both selectivity filters found uniquely in RSK1/2/4, the Cys at the end of the Gly-rich loop and the Thr gatekeeper (Cohen, Zhang, Shokat, & Taunton, 2005).

A cocrystal structure of the pyrazolopyrimidine PP1 bound to HCK (Schindler et al., 1999), a SRC-family kinase with a Thr gatekeeper but otherwise distantly related to RSK, inspired the design of a fluoromethylketone inhibitor, prosaically named **FMK** (Fig. 4.1B and C). We predicted that Cys436 in the RSK2 CTD would occupy a similar space as Val284 in HCK and would be within striking distance of a fluoromethylketone appended to C6 of a PP1-like pyrrolopyrimidine scaffold. A hydroxypropyl substituent at N7, predicted to be solvent accessible, would enable derivatization with fluorophores and affinity tags. Although fluoromethylketones have been used to target highly reactive catalytic cysteines in Cys proteases, their ability to covalently target noncatalytic cysteines had not been explored extensively.

FMK irreversibly inhibits RSK2 in biochemical (IC_{50} ~ 15 n*M*) and cellular assays (EC_{50} ~ 300 n*M*), and both CTD selectivity filters are required for potent inhibition (Cohen et al., 2005). Mutation of either Cys436 to Val or the gatekeeper Thr493 to Met confers resistance (biochemical IC_{50} > 3 μ*M*). Given that it has a similar kinase-recognition scaffold as PP1, which reversibly inhibits SRC and many other tyrosine kinases, we were surprised to find that **FMK** was extremely selective when tested against a large panel of kinases (unpublished results in collaboration with Eric Johnson, AbbVie). Out of 150 kinases tested, **FMK** inhibited only five with an IC_{50} below 5 μ*M*; only two kinases had submicromolar IC_{50}s (S6K1 ~ 0.5 μ*M* and PTK5 ~ 0.7 μ*M*). Contrary to a published report (Bain et al., 2007), **FMK** failed to inhibit LCK or SRC (IC_{50} > 5 μ*M*), whereas PP1 potently inhibits both kinases. The enhanced selectivity of **FMK** relative to PP1 suggests that the fluoromethylketone sterically clashes with most kinases that would otherwise accommodate the pyrrolopyrimidine scaffold.

We employed a similar strategy to target the structurally analogous cysteine in NEK2, a cell cycle-regulated kinase that localizes to centrosomes and is overexpressed in many cancers. A cocrystal structure of the oxindole SU11652 (related to the kidney cancer drug, sunitinib) bound to NEK2 suggested that the C5 position could project electrophiles to engage Cys22. Whereas the oxindole scaffold (including SU11652 and sunitinib) inhibited NEK2 weakly (IC_{50} > 5 μ*M*), appending a propynamide electrophile to C5 led to an irreversible NEK2 inhibitor that does not cross-react with PLK1, an essential cell cycle kinase with an equivalent cysteine (Henise & Taunton, 2011). Again, mutation of the key cysteine in NEK2 conferred resistance.

2.1.1 Applications of irreversible covalent kinase inhibitors

An advantage of irreversible (or slowly reversible; see below) covalent inhibitors that target a noncatalytic cysteine is the relative ease with which a cellular phenotype can be attributed to inactivation of the desired kinase, as opposed to inhibiting myriad off-target kinases. An inhibitor washout experiment is the simplest method: following acute treatment with a saturating amount of the inhibitor (e.g., 5 μ*M* **FMK**, 30–60 min incubation at 37 °C), cells are washed extensively (3 × 5 min with compound-free media) and the biological readout is assessed (e.g., RSK autophosphorylation in response to growth factor stimulation). Cellular phenotypes that persist after inhibitor washout may derive from irreversible inactivation of the desired kinase, whereas phenotypes that disappear after washout are likely due to reversible inhibition of irrelevant targets. Drug washout experiments may be confounded by rapid resynthesis of the kinase, giving a false-negative result; depending on the resynthesis rate, this is more likely to be an issue if the phenotype is scored many hours after the washout step (e.g., cell proliferation). In addition, some inhibitors are not easily washed out and may accumulate in cells at high concentrations. To control for this potential artifact and provide additional evidence linking a cellular phenotype to covalent targeting, we recommend synthesizing an isosteric version of the inhibitor that is not electrophilic. With **FMK**, for example, we synthesized the corresponding methyl ketone lacking the fluorine (unpublished results). With the NEK2 inhibitor, we added a methyl group to the electrophilic carbon of the propynamide (Henise & Taunton, 2011). The resulting "negative control" compounds are inactive versus RSK and NEK2, but they have nearly identical physical properties and are similar in their ability to reversibly engage kinases.

The most convincing way to demonstrate that a cellular phenotype results from covalent inactivation of the desired kinase is to "rescue" the phenotype by reconstituting cells with a resistant allele of the kinase. Typically, the key Cys is mutated to Val, Ala, or Ser. In many cases, it is possible to confer resistance to the electrophilic inhibitor by expressing the mutant kinase in cells that also express the endogenous wild-type kinase; we have used this approach with our RSK and NEK2 inhibitors (Doehn et al., 2009; Henise & Taunton, 2011). However, in certain cases, it may be difficult to achieve optimal expression levels of the transgene, or the inhibitor-bound endogenous kinase may act in a "dominant negative" fashion and interfere with the function of the ectopically expressed mutant kinase. In such cases, it may be necessary to use cells in which the endogenous kinase has been genetically deleted or knocked down. Using zinc finger nuclease technology, Frödin and colleagues engineered **FMK** resistance by introducing a C436V mutation into the endogenous RSK2 (also known as *RPS6KA3*) gene (Chen et al., 2011). Advances in CRISPR/Cas9 genome-editing technology will likely make this process much easier (Friedland et al., 2013; Wang et al., 2013).

2.2. Fluorescent and alkyne-tagged probes to quantify proteome-wide selectivity and RSK occupancy *in vivo*

To identify the intracellular targets of **FMK** and quantify RSK occupancy as a function of inhibitor concentration, we required tagged derivatives. We designed the following two probes for different applications (Cohen, Hadjivassiliou, & Taunton, 2007): (1) **FMK-PA**, a more potent, clickable derivative for assessing proteome-wide selectivity and (2) **FMK-BODIPY**, a cell-permeable fluorescent probe for convenient quantitation of RSK occupancy (Fig. 4.2). Based on the proposed binding mode of **FMK**, we

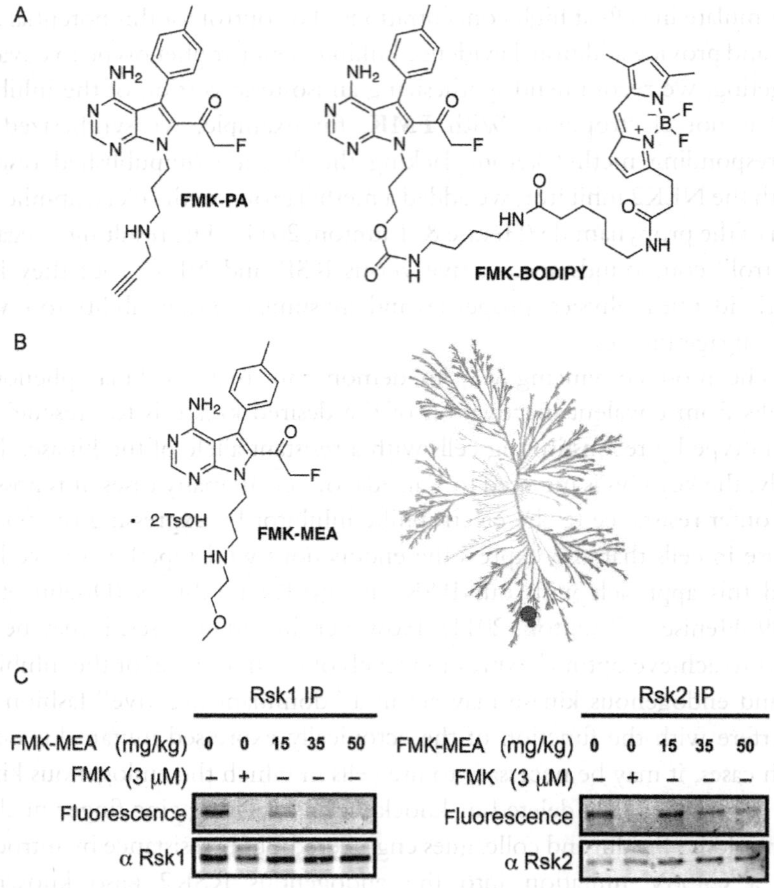

Figure 4.2 FMK derivatives used to elucidate pathophysiological roles of RSK. (A) Chemical structures of **FMK-PA** and **FMK-BODIPY**. (B) Kinome-wide selectivity of **FMK-MEA**, screened at 1 μ*M* versus 443 kinases (DiscoverX). Only RSK1 and RSK4 CTDs were bound >50% relative to DMSO controls. (C) Using **FMK-BODIPY** as a probe, RSK1/2 occupancy was assessed in cardiac tissue lysates derived from mice treated with the indicated dose of **FMK-MEA**. To visualize **FMK-BODIPY**-labeled RSK1 and RSK2, each kinase was separately immunoprecipitated with isoform-specific antibodies. A dose-dependent reduction in fluorescence indicates RSK occupancy by **FMK-MEA**. *Figure adapted with permission from Lippincott Williams and Wilkins/Wolters Kluwer Health: Circulation, Le et al. (2013).* (See the color plate.)

expected the hydroxypropyl group to be solvent exposed and able to accommodate various tags.

Substitution of the primary alcohol of **FMK** with propargylamine (via the di-Boc-protected mesylate) provided **FMK-PA**, amenable to bioorthogonal Cu(I)-mediated azide-alkyne cycloaddition, or click chemistry. **FMK-PA** was more potent than **FMK** in cellular assays, showing maximal inhibition of RSK autophosphorylation at 100 n*M*. After treating cells with **FMK-PA**, click chemistry with rhodamine-azide was performed on cell lysates. Only two rhodamine-labeled bands, corresponding to endogenous RSK1 and RSK2, were detected by SDS-PAGE and in-gel fluorescence scanning (Cohen et al., 2007). Thus, **FMK-PA** appears to exhibit extraordinary proteome-wide selectivity, discriminating between Cys436 in RSK1/2 and thousands of off-target cysteines.

We also synthesized the fluorescent derivative **FMK-BODIPY**, which albeit less potent than **FMK-PA**, is cell permeable and reasonably selective for RSK1/2 when added to intact cells (3–5 μ*M* in serum-free media) or cell lysates (Cohen et al., 2007). **FMK-BODIPY** is more convenient than **FMK-PA** for quantifying RSK1/2 occupancy by other RSK inhibitors, as there is no requirement for click chemistry (Serafimova et al., 2012). Moreover, for reasons that are not clear, detecting RSK1/2 with **FMK-PA** and click chemistry has proven difficult in certain contexts (e.g., lysates derived from mouse cardiac tissue). When assessing the pharmacodynamic potency of new RSK inhibitors in cells and mice, our "go-to" assay is to quantify RSK1/2 assay using **FMK-BODIPY** as the probe.

To elucidate the roles of RSK CTD activity in mouse disease models, we sought an **FMK** derivative with improved aqueous solubility. Based on the increased potency and solubility of **FMK-PA**, we synthesized the related methoxyethylamine derivative **FMK-MEA** (Fig. 4.2B). When screened against a 443-kinase panel at concentration of 1 μ*M* (DiscoveRX), **FMK-MEA** interacted with only two kinases, RSK1 and RSK4 CTD (Fig. 4.2B; RSK2 CTD was not tested). This level of selectivity is rarely achieved with noncovalent kinase inhibitors. The bis-tosylate salt of **FMK-MEA** dissolves in saline at >25 mg/mL, enabling convenient dosing by intraperitoneal injection or oral gavage ($C_{max} \sim 600$ n*M* in mice after 30 mg/kg oral dose, 38% bioavailability; unpublished results from Albany Molecular Research Inc.).

RSK activity is elevated in endothelial cells and cardiomyocytes in diabetic animals and has been implicated in heart disease exacerbated by diabetes. In a recent study, **FMK-MEA** was found to reduce endothelial dysfunction in two diabetic mouse models and was efficacious in an aggressive atherosclerosis model. We used **FMK-BODIPY** to demonstrate a correlation between RSK1/2 occupancy and a dose-dependent decrease in endothelial dysfunction readouts in mice treated with **FMK-MEA** (Le et al., 2013). Although **FMK-BODIPY**-labeled RSK1/2 is easily detected in lysates derived from mouse lung, liver, and brain tissue, the signal to background is low in heart lysates, necessitating immunoprecipitation with anti-RSK1/2 antibodies (Fig. 4.2C). A protocol for determining RSK1/2 occupancy in mouse cardiac tissue lysates using **FMK-BODIPY** is described below. The protocol can be used with fluorescent occupancy probes directed against other targets that are difficult to detect by gel electrophoresis of whole-cell lysates, as long as specific antibodies are available for the immunoprecipitation step. Relative to streptavidin affinity purification with biotinylated probes followed by Western blot detection, this protocol is simpler and easier to implement.

2.2.1 Assessing RSK1/2 occupancy after dosing mice with FMK-MEA

1. Administer **FMK-MEA** (or other RSK CTD inhibitors) or vehicle (2–4 mice per arm) by intraperitoneal injection or oral gavage; after 2–4 h, sacrifice the mice according to a protocol preapproved by your Institutional Animal Care and Use Committee. Harvest the desired tissues and freeze them immediately in liquid nitrogen. A time course (e.g., 6/12/24 h after dosing) can be used to assess drug-target residence time and/or the rate of RSK resynthesis.
2. While the tissue is still slightly frozen, mince it into small pieces with a clean razor blade.
3. Place each tissue sample into a 2 mL tube. Add 0.75 mL PBS supplemented with protease and phosphatase inhibitors (Roche). *Note*: buffers with stronger buffering capacity (e.g., 50 m*M* Tris, pH 8.0) may work better than PBS, depending on the target and the probe.
4. Prepare lysates using a mechanical rotor-stator style tissue homogenizer (Janke & Kunkel/IKA T25-Ultra-Turrax Homogenizer), keeping the tubes on ice. Begin by homogenizing for 30 s at low speed, allowing samples to rest for 30 s, and then homogenizing at high speed for 30 s.
5. Transfer the crude lysates to 1.5 mL ultracentrifuge tubes. Centrifuge samples at 4 °C in a bench top ultracentrifuge for 1 h at 30,000 × *g* (Beckman Coulter Optima TLX Ultracentrifuge).
6. Transfer supernatants to new 1.5 mL tubes on ice and snap freeze in liquid nitrogen if desired.

7. Quantify protein concentration by Bradford assay, normalizing to the most dilute sample by adding cold PBS. Transfer 100 μL of the lysate to new tubes for labeling with the fluorescent occupancy probe, **FMK-BODIPY**.
8. Add **FMK-BODIPY** (5.25 μL of a 100 μ*M* DMSO stock; 5 μ*M* final concentration) and incubate for 1 h at room temperature.
9. Remove 40 μL from each sample and add 10 μL of 5 × Laemmli sample buffer (pre-IP lysate sample). *Note*: depending on the abundance of the target and the specificity and efficiency of fluorescent probe labeling, it may be possible to quantify target occupancy by SDS-PAGE analysis of the crude lysates, without the need for immunoprecipitation.
10. To the remaining 60 μL of the lysates treated with **FMK-BODIPY**, add 300 μL of cold PBS + 1% NP40. Keep samples on ice.
11. Add 10 μL of α-RSK1 and 10 μL of α-RSK2 antibodies (Santa Cruz Biotechnology, sc-231 and sc-9986). Place samples in a rotator at 4 °C for 2–4 h.
12. Add 30 μL of Protein G Dynabeads (Invitrogen 100-04D) to each tube, place in a rotator, and incubate overnight at 4 °C. *Note*: shorter incubation times may suffice.
13. Place tubes in magnetic racks to separate the beads. Remove 40 μL and quench with 5 × Laemmli sample buffer (post-IP lysate sample; Western blot analysis of pre- and post-IP lysate samples can be used to assess immunoprecipitation efficiency).
14. Discard the remaining supernatant and add 300 μL of cold PBS + 1% NP40 to wash. Place in a rotator for 5 min.
15. Place tubes in magnetic racks to separate beads, and discard the supernatant. Add 300 μL of cold PBS + 1% NP40, mix thoroughly, and transfer each sample to a new prechilled 1.5 mL Eppendorf tube. Place on a rotator at 4 °C for 5 min.
16. Wash once more with 300 μL of cold PBS + 1% NP40 for 5 min, separate the beads with a magnetic rack, and discard the supernatant.
17. To elute the proteins from the beads, add 50 μL of 2 × Laemmli sample buffer and 10 μL of freshly made 1 *M* DTT to the beads. Vortex briefly and heat for 1 min at 90 °C.
18. Resolve proteins by SDS-PAGE. Quantify BODIPY labeling of RSK1/2 on a Typhoon scanner (GE Healthcare). Transfer proteins from the gel to nitrocellulose to detect total RSK1/2 by Western blot (Fig. 4.2C).

3. TARGETING NONCATALYTIC CYSTEINES WITH REVERSIBLE COVALENT INHIBITORS

Covalent irreversible inhibitors are typically avoided in drug discovery because of the potential for nonspecific alkylation, protein haptenization, and unpredictable toxicity (Evans et al., 2004; Uetrecht, 2008). To minimize these concerns, electrophiles with attenuated reactivity, e.g. acrylamides, have been employed. However, the *in vivo* targeting specificity of acrylamide-based kinase inhibitors has not been established, and thus far, such drugs have only been developed to treat advanced cancer. As an alternative approach, we sought to target noncatalytic cysteines with reversible covalent inhibitors. In this case, reactions with off-target cysteines would be transient and ideally less likely to exert idiosyncratic toxic effects. By contrast, the cooperative formation of specific covalent and noncovalent interactions with the desired target would result in high-affinity binding and slow dissociation kinetics.

In this section, we describe our recent efforts to tune the reactivity of Michael acceptors for the design and discovery of reversible covalent inhibitors. Although our work has focused on kinases, this strategy should be applicable to other targets with druggable cysteines.

3.1. Reversible Michael acceptors for cysteine-targeting applications

To explore the possibility of trapping cysteine thiols with alternative electrophiles, we synthesized several Michael acceptors activated by one or two electron-withdrawing groups (EWGs) attached to the incipient α-carbanion (Serafimova et al., 2012). As anticipated, α-cyanoacrylamides (with two EWGs) reacted with simple thiols like GSH more rapidly than acrylamides or acrylonitriles. What was unexpected was that the thiol adducts could not be isolated and instantaneously reverted back to the cyanoacrylamides upon dilution. The faster thiol elimination rate relative to the corresponding acrylamide adduct makes sense given the increased carbon acidity of the cyanoacrylamide adduct. However, the magnitude of the adduct's kinetic instability at physiological pH was unanticipated. Based on these kinetic properties, we reasoned that cyanoacrylamides could be exploited to yield cysteine-targeted, reversible covalent inhibitors with increased selectivity relative to irreversible inhibitors. These concepts led to the discovery of ultra-selective cyanoacrylamide-based RSK inhibitors with picomolar

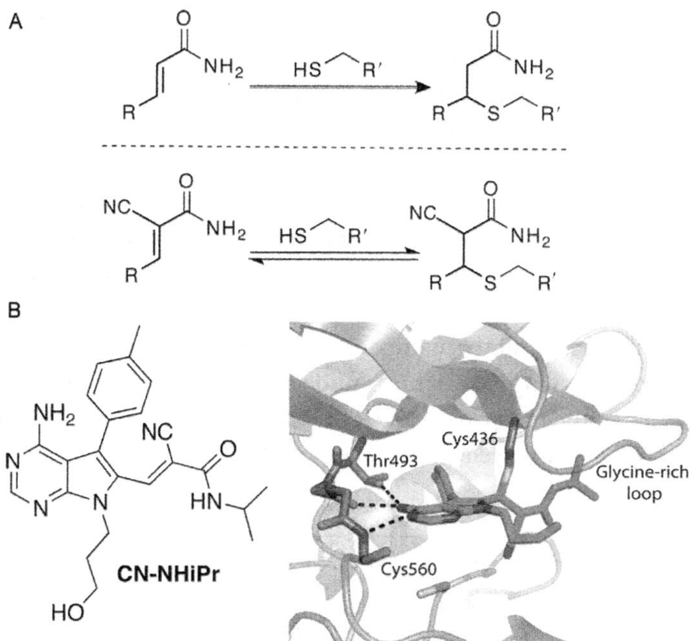

Figure 4.3 Reversible targeting of noncatalytic cysteines with cyanoacrylamides. (A) Cyanoacrylamides, unlike acrylamides, form rapidly reversible adducts with thiols. (B) *N*-isopropyl cyanoacrylamide variant of **FMK** is a potent, selective, and reversible RSK inhibitor with slow dissociation kinetics. A cocrystal structure of a *tert*-butyl cyanoacrylate derivative bound to RSK2 shows a network of noncovalent interactions that cooperatively stabilize the covalent complex. Upon unfolding of RSK2, the covalent bond is rapidly reversed. *Adapted with permission from Serafimova et al. (2012). Copyright Nature Publishing Group, 2012.* (See the color plate.)

affinity and slow off-rates, the structural basis of which was revealed by X-ray crystallography (Fig. 4.3). Because covalent bond formation is under thermodynamic control and requires multiple specific noncovalent interactions to stabilize the covalent complex, the likelihood of forming spurious adducts with off-target proteins is reduced relative to irreversible electrophiles like fluoromethylketones and acrylamides.

3.2. Electrophilic fragment-based ligand discovery with cyanoacrylamides

The intrinsically reversible, yet energetically favorable nature of the cysteine/cyanoacrylamide reaction suggested that this "sweet spot" in electrophile space could form the basis of a fragment screening library for cysteine-containing targets (Miller, Paavilainen, Krishnan, Serafimova, & Taunton,

2013). The approach is conceptually related to the disulfide fragment tethering technology developed by Erlanson et al. (2000), but with critical differences: (1) unlike most disulfides, cyanoacrylamides can be used in cells and animals, and (2) the amide group can serve as a medicinal chemistry handle for optimizing affinity and selectivity. In developing a cyanoacrylamide fragment screening approach, we exploited both of these advantages. We tested ten cyanoacrylamide fragments against three cysteine-containing kinases (RSK2, NEK2, and PLK1) under stringent conditions (10 m*M* GSH and 0.1 m*M* ATP). Even with such a small screening set, we identified unique inhibitors for all three kinases with potencies in the low micromolar to submicromolar range. Structure-guided merging of two of the initial fragment hits led to the first pan-MSK/RSK CTD inhibitor, with subnanomolar affinity and >500-fold selectivity over NEK2 and PLK1 (Fig. 4.4).

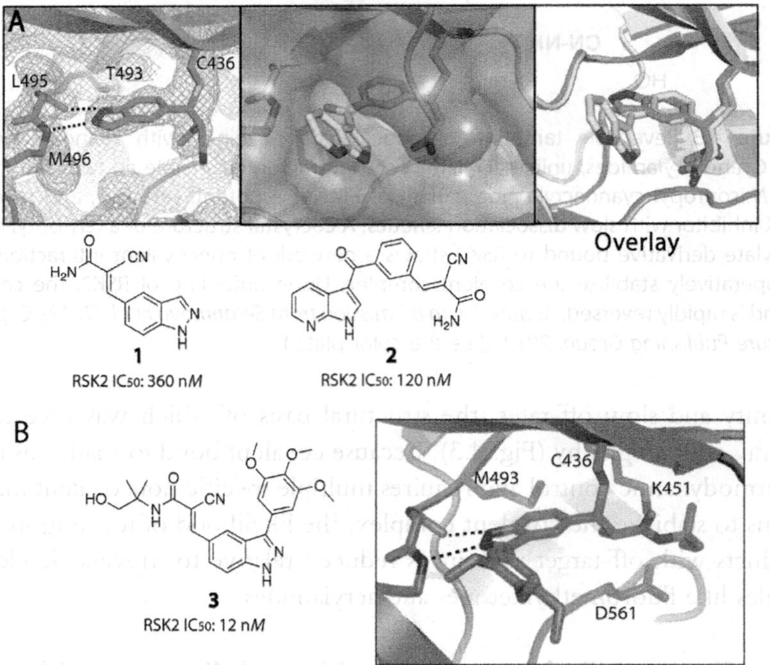

Figure 4.4 Cyanoacrylamide fragment screening identifies pan-MSK/RSK inhibitors. (A) Using a RSK2 kinase assay, we screened a panel of 10 cyanoacrylamide fragments (up to 300 μ*M*) and discovered **1** and **2** as the most potent hits. We solved cocrystal structures of **1** and **2** bound RSK2 (PDB codes: 4JG6 and 4JG7). An overlay of both structures suggested the design of 3-aryl indazole variants. (B) Structure-guided optimization led to 3-aryl indazole **3** (RMM-46), which potently and selectively blocks MSK and RSK signaling in cells. *Adapted with permission from Miller et al. (2013). Copyright 2013, American Chemical Society.* (See the color plate.)

3.2.1 Assembling and screening a cyanoacrylamide fragment library

Cyanoacrylamides are easy to synthesize in one step via Knoevenagel condensation of 2-cyanoacetamide with an aldehyde fragment. Searching the ZINC database (Irwin, Sterling, Mysinger, Bolstad, & Coleman, 2012) revealed over 12,000 commercially available aldehyde fragments (MW <250 Da). We typically mix aryl or heteroaryl aldehydes (0.3 mmol) with one equivalent each of 2-cyanoacetamide and piperidine in THF (~100 m*M*) for 24–48 h. The pure cyanoacrylamide often precipitates directly from the reaction mixture and can be isolated as a pure solid by brief centrifugation onto a 0.22 μm filter unit, followed by washing the retained solid with THF. For cyanoacrylamides that do not precipitate, we recommend preparative TLC as an efficient purification method. While cyanoacrylamides are often found in commercial small-molecule screening libraries, they tend to be larger (>300 Da) and have large hydrophobic substituents on the amide nitrogen. Such compounds may have lower solubility, which can lead to screening artifacts. Thus far, we have only used unsubstituted (primary) amides derived from 2-cyanoacetamide for initial fragment screens; we then incorporate amide substituents during the subsequent hit optimization phase. This latter modification step may be critical for certain cellular assays, as we have found that primary cyanoacrylamides are more prone to hydrolysis (via retro-Knoevenagel reaction to give the aldehyde and 2-cyanoacetamide) than secondary or tertiary cyanoacrylamides. Cyanoacrylate esters are even less stable.

Nearly all of the cyanoacrylamides that we have tested react reversibly with simple thiols like GSH and β-mercaptoethanol (BME). Monitoring the thermodynamics and kinetics of these reactions provides a simple approximation of an electrophile's intrinsic reactivity toward cysteine residues in unstructured regions of proteins. One way to compare the intrinsic reactivity of different cyanoacrylamides is to measure the equilibrium dissociation constant (K_d) by titrating GSH or BME in buffered solution. The reaction is conveniently monitored by UV–vis spectroscopy, as most β-aryl or -heteroaryl cyanoacrylamides have a strong absorption band (300–400 nm) that is disrupted upon thiol conjugate addition (Fig. 4.5A). The K_d can be determined fitting the titration data to a one-site binding model using GraphPad Prism or similar software (Fig. 4.5B). Reversibility can be demonstrated by diluting the thiol/cyanoacrylamide adduct into buffer that either lacks or contains excess thiol and monitoring recovery of the cyanoacrylamide UV absorption band (Fig. 4.5C). For most thiol/cyanoacrylamide adducts, a new equilibrium is established within seconds upon dilution. Cyanoacrylamide/thiol equilibria can also be characterized by ^{1}H NMR.

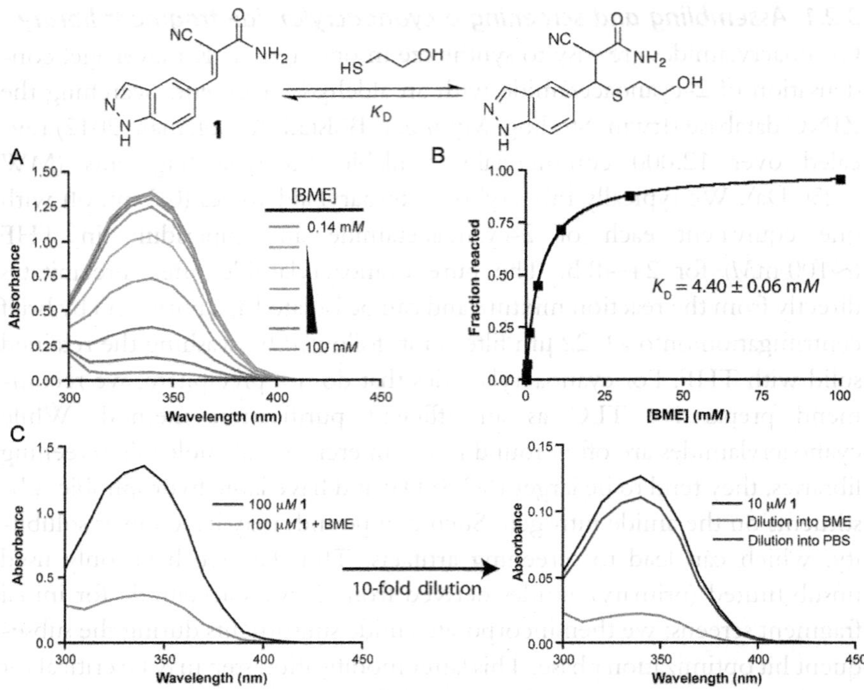

Figure 4.5 Characterization of thiol conjugate addition reactions with cyanoacrylamides. (A) Reaction of cyanoacrylamide **1** with BME is accompanied by a decrease in the 340 nm absorption band. (B) The K_d is determined by fitting the titration data to a simple one-site binding model. (C) Dilution of the BME/cyanoacrylamide adduct into buffer lacking BME establishes a new equilibrium favoring the cyanoacrylamide. (See the color plate.)

Depending on the cyanoacrylamide, K_d values with GSH or BME typically range from 1 to 100 m*M*. Assuming millimolar GSH concentrations in cells, a cyanoacrylamide with a K_d of <10 m*M* will be significantly bound to GSH, yet it will still be available to engage the desired target due to rapid β-elimination from GSH and dynamic equilibration. In our initial fragment screen, there was no correlation between kinase inhibition potency and BME reactivity as determined by K_d measurements (unpublished results). This further supports our conclusion that increased potency derives from specific noncovalent interactions involving the unique β-substituent of the cyanoacrylamide, in addition to the covalent bond formation. To expand the noncovalent recognition potential of the electrophile, we have begun to explore alternative Michael acceptors for electrophilic fragment screens and structure-based inhibitor design (Krishnan et al., 2014). Given the challenge of predicting whether a new electrophilic chemotype will undergo reversible thiol-addition reactions, we routinely measure the intrinsic reactivity and reversibility toward BME or GSH using the methods described above.

4. SEMIPROMISCUOUS COVALENT INHIBITORS AS CHEMOPROTEOMIC PROBES

Quantifying drug/target engagement in living cells is a challenging problem for which new chemical methods are needed (Moellering & Cravatt, 2012). This issue is especially critical for kinase inhibitors, which may interact with any one of 500 structurally related "off-target" kinases. High-throughput assays with purified proteins often fail to recapitulate the state of endogenous kinases in their native subcellular milieu, in which dynamic regulatory interactions modulate their affinity for ATP and inhibitors. Powerful technologies have been developed for quantifying interactions with hundreds of endogenous kinases, but these probes only work in dilute cell lysates (Bantscheff et al., 2007; Patricelli et al., 2011). We were therefore motivated to develop complementary probes for quantifying inhibitor/kinase engagement in living cells.

4.1. Identification of new therapeutic kinase targets with a semipromiscuous inhibitor

We recently designed a chemoproteomic probe based on hypothemycin, an electrophilic polyketide natural product (Fig. 4.6), for the purpose of identifying and prioritizing potential therapeutic targets in the disease-causing parasite, *Trypanosoma brucei* (Nishino et al., 2013). Hypothemycin had

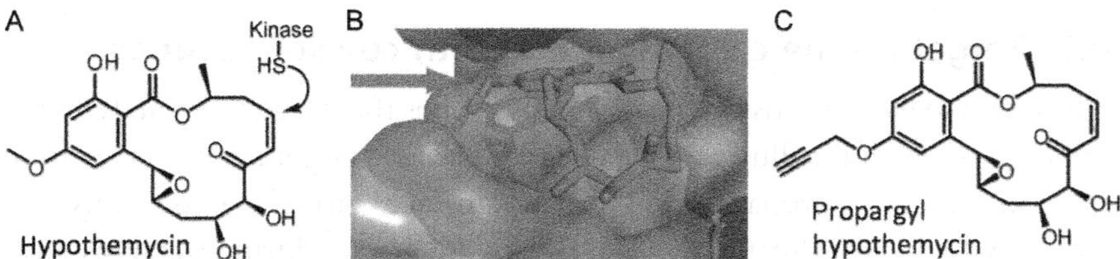

Figure 4.6 Semipromiscuous occupancy probe for CDXG kinases. (A) Hypothemycin reacts with the cysteine in CDXG kinases via conjugate addition to the *cis*-enone. (B) Cocrystal structure of hypothemycin bound to ERK2 (PDB code: 3C9W) suggests the methyl ether as a derivatization site for an alkyne tag. (C) Chemical structure of propargyl-hypothemycin, prepared from desmethyl-hypothemycin and propargyl bromide. (See the color plate.)

previously been shown to covalently inactivate several human kinases bearing a cysteine adjacent to the conserved DXG motif, albeit with widely varying potency. In many eukaryotes, including humans and trypanosomes, such "CDXG kinases" comprise 10–20% of the kinome. We found that hypothemycin potently kills *T. brucei* parasites. Twenty-one of the 182 protein kinases in *T. brucei* have a CDXG motif and were therefore strong candidates for mediating hypothemycin's trypanocidal effects.

To identify the relevant target(s), we synthesized a propargyl ether derivative of hypothemycin (guided by a cocrystal structure) for quantitative chemoproteomic experiments. *T. brucei* cell lysates were pretreated with increasing concentrations of hypothemycin, followed by treatment with the propargyl-hypothemycin probe, click chemistry with biotin-azide, streptavidin affinity purification, and mass spectrometry analysis. We used isobaric mass spectrometry tags (Ross et al., 2004) to quantify the relative amount of every protein identified in the streptavidin eluates as a function of increasing hypothemycin concentration. This analysis revealed that propargyl-hypothemycin specifically labeled 11 CDXG kinases (without labeling any non-CDXG kinases), yet only three kinases were competed with nanomolar concentrations of hypothemycin. A similar quantitative occupancy assay in intact parasites revealed that only TbCLK1, a previously unstudied CLK-family kinase, was fully engaged by cytotoxic concentrations of hypothemycin. RNAi knockdown demonstrated that TbCLK1 (but not the closely related paralog, TbCLK2) is essential for *T. brucei* viability. Hence, this study led to the discovery of a new potential therapeutic target for African sleeping sickness, as well as a chemoproteomic probe for quantifying inhibitor occupancy of TbCLK1 and other CDXG kinases in living cells.

4.2. Targeting the catalytic lysine with covalent probes

Thus far, we have discussed the design of probes that target noncatalytic cysteines. To develop cellular occupancy probes that target a broader swath of the kinome than propargyl-hypothemycin, we turned to the conserved lysine, which coordinates the ADP-leaving group during catalysis (e.g., K295 in human SRC, Fig. 4.7A). While this lysine does not act as a catalytic nucleophile in the phospho-transfer step, its reactivity toward certain electrophiles may be enhanced by electrostatic interactions with conserved Asp and Glu residues; the latter interactions are observed in many but not all kinase structures. Targeting the catalytic Lys in kinases without targeting

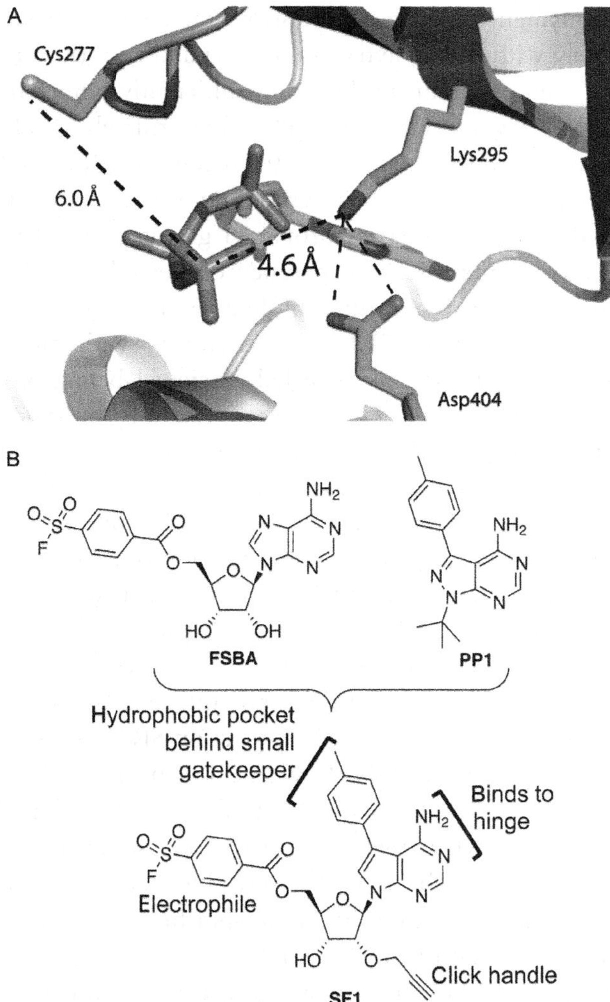

Figure 4.7 Design of a semipromiscuous, lysine-targeted kinase probe. (A) Structure of SRC bound to AMP-PNP (PDB code: 2SRC), showing distances from the terminal phosphate to the catalytic Lys295 and nonconserved Cys277. (B) Design of a lysine-targeted sulfonyl fluoride probe, **SF1**, by merging FSBA and PP1. *Adapted with permission from Gushwa et al. (2012). Copyright 2012, American Chemical Society.* (See the color plate.)

thousands of other lysines frequently found on protein surfaces presents a potential chemoselectivity challenge, and few Lys-targeted kinase probes have been reported that are suitable for interrogating kinase occupancy in cells. Nevertheless, ATP-biotin and ADP-biotin acylate lysine residues in kinases and other ATP binding proteins in gel-filtered cell lysates and are useful as occupancy probes (Patricelli et al., 2011).

Flurosulfonylbenzoyl adenosine (FSBA, Fig. 4.7B) has been reported to react selectively with the catalytic Lys of several kinases (Kamps, Taylor, & Sefton, 1984), although this probe is not sufficiently potent to be used in cells. Our strategy was to retain the sulfonyl fluoride electrophile of FSBA while increasing the intrinsic affinity of the recognition scaffold by incorporating a *p*-tolyl-substituted pyrrolopyrimidine, similar to the promiscuous tyrosine kinase inhibitor PP1. A propargyl group attached to the ribose 2′-OH fortuitously increased the potency of the sulfonyl fluoride probe **SF1** toward all kinases tested (unpublished results). After brief treatment of intact human cells with **SF1**, click chemistry with biotin-azide and streptavidin affinity purification revealed efficient labeling of endogenous SRC-family kinases. By replacing the benzoyl sulfonyl fluoride of **SF1** with a vinylsulfonate, we generated a second, more selective probe that targets a proximal cysteine (Fig. 4.7A) found in only three SRC-family kinases (e.g., SRC, but not LCK, LYN, HCK, or FYN), in addition to FGFR1–4 and a few other kinases (Gushwa, Kang, Chen, & Taunton, 2012). Other groups have developed useful irreversible inhibitors that target this cysteine (Kwarcinscki, Fox, Steffey, & Soellner, 2012; Zhou et al., 2010).

Using **SF1** to probe kinase occupancy in living cells, we discovered that the recently approved drug ponatinib potently engages the SRC-family kinase, LCK, while having no apparent effect on SRC or FYN. Given that all three kinases were previously shown to be sensitive to low nanomolar concentrations of ponatinib in biochemical assays, these results illustrate potential differences when measuring native kinase engagement in cells. In our initial study with **SF1**, Western blot detection with specific antibodies enabled quantitation of affinity-purified, probe-labeled kinases. Because antibodies against many human kinases are commercially available, this method provides a convenient alternative to instrument-intensive quantitation by mass spectrometry. Nevertheless, quantitation by Western blotting is limited in terms of the number of kinases that can be analyzed in a single experiment. In unpublished work, we have used **SF1** in combination with quantitative mass spectrometry to identify a large set of *T. brucei* kinases, only a few of which are labeled by propargyl-hypothemycin and none of which are tyrosine kinases. Many of these kinases are essential for viability and hence, potential therapeutic targets. Given that little is known about their upstream activators or downstream substrates, quantitative occupancy assays with probes such as propargyl-hypothemycin and **SF1** are among the best ways to demonstrate on-target cellular effects of new inhibitors. We anticipate that **SF1** and related sulfonyl fluoride probes will have broad utility for kinase occupancy studies in human cells as well as disease-causing parasites.

5. CONCLUSIONS AND FUTURE DIRECTIONS

Understanding the intricate roles of protein kinases in cellular physiology requires the development of potent and selective small-molecule probes. Moreover, protein kinases remain an important and ever-expanding class of therapeutic targets for the treatment of cancer and other serious diseases. Although genetic gain- and loss-of-function studies can provide important clues about which kinases to target for a particular disease state, selective inhibitors are absolutely essential for early-stage target validation studies. Covalent targeting of noncatalytic cysteines has emerged as an efficient strategy for obtaining selective kinase inhibitors that exhibit prolonged target occupancy *in vivo*. Indeed, covalent kinase inhibitors often confer sustained inactivation of the desired target even when their pharmacokinetic properties are suboptimal. In this chapter, we have presented our laboratory's approach to the design of covalent probes and their application to the study of kinase targets for which there is minimal genetic validation in disease models. For example, our RSK1/2/4 inhibitor, **FMK-MEA**, has revealed a previously unknown role for RSK hyperactivity in atherosclerosis and endothelial dysfunction (Le et al., 2013). Given that **FMK-MEA** targets the C-terminal kinase domain of three RSK paralogs, while sparing the N-terminal kinase domain, modeling its pharmacology with a purely genetic approach would be challenging.

The Protein Data Bank contains thousands of kinase/ligand cocrystal structures that reveal the precise location of potentially nucleophilic side chains relative to the bound ligands. For kinases and/or ligands whose structures have not been determined, the approximate locations of nucleophilic side chains (e.g., Cys or Lys) can be predicted from homology models and ligand docking. Knowledge of these distances and vectors—between a bound ligand and a noncatalytic Cys or catalytic Lys—should enable the design of either highly selective or semipromiscuous covalent probes against many different kinases. One of our major goals for the future is to identify novel electrophilic functional groups with the optimal balance of reactivity and stability to target cysteine, lysine, and other potentially nucleophilic side chains in a selective, site-directed manner.

REFERENCES

Bain, J., Plater, L., Elliot, M., Shpiro, N., Hastie, C. J., Mclauchlan, H., et al. (2007). The selectivity of protein kinase inhibitors: A further update. *The Biochemical Journal, 408*, 297–315.

Bantscheff, M., Eberhard, D., Abraham, Y., Bastuck, S., Boesche, M., Hobson, S., et al. (2007). Quantitative chemical proteomics revels mechanisms of action of clinical ABL kinase inhibitors. *Nature Biotechnology*, *25*, 1035–1044.

Barf, T., & Kaptein, A. (2012). Irreversible protein kinase inhibitors: Balancing the benefits and risks. *Journal of Medicinal Chemistry*, *55*, 6243–6262.

Buzko, O., & Shokat, K. M. (2002). A kinase sequence database: Sequence alignments and family assignment. *Bioinformatics*, *18*, 1274.

Chen, F., Pruett-Miller, S. M., Huang, Y., Gjoka, M., Duda, K., Taunton, J., et al. (2011). High-frequency genome editing using ssDNA oligonucleotides with zinc-finger nucleases. *Nature Methods*, *8*, 753–755.

Cohen, M. S., Hadjivassiliou, H., & Taunton, J. (2007). A clickable inhibitor reveals context-dependent autoactivation of p90 RSK. *Nature Chemical Biology*, *3*, 156–160.

Cohen, M. S., Zhang, C., Shokat, K. M., & Taunton, J. (2005). Structural bioinformatics-based design of selective, irreversible kinase inhibitors. *Science*, *308*, 1318–1321.

Cuello, F., Snabaitis, A. K., Cohen, M. S., Taunton, J., & Avkiran, M. (2006). Evidence for direct regulation of myocardial Na+/H$^+$ exchanger isoform 1 phosphorylation and activity by 90-kDa ribosomal S6 kinase (RSK): Effects of the novel and specific RSK inhibitor fmk on responses to alpha1-adrenergic stimulation. *Molecular Pharmacology*, *71*, 799–806.

Dalby, K. N., Morrice, N., Caudwell, F. B., Avruch, J., & Cohen, P. (1998). Identification of regulatory phosphorylation sites in mitogen-activated protein kinase (MAPK)-activated protein kinase-1a/p90rsk that are inducible by MAPK. *The Journal of Biological Chemistry*, *273*, 1496–1505.

David, J., Mehic, D., Bakiri, L., Schilling, A. F., Mandic, V., Priemel, M., et al. (2005). Essential role of RSK2 in c-Fos–dependent osteosarcoma development. *The Journal of Clinical Investigation*, *115*, 664–672.

Doehn, U., Hauge, C., Frank, S. R., Jensen, C. J., Duda, K., Nielsen, J. V., et al. (2009). RSK is a principal effector of the RAS-ERK pathway for eliciting a coordinate promotile/invasive gene program and phenotype in epithelial cells. *Molecular Cell*, 511–522.

Erlanson, D. A., Braisted, A. C., Raphael, D. R., Randal, M., Stroud, R. M., Gordon, E. M., et al. (2000). Site-directed ligand discovery. *Proceedings of the National Academy of Sciences of the United States of America*, *97*, 9367–9372.

Evans, D. C., Watt, A. P., Nicoll-Griffith, D. A., & Baillie, T. A. (2004). Drug–protein adducts: An industry perspective on minimizing the potential for drug bioactivation in drug discovery and development. *Chemical Research in Toxicology*, *17*, 3–16.

Fisher, T. L., & Blenis, J. (1996). Evidence for two catalytically active kinase domains in pp90rsk. *Molecular and Cellular Biology*, *16*, 1212–1219.

Friedland, A. E., Tzur, Y. B., Esvelt, K. M., Colaiácovo, M. P., Church, G. M., & Calarco, J. A. (2013). Heritable genome editing in *C. elegans* via a CRISPR-Cas9 system. *Nature Methods*, *10*, 741–743.

Frödin, M., Jensen, C. J., Merienne, K., & Gammeltoft, S. (2000). A phosphoserine-regulated docking site in the protein kinase RSK2 that recruits and activates PDK1. *The EMBO Journal*, *19*, 2924–2934.

Fry, D. W., Bridges, A. J., Denny, W. A., Doherty, A., Greis, K. D., Hicks, J. L., et al. (1998). Specific, irreversible inactivation of the epidermal growth factor receptor and erbB2, by a new class of tyrosine kinase inhibitor. *Proceedings of the National Academy of Sciences of the United States of America*, *95*, 12022–12027.

Gushwa, N. N., Kang, S., Chen, J., & Taunton, J. (2012). Selective targeting of distinct active site nucleophiles by irreversible SRC-family kinase inhibitors. *Journal of the American Chemical Society, 134*, 20214–20217.

Hauge, C., & Frödin, M. (2006). RSK and MSK in MAPK signaling. *Journal of Cell Science, 119*, 3021–3023.

Henise, J., & Taunton, J. (2011). Irreversible Nek2 kinase inhibitors with cellular activity. *Journal of Medicinal Chemistry, 54*, 4133–4146.

Irwin, J. J., Sterling, T., Mysinger, M. M., Bolstad, E. S., & Coleman, R. G. (2012). ZINC: A free tool to discover chemistry for biology. *Journal of Chemical Information and Modeling, 52*, 1757–1768.

Kamps, M. P., Taylor, S. S., & Sefton, B. M. (1984). Direct evidence that oncogenic tyrosine kinases and cyclic AMP-dependent protein kinase have homologous ATP-binding sites. *Nature, 310*, 589–592.

Kang, S., Elf, S., Lythgoe, K., Hitosugi, T., Taunton, J., Zhou, W., et al. (2010). p90 ribosomal S6 kinase 2 promotes invasion and metastasis of human head and neck squamous cell carcinoma cells. *The Journal of Clinical Investigation, 120*, 1165–1177.

Krishnan, S., Miller, R. M., Tian, B., Mullins, R. D., Jacobson, M. P., & Taunton, J. (2014). Design of reversible, cysteine-targeted Michael acceptors guided by kinetic and computational analysis. *Journal of the American Chemical Society, 136*, 12624–12630.

Kwarcinscki, F. E., Fox, C. C., Steffey, M. E., & Soellner, M. B. (2012). Irreversible inhibitors of c-Src kinase that target a nonconserved cysteine. *ACS Chemical Biology, 7*, 1910–1971.

Le, N. T., Heo, K. S., Takei, Y., Lee, H., Woo, C. H., Chang, E., et al. (2013). A crucial role for p90RSK-mediated reduction of ERK5 transcriptional activity in endothelial dysfunction and atherosclerosis. *Circulation, 187*, 486–499.

Leproult, E., Barluenga, S., Moras, D., Wurtz, J. M., & Winssinger, N. (2011). Cysteine mapping in conformationally distinct kinase nucleotide binding sites: Application to the design of selective covalent inhibitors. *Journal of Medicinal Chemistry, 54*, 1347–1355.

Liu, Y., Bishop, A., Witucki, L., Kraybill, B., Shimizu, E., Tsien, J., et al. (1999). Structural basis for selective inhibition of Src family kinases by PP1. *Chemistry and Biology, 6*, 671–678.

Liu, Q., Sabnis, Y., Zhao, Z., Zhang, T., Buhrlage, S. J., Jones, L. H., et al. (2013). Developing irreversible inhibitors of the protein kinase cysteinome. *Chemistry and Biology, 20*, 146–159.

Manning, G., Whyte, D. B., Martinez, R., Hunter, T., & Sudarsanam, S. (2002). The protein kinase complement of the human genome. *Science, 298*, 1912–1934.

Miller, R. M., Paavilainen, V. O., Krishnan, S., Serafimova, I. M., & Taunton, J. (2013). Electrophilic fragment-based design of reversible covalent kinase inhibitors. *Journal of the American Chemical Society, 135*, 5298–5301.

Moellering, R. E., & Cravatt, B. F. (2012). How chemoproteomics can enable drug discovery and development. *Chemistry and Biology, 19*, 11–22.

Nishino, M., Choy, J. W., Gushwa, N. N., Oses-Prieto, J. A., Koupparis, K., Burlingame, A. L., et al. (2013). Hypothemycin, a fungal natural product, indentifies therapeutic targets in *Trypanosoma brucei*. *eLife, 2*, 1–15.

Pan, Z., Scheerens, H., Li, S. J., Schultz, B. E., Sprengeler, P. A., Burrill, L. C., et al. (2007). Discovery of selective irreversible inhibitors for Bruton's tyrosine kinase. *ChemMedChem, 2*, 58–61.

Patricelli, M. P., Nomanbhoy, T. K., Wu, J., Brown, H., Zhou, D., Zhang, J., et al. (2011). In situ kinase profiling reveals functionally relevant properties of native kinases. *Chemistry and Biology, 18*, 699–710.

Romeo, Y., Zhang, X., & Roux, P. P. (2012). Regulation and function of the RSK family of protein kinases. *The Biochemical Journal, 441*, 553–569.

Ross, P. L., Huang, Y. N., Marchese, J. N., Williamson, B., Parker, K., Hattan, S., et al. (2004). Multiplexed protein quantitation in *Saccharomyces cerevisiae* using amine-reactive isobaric tagging reagents. *Molecular & Cellular Proteomics, 3*, 1154–1169.

Sapkota, G. P., Cummings, L., Newell, F. S., Armstrong, C., Bain, J., Frödin, M., et al. (2007). BI-D1870 is a specific inhibitor of the p90 RSK (ribosomal S6 kinase) isoforms in vitro and in vivo. *The Biochemical Journal, 401*(I), 29–38.

Schindler, T., Sicheri, F., Pico, A., Gazit, A., Levitzki, A., & Kuriyan, J. (1999). Crystal structure of Hck in complex with a Src family-selective tyrosine kinase inhibitor. *Molecular Cell, 3*, 639–648.

Serafimova, I. M., Pufall, M. A., Krishnan, S., Duda, K., Cohen, M. S., Maglathlin, R. L., et al. (2012). Reversible targeting of noncatalytic cysteines with chemically tuned electrophiles. *Nature Chemical Biology, 8*, 471–476.

Shahbazian, D., Roux, P. P., Mieulet, V., Cohen, M. S., Raught, B., Taunton, J., et al. (2006). The mTOR/PI3K and MAPK pathways converge on eIF4B to control its phosphorylation and activity. *The EMBO Journal, 25*, 2781–2791.

Singh, J., Dobrusin, E. M., Fry, D. W., Haske, T., Whitty, A., & McNamara, D. J. (1997). Structure-based design of a potent, selective, and irreversible inhibitor of the catalytic domain of the erbB receptor subfamily of protein tyrosine kinases. *Journal of Medicinal Chemistry, 40*, 1130–1135.

Singh, J., Petter, R. C., Baillie, T. A., & Whitty, A. (2011). The resurgence of covalent drugs. *Nature Reviews Drug Discovery, 10*, 307–317.

Smolen, G. A., Zhang, J., Zubrowski, M. J., Edelman, E. J., Luo, B., Yu, M., et al. (2010). A genome-wide RNAi screen identifies multiple RSK-dependent regulators of cell migration. *Genes & Development, 24*, 2654–2665.

Solca, F., Dahl, G., Zoephel, A., Bader, G., Sanderson, M., Klein, C., et al. (2012). Target binding properties and cellular activity of afatinib (BIBW 2992), an irreversible ErbB family blocker. *The Journal of Pharmacology and Experimental Therapeutics, 343*, 342–350.

Takahashi, E., Abe, J., Gallis, B., Aebersold, R., Spring, D. J., Krebs, E. G., et al. (1999). p90(RSK) is a serum-stimulated $Na+/H^+$ exchanger isoform-1 kinase. Regulatory phosphorylation of serine 703 of $Na+/H^+$ exchanger isoform-1. *The Journal of Biological Chemistry, 274*, 20206–20214.

Uetrecht, J. (2008). Idiosyncratic drug reactions: Past, present, and future. *Chemical Research in Toxicology, 21*, 84–92.

Wang, H., Yang, H., Shivalila, C. S., Dawlaty, M. M., Cheng, A. W., Zhang, F., et al. (2013). One-step generation of mice carrying mutations in multiple genes by CRISPR/Cas-mediated genome engineering. *Cell, 153*, 910–918.

Zhou, W., Hur, W., McDermott, U., Dutt, A., Xian, W., Ficarro, S. B., et al. (2010). A structure-guided approach to creating covalent FGFR inhibitors. *Chemistry and Biology, 17*, 285–295.

CHAPTER FIVE

The Resistance Tetrad: Amino Acid Hotspots for Kinome-Wide Exploitation of Drug-Resistant Protein Kinase Alleles

Fiona P. Bailey*, Veselin I. Andreev, Patrick A. Eyers*,[1]
*Department of Biochemistry, Institute of Integrative Biology, University of Liverpool, Liverpool, United Kingdom
[1]Corresponding author: e-mail address: patrick.eyers@liverpool.ac.uk

Contents

Methods in Enzymology, Volume 548
ISSN 0076-6879
http://dx.doi.org/10.1016/B978-0-12-397918-6.00005-7

Abstract

Acquired resistance to targeted kinase inhibitors is a well-documented clinical problem that is potentially fatal for patients to whom a suitable back-up is not available. However, protein kinase alleles that promote resistance to inhibitors can be exploited experimentally as gold-standards for "on"- and "off"-target validation strategies and constitute a powerful resource for assessing the ability of new or combined therapies to override resistance. Clinical resistance to kinase inhibitors is an evident in all tyrosine kinase-driven malignancies, where high rates of mutation drive tumor evolution toward the insidious drug-resistant (DR) state through a variety of mechanisms. Unfortunately, this problem is likely to intensify in the future as the number of target kinases, approved inhibitors, and clinical indications increase. To empower the analysis of resistance in kinases, we have validated a bioinformatic, structural, and cellular workflow for designing and evaluating resistance at key mutational hotspots among kinome members. In this chapter, we discuss how mutation of amino acids in the gatekeeper and hinge-loop regions (collectively termed the "resistance tetrad") and the DFG motif represent an effective approach for generating panels of DR kinase alleles for chemical genetics and biological target validation.

1. INTRODUCTION

Protein kinases regulate a myriad of fundamental cellular events, and in light of their inappropriate regulation in human diseases represent highly attractive pharmacological targets. Despite problems with clinical resistance, the remarkable success of first-generation targeted inhibitors such as Imatinib spurred-on an initially skeptical industry (Cohen, 2002), and the protein kinase superfamily now represents the second most targeted family of proteins after G-protein coupled receptors (Cohen & Alessi, 2013). However, the presence of preexisting somatic mutations, coupled with the seemingly inevitable appearance of acquired drug-resistant (DR) populations after conventional cycles of drug exposure, suggests that screening and medicinal chemistry efforts must be focused at an early stage to both predict and target mutant kinases and the reprogrammed signaling networks that rapidly evolve in cancer cells. Resistance to kinase inhibitors occurs via multiple mechanisms. These include, but are not limited to, changes in drug metabolism and efflux, genetic and epigenetic events leading to kinase overexpression or hyperactivation and compensatory signaling through various mechanisms, which reactivate signals normally transduced by target kinase(s). Below, we focus on how specific patterns of mutation in the kinase drug-binding site, which impose a measurable decrease in binding affinity and promote direct drug resistance in cells, can be exploited experimentally to directly validate the targets of multiple kinase inhibitors, providing preclinical and clinical examples. In other words, the nearly has crept in and is wrong.

2. PROTEIN KINASES AND KINASE INHIBITORS

Eukaryotic protein kinases share a conserved catalytic domain (Manning, Whyte, Martinez, Hunter, & Sudarsanam, 2002), which adopts different conformers corresponding to catalytically "active" and "inactive" forms (Endicott, Noble, & Johnson, 2012). The canonical kinase domain is composed of an N-terminal N-lobe and a predominantly α-helical C-terminal C-lobe, which both contribute clusters of conserved amino acids that interact with ATP and substrates (Hanks & Hunter, 1995). The two domains articulate through a hinge region, which provides a high-affinity binding site for adenine and forms stereotypical hydrogen bonds with synthetic ATP-competitive kinase inhibitors. The N-lobe contains a flexible Gly-rich (G) loop, which positions amino acid side chains required for nucleotide and inhibitor binding. Interspersed conserved and nonconserved amino acids line the ATP site, endowing kinases with differing levels of discriminatory power toward chemical inhibitors (Fedorov et al., 2007; Huang, Zhou, Lafleur, Nevado, & Caflisch, 2010; Knight & Shokat, 2005; Sheinerman, Giraud, & Laoui, 2005). Mutations at several common sites, including the catalytic αC-helix and the Gly-rich loop (G-loop, also called the P-loop), perturb the binding affinity of small molecule inhibitors and can drive drug resistance *in vivo* (Barouch-Bentov & Sauer, 2011; Bikker, Brooijmans, Wissner, & Mansour, 2009; Burkard & Jallepalli, 2010; Krishnamurty & Maly, 2010; Vieth, Sutherland, Robertson, & Campbell, 2005). In this chapter, we discuss how rational mutagenesis of amino acids that often project into inhibitor-occupied space (Table 5.1), provide simple experimental opportunities for kinome-wide deployment of DR alleles.

3. PROTEIN KINASE INHIBITORS

Depending on their biochemical mode of action, small molecule protein kinase inhibitors can be classified into two basic classes (Fig. 5.1), although allosteric and covalent compounds (e.g., Ibrutinib, which target Cys residues) are likely to suffer from the same issues of drug resistance as these compounds. Due to the similarity of the catalytically active kinase fold targeting or stabilizing catalytically inactive confirmations provides an

Table 5.1 Resistance mutations in Gly-rich loop, resistance tetrad, and the "DFG" motif in protein kinases

Name/position of residue	Common mutations	Selected references
G-loop (position +5 hydrophobic) in Gly-X-Gly-X-Tyr/ Phe-Gly	Phe, Tyr, or His	Azam, Latek, and Daley (2003), Barouch-Bentov and Sauer (2011), Heuckmann et al. (2011), Roumiantsev et al. (2002), and Shah et al. (2002)
Gatekeeper (tetrad position 0)	Met, Thr, Leu, Phe, Val, Ile, Gln, Ala, Gly	Bishop et al. (2000), Blencke et al. (2004), Carter et al. (2005), Clark et al. (2012), Eyers, Craxton, Morrice, Cohen, and Goedert (1998), Gorre et al. (2001), Huang et al. (2010), Kwiatkowski et al. (2010), Littlefield, Moasser, and Jura (2014), Schmidt, Budirahardja, Klompmaker, and Medema (2005), and Zhang et al. (2011)
Hydrophobic (tetrad position +2)	His, Gly	Azam et al. (2003), Balzano, Santaguida, Musacchio, and Villa (2011), Girdler et al. (2008), and Heuckmann et al. (2011)
Specificity surface (tetrad position +6)	Leu, Val, Arg, Gly	Awad, Engelman, and Shaw (2013), Balzano et al. (2011), Girdler et al. (2008), Scutt et al. (2009), and Wacker, Houghtaling, Elemento, and Kapoor (2012)
Specificity surface (tetrad position +7)	Asp, Glu, Asn, Ser, Thr	Aliagas-Martin et al. (2009), Heuckmann et al. (2011), and Sloane et al. (2010)
−1 and +1 relative to DFG motif	Thr, Trp	Deng et al. (2011), Girdler et al. (2006), and Nichols et al. (2009)

Selected activity-preserving mutations that are known to induce biochemical resistance to kinase inhibitors at the resistance tetrad loci are shown. To create a complete kinase resistance profile, amino acid saturation mutagenesis (to include all 20 side chains) at each of these loci is preferable, and the subsequent mutant kinase library should then be analyzed for preservation of catalytic function, ATP binding, and drug resistance using the procedures described below. Publications describing clinical and experimental analysis of pertinent drug-resistant mutants, especially those with comprehensive methods sections, are provided.

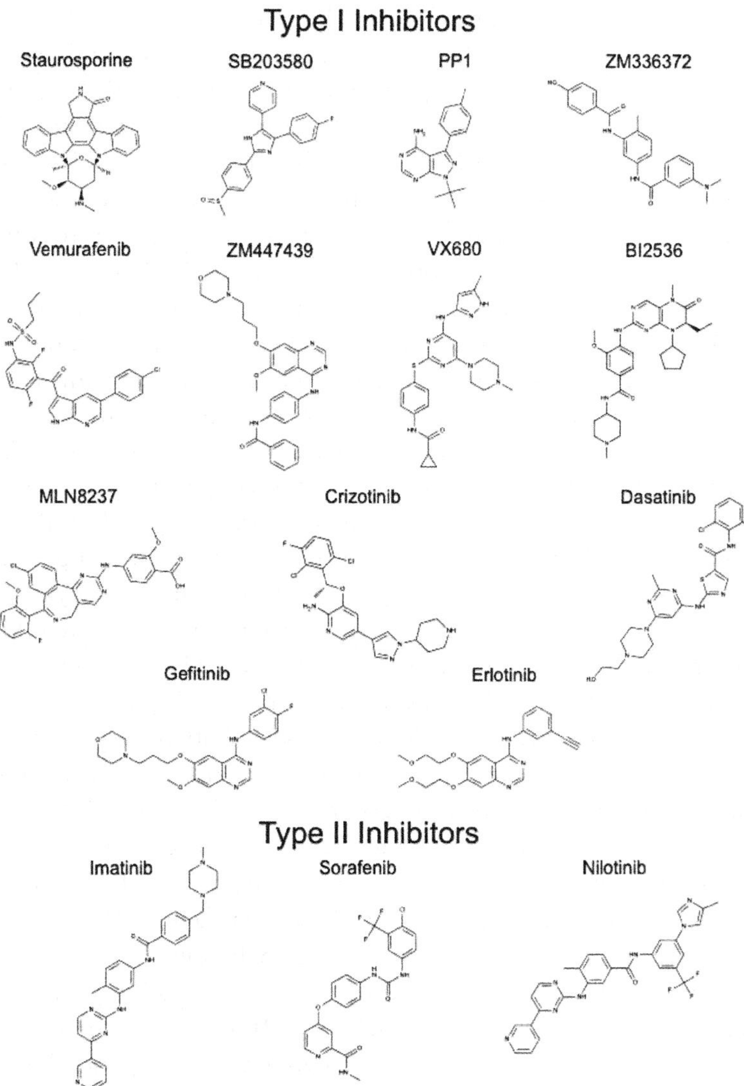

Figure 5.1 Chemical structure of selected type I and type II kinase inhibitors (Davis et al., 2011). With the possible exception of Staurosporine, chemical genetic approaches employing drug-resistant kinase alleles have been successfully exploited to validate on-target biological effects of each of these compounds. Even Staurosporine has recently succumbed to chemical biology, with modified versions (staralogs) being of particular interest as much more specific inhibitors of analog-sensitive kinases (Lopez et al., 2013).

opportunity for the design of specific inhibitors (Davis et al., 2011; Liu & Gray, 2006; Wodicka et al., 2010). However, many compounds employed in laboratory and clinical settings were identified from binding or enzyme assays employing active kinases, and are termed type I inhibitors. These compounds form hydrogen bonds with the main chain amide backbone of the hinge region, exhibiting high affinity for the (conserved) active conformation (Zuccotto, Ardini, Casale, & Angiolini, 2010). Moreover, they do not impose "inactive" conformations across the DFG motif, which in the active "DFG-in" state orients the Asp residue that stabilizes Mg^{2+} and ATP. The kinome-wide promiscuity of ATP-competitive inhibitors such as Staurosporine (Fig. 5.1) implies that its target site is spatially similar in most human kinases, which raised (largely unfounded) fears about the feasibility of creating highly specific inhibitors (Cohen, 2002; Knight & Shokat, 2005). Fortunately, many compounds acquire additional specificity through interaction with binding pockets and specificity surfaces created by the three-dimensional distribution of amino acid side chains in their target kinases. The "gatekeeper" amino acid (Fig. 5.2A) guards one such hydrophobic pocket, and is a very significant selectivity determinant in protein kinases, as determined in part by its common mutation and the promotion of drug resistance. For example, the type I inhibitors SB203580 (Cuenda et al., 1995) and PP1 (Hanke et al., 1996) exhibit selectivity toward multiple catalytically active protein kinases, with a preference for kinases possessing small residues at the gatekeeper locus. In contrast, although type II inhibitors can interact with both hinge and gatekeeper, they stabilize an inactive "DFG-out" confirmation, which is incompatible with catalysis. For example the clinical inhibitor Imatinib binds to ABL, facilitating an $\sim$180° rotation of the DFG motif that prevents ATP binding (Schindler et al., 2000). Such rearrangement of the Phe residue uncovers a hydrophobic pocket, which is characteristically occupied by such type II inhibitors. In contrast to ATP-competitive inhibitors, type III compounds, exemplified by MEK1 inhibitors such as PD318088 (Ohren et al., 2004) target nonconserved regions adjacent to the ATP site, and often require atypical amino acid mutations to provoke drug resistance (Emery et al., 2009).

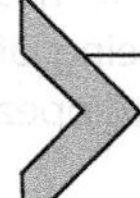

4. SCREENING APPROACHES TO DECIPHER PROTEIN KINASE-INHIBITOR SPECIFICITY

For inhibitors binding in the ATP-binding site, target quantification and validation represent a major challenge given the vast complexity of the

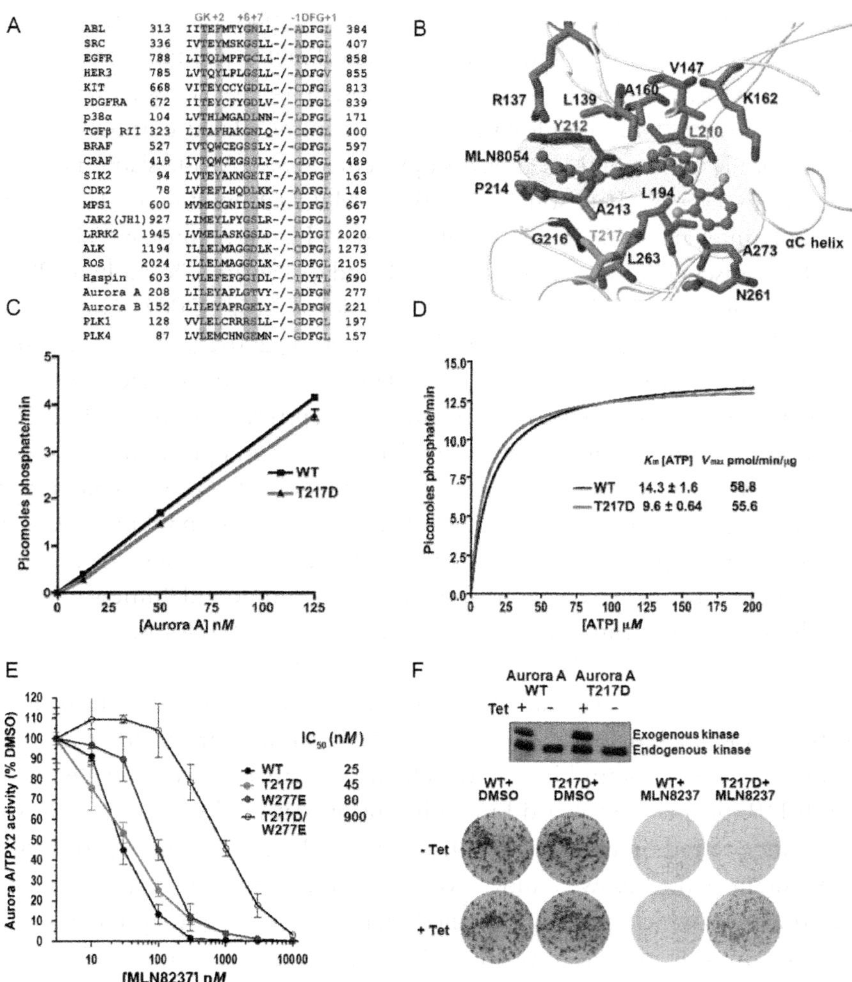

Figure 5.2 Experimental procedure for evaluating key drug-resistance loci in protein kinases. (A) Alignment of the resistance tetrad and "DFG" region of multiple kinases that have been analyzed using drug-resistance approaches, and for which drug-resistant alleles are readily available. Key side chains highlighted in kinases targeted for drug-resistance approaches are the gatekeeper residue (red), the +2 hydrophobic (green), the +6 (blue), the +7 (orange), and the DFG −1, DFG, and DFG +1 residues (pink). A similar color scheme is employed in (B), demonstrating the binding mode of MLN8054 in Aurora A (PDB ID: 2X81), with the resistance tetrad highlighted. (C) Kinetic analysis of WT and T217D Aurora A mutant (D) K_M[ATP] values for WT and T217D Aurora A. (E) Aurora A/TPX2 drug resistance toward MLN8227 induced by rationally designed mutations *in vitro*. (F) Tet-inducible overexpression of WT or DR (T217D) Aurora A analyzed by Western blotting (top panel) and on-target validation of MLN8237 cytotoxicity using a cellular assay (lower panels). *Panels (C)–(F) are reproduced from Sloane et al. (2010) with permission.* (See the color plate.)

~2000 member human "purinome" (Haystead, 2006), only approximately a quarter of which are protein kinases. A logical method to rationalize kinase specificity (or, viewed conversely, kinase resistance) is via high-throughput screening approaches employing soluble or immobilized compounds and large, ideally native, and assemblies of protein kinases (Anastassiadis, Deacon, Devarajan, Ma, & Peterson, 2011; Bain, McLauchlan, Elliott, & Cohen, 2003; Bain et al., 2007; Bamborough, Drewry, Harper, Smith, & Schneider, 2008; Bantscheff et al., 2007; Davies, Reddy, Caivano, & Cohen, 2000; Davis et al., 2011; Fabian et al., 2005; Fedorov et al., 2007; Godl et al., 2003; Miduturu et al., 2011; Patricelli et al., 2011, 2007; Vieth et al., 2005). Screening provides an invaluable source of mineable information for the logical design of DR kinase mutants from an inhibitor-centric perspective, although all databases inevitably suffer from incomplete kinome and purinome coverage. Moreover, they are currently unable to categorically prove whether an intended "on-target" or unintended "off-target" effect of an inhibitor is really responsible for phenotypic or clinical outcomes in a responsive biological system. This has led to concerns about a lack of compound "fitness" and the publication of guidelines to help standardize kinase-inhibitor exploitation in research (Cohen, 2010; Workman & Collins, 2010).

In order to achieve our experimental goal of target validation using drug-resistance kinase alleles, we established a simple approach for targeted mutagenesis of protein kinases, which aims to create a functional enzyme with DR properties toward kinase inhibitors caused by mutations in the resistance tetrad or around the "DFG" motif (Fig. 5.2). These procedures are described below, and alongside kinome-wide quantification approaches, which provide detailed snapshots of the adaptive kinome reprogramming that accompanies drug resistance (Cooper et al., 2013; Duncan et al., 2012), have the potential to improve our understanding of drug resistance.

5. RANDOM AND DIRECTED MUTAGENESIS APPROACHES REVEAL COMMON RESISTANCE MECHANISMS

RT-PCR analysis of samples from relapsed (DR) chronic myeloid leukemia patients treated with Imatinib reveals prevalent classes of mutations in the catalytic domain of BCR-ABL, including Y253F/E255K (G-loop) and T315I (gatekeeper) substitutions (Gorre et al., 2001; Roumiantsev et al., 2002). Molecular modeling demonstrates that the Thr 315 hydroxyl group forms an H-bond with Imatinib, and that an Ile gatekeeper swop abolishes this interaction by inducing a steric clash with the drug. A second "cluster" of mutations, which do not directly contact the drug, appear to destabilize the "DFG-out" conformation and thus decrease affinity toward type II inhibitors such as Imatinib (Shah et al., 2002). Importantly, the resistance of Y253F (but not T315I) mutants can be overcome clinically with second-line therapeutics such as Dasatinib (Shah et al., 2004) and T315I mutant ABL can be targeted with Ponatinib, which has the advantage of inhibiting multiple DR ABL mutants (Zhou et al., 2011), or compounds such as VX680, originally described as a pan-Aurora kinase inhibitor (Carter et al., 2005; Tyler, Shpiro, Marquez, & Eyers, 2007; Young et al., 2006). Interestingly, the BCR-ABL gatekeeper +2 amino acid (F317) also engages in aromatic interactions with Dasatinib (Tokarski et al., 2006), and a F317L mutation has been detected in cancer patients unresponsive to Dasatinib treatment. Mutation at the +6 position (G321) also renders BCR-ABL refractory to Imatinib inhibition (Balzano et al., 2011; Melo & Chuah, 2007).

To help discover potential mechanisms of drug resistance in oncogenic kinases, a DNA repair pathway defective *Escherichia coli* strain has been exploited to randomly mutagenize BCR-ABL; other protein kinases can be assessed in a similar manner (Azam, 2012; Azam et al., 2003; Cools et al., 2004; Ray, Cowan-Jacob, Manley, Mestan, & Griffin, 2007). Indeed, the recovery of Imatinib-resistant clones from transfected model Ba/F3 cells (including cancer-associated T315I ABL mutants), confirms the general applicability of such predictive approaches for assessing drug-resistance mutations in oncogenic kinases, which can subsequently be employed as targets for second-line therapeutics (Carter et al., 2005; Zhou et al., 2011). In addition, random mutagenesis using error-prone PCR or exposure to the mutagen *N*-ethyl-*N*-nitrosourea has uncovered gatekeeper-mediated resistance of EML4-ALK to Crizotinib in Ba/F3 cells (Zhang et al., 2011); BCR-ABL mutants resistant to Imatinib, Dasatinb, and/or Nilotinib (Bradeen et al., 2006); and DR oncogenic V617F JAK2 mutations (Deshpande et al., 2012; Marit et al., 2012). Finally, type III phosphoinositide-4-kinase (PI4K) mutagenesis (Balla et al., 2008) or *in vivo* screening around the affinity pocket in the oncogenic PI3K isoform p110α, discloses mutations imparting lipid kinase-inhibitor resistance (Zunder, Knight, Houseman, Apsel, & Shokat, 2008).

6. A GENERAL PROCEDURE FOR DIRECTED (NONRANDOM) MUTAGENESIS OF dsDNA PLASMIDS

- Clone protein kinase into plasmid vector and isolate from methylation-competent host. Validated vectors for prokaryotic expression and affinity purification include Novagen's pET (6-His), Pharmacia's pGEX (GST), or NEB's pMal (Maltose-binding protein) vectors. pBAC-2cp (Novagen) is suitable for kinase expression using the Baculovirus Sf9 system, and pCDNA5 FRT vectors (Invitrogen) have been successfully employed for the creation of stable cell lines expressing kinases (see below).
- Design complimentary ~36 bp primer pairs, with the mutated codon positioned near the center. Aim for a GC content of >50%, and ideally end primers in G or C. We do not find that PAGE purification of commercial primers is necessary for highly efficient (>95%) mutagenesis.
- Dissolve primers in sterile H_2O at a concentration of 0.5 mg/ml and store at −20 °C.
- Assemble PCR reaction containing 5–50 ng kinase template, 500 ng of each primer, 1.5 m*M* $MgSO_4$, 200 μ*M* dNTPs, and 2.5 U of DNA polymerase. Validated thermostable DNA polymerases include PfuTurbo (Stratagene), KOD (Novagen), and DeepVent (NEB).
- Cycle for ~2 min per kb plasmid length (e.g., 20 min for a 10 kb plasmid), with an annealing temperature of between 50 and 60 °C for 18 cycles.
- Incubate PCR reaction with 10 U Dpn1 (NEB) at 37 °C for 1 h to the digest WT template.
- Confirm amplified PCR product on TAE-Agarose gel 1% (w/v) and immediately transform 1/20th of the PCR reaction into competent bacteria, e.g., XL-1 or Top10′ ($\sim 1\times10^8$–1×10^9 cfu/μg), culture on LB Agar plates containing selective antibiotic (typically 50 μg/ml kanamycin or 100 μg/ml carbenicillin) and culture overnight at 37 °C.
- Pick several colonies, grow overnight in 3 ml LB + antibiotic, and purify plasmid by standard procedures. We recommend Qiagen™ for purification of high quality and transfectable plasmids.
- Sequence complete coding region to confirm specific mutagenesis.
- For high-level expression of recombinant kinases for assay, transform purified plasmid into bacterial strain BL21 (DE3) pLysS (Novagen).
- For stable isogenic human cell production, transform plasmid prior to antibiotic selection (see below).

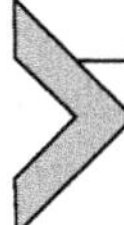

7. THE RESISTANCE TETRAD POSITION 0: THE GATEKEEPER RESIDUE

The gatekeeper residue lies adjacent to the N^6 atom of the purine ring, and controls access to a hydrophobic pocket proximal to the canonical ATP-binding site (Dar & Shokat, 2011; Zuccotto et al., 2010). All human kinases encode gatekeeper residues larger than Gly and Ala, with a propensity toward either Ser/Thr (small) or Met/Phe/Leu (bulky) side chains (Fig. 5.3). In the 1990s, it became clear that substitution of the bulky Ile

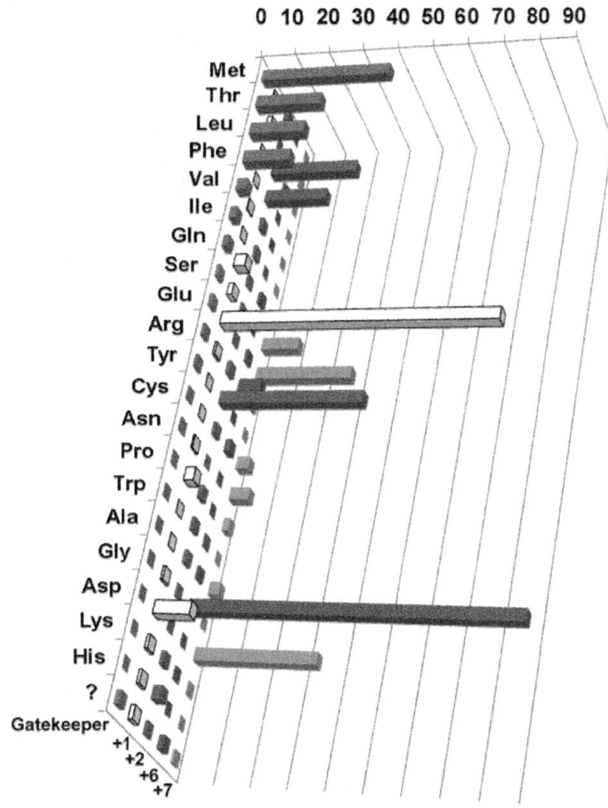

Figure 5.3 Frequency distribution (%) of amino acids found within the resistance tetrad. Gatekeeper, +1 (commonly an Asp or Glu residue), +2, +6, and +7 amino acid statistics from the human kinome are plotted as percentage of total (Manning et al., 2002). The most common residues at the gatekeeper residue are Met, Thr, Leu, Phe, and Val, whereas Tyr, Leu, Phe, and His predominate at the +2 position. Among benchmark AGC, CAMK, and TKs, the +6 position is very commonly Gly, while Asp, Glu, and Ser prevail at +7. Many amino acids are rare or absent in the resistance tetrad, helping simplify the choice of initial logical mutations for developing drug-resistance alleles. (See the color plate.)

gatekeeper of SRC with a smaller side chain endows it with unique properties (Shah, Liu, Deirmengian, & Shokat, 1997). Consistent with its role in accommodating bulky ATP analogs, the SRC gatekeeper (I338) also controls inhibition by PP1 and SB203580 in an amino acid side chain volume-dependent manner (Liu et al., 1999). Interestingly, the replacement of a Phe for an Ala or Gly gatekeeper in unrelated kinases sensitizes these kinases to PP1 and N^6-modified PP1 variants (Liu et al., 1999), and this technique has since been expanded kinome wide (Bishop et al., 2000; Blencke et al., 2004; Hegarat et al., 2011). Many, but not all, protein kinases are tolerant to such changes, and inspection of the amino acid at this position, combined with structural and frequency analysis (Figs. 5.3 and 5.4), provides a rationale for cross-kinome exploitation of kinases bearing mutant gatekeeper residues that can be screened toward any number of compounds.

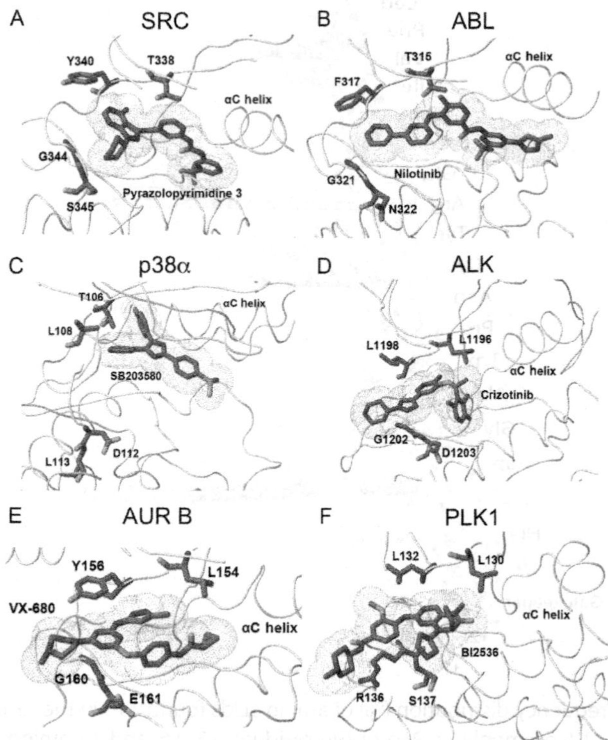

Figure 5.4 Structural analysis of the resistance tetrad in diverse human kinases. The cocrystal structures of (A) SRC and a PP1 analog (PDB ID: 3EL7), (B) ABL and Nilotinib (PDB ID: 3CS9), (C) p38α and SB203580 (PDB ID: 1A9U), (D) ALK and Crizotinib (PDB ID: 2XPT), (E) Aurora B and VX680 (PDB ID: 4AF3), and (F) Plk1 and BI2536 (PDB ID: 2RKU). Amino acids corresponding to the resistance tetrad are indicated for each kinase. (See the color plate.)

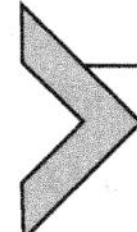

8. SB203580: A PARADIGM FOR GATEKEEPER-MEDIATED DRUG RESISTANCE FROM TEST TUBE TO MOUSE

The target specificity of all kinase inhibitors remains unproven until drug-resistance data provide unequivocal evidence to the contrary. A case-in-point is the pyridinyl imidazole inhibitor SB203580, which inhibits p38α MAPK, but does not affect closely related MAPKs such as ERK or JNK, despite very high-sequence conservation (Cuenda et al., 1995). With the subsequent discovery of additional p38 MAPK family members (including p38γ and p38δ) the specificity of SB203580 for p38α/β was traced to occupancy of the 4-fluorophenyl moiety in the hydrophobic pocket lying adjacent to the Thr residue in p38α/β, which is blocked by a bulkier residue in all other (SB203580-resistant) MAPKs (Eyers et al., 1998). Interestingly, many kinases with small gatekeepers (e.g., RAF kinases and TGFβ receptors) also exhibit sensitivity to type I inhibitors such as PP1 and SB203580 that target the hydrophobic pocket (Blencke et al., 2004; Eyers et al., 1998; Godl et al., 2003; Hall-Jackson, Goedert, Hedge, & Cohen, 1999). The relationship between gatekeeper volume and sensitivity was subsequently exploited to investigate cellular drug resistance, through ectopic overexpression of a DR T106M p38α mutant, permitting validation of "on-target" p38α substrates in cells (Eyers et al., 1998; Kovarik et al., 1999). Interestingly, the unrelated CRAF inhibitor ZM336372 also inhibits p38α/β, and T106M mutations also render p38α/β insensitive to this compound, suggesting a similar gatekeeper-driven binding mode (Hall-Jackson, Eyers, et al., 1999). Notably, the (as then) unexplained activation of the RAF kinases by SB203580 was validated as a *bona fide* "off-target" effect independent of p38α inhibition (Eyers, van den, Quinlan, Goedert, & Cohen, 1999). Remarkably, SB203580 and ZM336372 both stimulate RAF activity in cells (Hall-Jackson, Goedert, et al., 1999), and *in vivo* chemical genetics proves that the BRAF inhibitor Vemurafenib paradoxically activates (rather than inhibits) the MAPK pathway through direct binding to BRAF in cells (Poulikakos, Zhang, Bollag, Shokat, & Rosen, 2010). Rodent models of inhibitor resistance provide a powerful opportunity to validate compound selectivity and "on-target" pharmacology in nonhuman systems. For example, DR p38α expression diminishes the inhibitory effect of SB203580 on TNFα (but not IL-10) induction in macrophages, validating "on"- and "off"-target activities attributable to distinct SB203580-sensitive kinases (Godl et al., 2003; Guo, Gerl, & Schrader, 2003). Genetic "knock-in" of p38α and p38β T106M mutants also validate the specific contribution of each kinase to the TNFα response *in vivo* (O'Keefe et al., 2007). In addition to their unfortunate appearance in clinical subjects, DR kinase alleles have also proven useful for validating the targets of small molecule inhibitors in cancer models (Burkard & Jallepalli, 2010) and the function of SIK2 in the inflammatory system (Clark et al., 2012). For example, the constitutively active BRAF V600E oncogenic mutant is a proliferative driver in melanoma, and mutation of the Thr gatekeeper to Asn induces drug resistance to multiple inhibitors *in vitro*, although not yet *in vivo*, including the clinically approved BRAF inhibitor Vemurafenib and the multikinase inhibitor Sorafenib. Interestingly, Vemurafenib fails to prevent tumor growth in mice bearing T279N BRAF-driven tumors, whereas Sorafenib remains effective (Whittaker et al., 2010). This elegant approach validates BRAF (and no other kinase) as the cytotoxic Vemurafenib target in a melanoma model and highlights a BRAF-independent target for Sorafenib, consistent with the weak affinity of this compound toward native BRAF in cells (Patricelli et al., 2011). Interestingly, Sorafenib has also been reported to be an antagonist of 5-hydroxytryptamine (5HT) receptors (Lin et al., 2012), highlighting how chemical genetic approaches that tease apart "on"- and "off"-target effects of multitargeted agents have the potential to be such powerful validation tools.

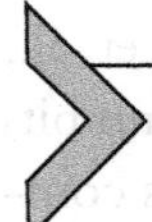

9. EXPANDING THE RESISTANCE TETRAD: +2 (HYDROPHOBIC) AND +6/+7 SPECIFICITY SURFACES IN KINASES

Aurora and Polo-like kinases (Plks) are oncological drug targets whose elevated expression in tumors has prompted the preclinical and clinical assessment of multiple inhibitors. Interestingly, and in contrast to many clinically targeted kinases, none of these kinases possesses a small gatekeeper residue (Fig. 5.2A), suggesting that type I inhibitors must achieve specificity and potency through different interfaces. The cellular target of the experimental Aurora inhibitor ZM447439 was validated as Aurora B, through retrieval of inhibitor-resistant Aurora B Y156H (+2 position) and G160V (+6 position) alleles from hyper-mutagenic (HCT-116) colon cancer cell lines selected after chronic ZM447439 exposure (Girdler et al., 2008). That Aurora B is the only phenotypic target of this compound was confirmed by overexpression of each mutant (in the presence of endogenous Aurora B), which restored the phosphorylation of histone H3 on S10 and promoted cell viability in the presence of ZM447439. Interestingly, these acquired mutations differ from classical gatekeeper mutants, although Y156 and G160 lie in close proximity to ZM447439, suggesting that their mutation generates a steric clash without directly affecting ATP binding or catalytic activity. Indeed, the expression of a DR G160L Aurora B allele also reverses phenotypic effects induced by the equipotent Aurora A and B inhibitor VX680 and restores clonogenic potential in the presence of the drug, directly validating Aurora B (rather than Aurora A) as the antiproliferative target (Scutt et al., 2009). Using an inhibitor resistance screen for Aurora A in the presence of its physiological activator TPX2 (Eyers, Churchill, & Maller, 2005) the hinge-loop residue G215, corresponding to Aurora B G160, emerged as a critical requirement for VX680 binding because Aurora A G216 substitutions induced volume-dependent VX680 resistance *in vitro* and phenotypic reversal of VX680-dependent mitotic defects. The ability to tease apart the *simultaneous* effects of VX680 on two separate kinases and unequivocally map the cytotoxic target demonstrates the immense power of chemical genetics for target validation. Indeed, to explain the specificity of a 2,4-bisanilinopyrimidine inhibitor class for Aurora A (but not Aurora B) the +7 residue of the Aurora kinase resistance tetrad, which is a Thr in Aurora A and a Glu in Aurora B (Fig. 5.2A) was examined. Remarkably, a simple swap exchanges inhibitor susceptibility (Aliagas-Martin et al., 2009; Dodson et al., 2010) and T217D/E Aurora A alleles induce cellular resistance toward MLN8054 and MLN8237 (Alisertib), validating Aurora A as the authentic cytotoxic target for this late stage clinical agent (Sloane et al., 2010).

10. THE RESISTANCE TETRAD IS A SELECTIVITY FILTER APPLICABLE FOR KINOME-WIDE DRUG-RESISTANCE STUDIES

Encouraged by the +6 amino acid driven mechanism of VX680 resistance in Aurora kinases, this residue was also investigated in Plk1 as a resistance determinant toward BI2536, which inhibits Plk1–3, but not Plk4 (Kothe et al., 2007). Arg 136 (conserved in Plk1–3) lies adjacent to BI2536, and mutating this residue to Gly renders Plk1 resistant toward BI2536, while retaining sensitivity to the unrelated inhibitor GW843682X, permitting its exploitation for chemical genetic target validation (Scutt et al., 2009). Significantly, this experimental R136G Plk1 mutant was also discovered *in vivo* using whole transcriptome approaches in cancer cells that had acquired resistance to BI2536 (Wacker et al., 2012). Interestingly, Plk4 contains a Gly at the +6 tetrad position (Fig. 5.2A), and is sensitive to both VX680 and MLN8054 (neither of which inhibit the Arg-containing Plk1–3), whereas a G95R mutation renders Plk4 resistant to both drugs, confirming the importance of the +6 Gly residue for sensitivity (Sloane et al., 2010). Recent clinical data from a metastatic lung adenocarcinoma also confirms that the +6 Gly side chain of the CD74-ROS1 fusion kinase mediates sensitivity to Crizotinib, since a G2032R mutation induces marked resistance *in vitro* and *in vivo*. To test the generality of hinge-loop region resistance determinants, Balzano and coworkers recently exploited *in vitro* kinase assays to explore effects of mutating the hinge-loop +2 or +6 residues in unrelated kinases, namely ABL/SRC (TK group), CK1δ (CK1 group), PAK5 (STE group), PHKγ (CAMK group), and the atypical kinase Haspin (Balzano et al., 2011). Remarkably, F317H ABL mutants are resistant to VX680, Imatinib, Dasatinib, and Nilotinib validating the +2 side chain as a general resistance determinant. In a similar fashion, Y340H/A substitutions renders SRC resistant to inhibition by Staurosporine, PP2 and VX680, and a F525L mutation imparts resistance of PAK5 toward Staurosporine. Gly +6 substitutions in PAK5 and CK1δ impart resistance toward both D4476 and Staurosporine, while PHKγ G110N/E mutants are Staurosporine resistant. Strikingly, a 5-iodotubericidin resistant G609D Haspin mutant was employed in cells to validate the on-target inhibition of histone H3 T3 phosphorylation (Balzano et al., 2011). Interestingly, Crizotinib-resistance in the oncogenic EML4-ALK kinase can readily be induced by cellular +2 (L1198P) or +7 (D1203N) tetrad mutations (Heuckmann et al., 2011). Finally, by combining logical mutations in the hinge region of the cyclin-dependent kinases (CDKs), DR alleles of Cdk2 have been designed to confirm a rate-limiting role for this kinase in DNA replication in interphase *Xenopus laveis* extracts (Echalier et al., 2012). Taken together, these findings extol the virtues of exploiting modification of selectivity filtering amino acids found in the resistance tetrad, which provide a blue-print for facile kinome-wide validation approaches.

11. ENGINEERING AND ANALYSIS OF LOGICALLY DESIGNED DRUG-RESISTANCE MUTATIONS

Inspection of aligned human protein kinase families (Manning et al., 2002) provides an excellent starting point for evaluating how the design of resistance-inducing mutations can be matched to the plasticity of the ATP-binding site, thus maximizing the chances of simultaneously preserving function and inducing drug resistance. We employ web-based kinome bioinformatics, using datasets retrieved from www.kinase.com, www.uniprot.org, or www.shokatlab.ucsf.edu/, to align the catalytic domain of pertinent kinases and reveal amino acids profiles in the resistance tetrad and DFG motif (Fig. 5.2A). Figure 5.3 presents the frequency and identity of amino acids in the gatekeeper (allotted position 0), +2 (hydrophobic), +6 (often Gly), and +7 (often small hydrophilic or negatively charged) tetrad loci in human kinases, revealing a very strong preference for distinct amino acids at each of these loci. By employing kinomics, logical mutations can therefore be introduced at the tetrad positions of any kinase and their effects on activity and drug resistance evaluated. In addition, structural biology and modeling have a critical role to play in designing specific drug-resistance alleles, and for many (but not all) compounds, the inspection of a crystal structure of a kinase-inhibitor complex reveals the binding mode, and is indispensable for defining how selectivity-determining tetrad surfaces engage with compounds (Fig. 5.4). For kinase/inhibitor pairs where unique or novel interaction sites are suspected, nontetrad mutations also permit chemical genetic validation of inhibitors (Azam et al., 2003; Burkard, Santamaria, & Jallepalli, 2012; Emery et al., 2009; Fitzgerald et al., 2003; Green et al., 2008; Wacker et al., 2012). Saturation mutagenesis is indispensable to fully explore the potential for drug resistance at different kinase loci, especially if inhibitor binding occurs independent of the Gly-rich loop and/or resistance tetrad, where stereotypical and well-tolerated replacements are readily made. In many cases, the exploitation of published SAR and kinase specificity data can also be applied on a case-by-case basis to the design of logical mutations in the tetrad, or at kinase-specific amino acids. For example, substitution of the unique Trp residue at the DFG+1 position in Aurora kinases (Fig. 5.2) has been employed for chemical genetic validation of Aurora A kinases (Girdler et al., 2006; Sloane et al., 2010) and a mutation at the DFG−1 position (A2016T) creates a DR LRRK2 mutant with wild-type activity (Nichols et al., 2009). Finally, "atypical" Gln gatekeeper substitutions create DR M602Q TTK/Mps1 (Jemaa et al., 2012; Kwiatkowski et al., 2010; Schmidt et al., 2005) or DR T96Q SIK2 (Clark et al., 2012) mutants for "on-target" validation approaches. We think it likely that further mutations that induce resistance in the DFG region of kinases will be identified, although as with "activating" oncogenic gatekeeper mutants, some thought as to their regulation should be given, especially since mutations at the DFG+1 motif might also change substrate specificity markedly (Chen et al., 2014), violating a central tenet needed for successful exploitation of DR alleles of kinases.

12. Analysis of Inhibitor Resistance Toward WT and DR Mutants *In Vitro*

To monitor unanticipated or unwanted effects of drug-resistance mutations on the kinetics of mutant kinases, the DR kinase should be compared with the WT kinase with respect to intrinsic activity (Fig. 5.2C) and the ability to transphosphorylate common substrates (Balzano et al., 2011; Carter et al., 2005; Eyers et al., 1998; Scutt et al., 2009; Sloane et al., 2010). Standard assay techniques often employ catalytically active kinase populations, which are targeted most efficiently by type I inhibitors, although type II inhibitors often have sufficient affinity for "active" kinases to permit enzymatic assay for resistance-validation purposes (Balzano et al., 2011). Proprietary approaches to assess type I- and II-inhibitor bindings are also suited for probing drug resistance (Carter et al., 2005; Davis et al., 2011; Fedorov et al., 2007). Drug resistance can be quantified *in vitro* using phosphotransferase assays and calculation of IC_{50} values for WT and DR kinases (Fig. 5.2E). Protein kinases can be assayed in multiple formats, although we use medium-throughput systems employing peptide mobility (EZReader II, Perkin Elmer) or phosphorylation of basic substrates (e.g., histones or MBP), which permit [^{32}P]-binding and quantification using phosphocellulose (p81) filter papers.

- Purify and determine specific activity of WT and DR kinase toward a common substrate and confirm that mutation supports catalytic activity.
- Some kinases need to be activated by other kinases (e.g., MAPKs) or are most appropriately assayed after reconstitution with physiological subunits (e.g., CDKs and Aurora kinases) prior to analysis of drug resistance.
- Employ a final [$\gamma-^{32}$P]ATP concentration (~500 cpm/pmol) close to the calculated K_M value for ATP (as determined below).
- To evaluate drug resistance, perform inhibitor assays using WT and DR kinases. A 5-log dilution series of inhibitor concentration in triplicate is suggested. We perform our kinase assays in a standard buffer, 25 m*M* Tris (pH 7.4), 100 m*M* NaCl, 1 m*M* DTT, 10 m*M* $MgCl_2$.
- As a control, use 1% (v/v) DMSO solvent, which does not inhibit the activity of most kinases investigated by >10% at this concentration.
- Calculate kinase activity as a function of picomoles phosphate transferred per minute in the presence and absence of inhibitors.
- Employ software to plot data and calculate IC_{50} values.

- A minimum IC_{50} resistance ratio (DR:WT kinase) of two- to fivefold (e.g., T217D in Fig. 5.2E) provides reasonable evidence that a drug-resistance allele should be assessed in cells, although 1–4 log orders of resistance ratios are often evident in engineered DR kinases.

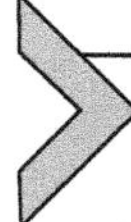

13. ONCOGENIC GATEKEEPER MUTATIONS: UNANTICIPATED MECHANISMS OF GATEKEEPER RESISTANCE MERIT BIOCHEMICAL SCRUTINY OF DR MUTANTS

Tyrosine kinases represent paradigms for understanding gatekeeper-mediated drug resistance, and more recently targeted clinical kinases, such as the oncogenic EML4-ALK fusion, also evolve resistance-inducing gatekeeper mutations to approved drugs such as Crizotinib (Katayama et al., 2011). Activating mutations in the EGFR tyrosine kinase are also common drivers in lung cancers. For example, the clinically approved quinazoline inhibitors Gefitinib and Erlotinib target constitutively active L858R EGFR, but a T790M gatekeeper substitution renders the kinase resistant to both drugs. Remarkably, in direct-binding assays, a L858R/T790M double mutant does not exhibit a significant difference in inhibitor affinity when compared to EGFR using nonphysiological ATP concentrations (Yun et al., 2008). Furthermore in a physiological context irreversible (covalent), quinazoline inhibitors potently inhibit T790M mutants, even though these drugs engage in a similar set of interactions as Erlotinib and Gefitinib. Remarkably, the L858R substitution increases the K_M[ATP] value compared to WT EGFR, whereas an additional gatekeeper (T790M) mutation reduces it again by an order of magnitude. In a remarkable twist, crystallographic analysis reveals that the "drug-resistant" met EGFR gatekeeper assumes a rotamer conformation that permits accommodation of inhibitors (Yun et al., 2008). Together, these observations argue against steric clash between the bulkier gatekeeper and inhibitor molecule (as is the case with Imatinib and BCR-ABL). Instead, the gatekeeper mediates a reduction in K_M[ATP], and serves as a general resistance mechanism in cells against any (reversible) ATP-competitive inhibitor with K_i insufficient to overcome the increased ATP affinity. Importantly, by screening focused library for T790M-specific inhibitors, covalent (irreversible) inhibitors, which exhibit high affinity for T790M EGFR, have also been reported (Ward et al., 2013; Zhou et al., 2009). These studies argue that drug-resistance alleles should be analyzed biochemically where possible, because gatekeeper and hinge mutations also have the ability to change oncogenic signaling potential (Chen et al., 2007; Skaggs et al., 2006) in addition to modulating drug binding. Indeed, gatekeeper mutations can be present in patients prior to therapy since hydrophobic (DR) substitutions in ABL and SRC appear to *enhance* catalytic activity, and potentially endowing tumors with a selective oncogenic growth advantage *in vivo* (Azam, Seeliger, Gray, Kuriyan, & Daley, 2008).

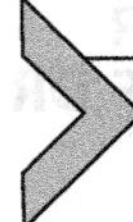

14. EVALUATION OF CATALYTIC BEHAVIOR AND K_M[ATP] VALUE FOR WT AND DR KINASE MUTANTS *IN VITRO*

To monitor, the effects of drug-resistance mutations on the Michaelis–Menten constant for ATP, WT, and DR kinases should be routinely compared using fixed concentrations of kinase/substrate over a range of high- and low-ATP concentrations (Fig. 5.2D). Some mutations also modulate ATP binding, potentially perturbing signaling functions in a cellular context (although this does not necessarily preclude them from cellular drug validation).

- Express tagged WT or DR kinase in an appropriate system (e.g., *E. coli*, Sf9, human HeLa, or HEK293 cell line).
- Affinity purify kinase using affinity tag, assay immediately, or dialyze into an all-purpose kinase buffer (25 m*M* HEPES (pH 7.4), 50 m*M* NaCl, 1 m*M* DTT) prior to storage at −80 °C.
- Assess specific activity of kinase in triplicate over a range of ATP concentrations, with Mg^{2+} ions fixed at 10 m*M*. Include no substrate and no ATP controls, to allow assay background to be calculated accurately. Most kinases phosphorylate substrates with K_M[ATP] values of ~10 μ*M*, although this varies widely. We typically assay final ATP concentrations of between 0.5 and 2 m*M* [$\gamma-{}^{32}$P]ATP, employing ATP with a specific activity of 500 cpm/pmol. A typical experimental range is: 0, 0.5, 1, 2, 5, 10, 20, 50, 100, 200, 500, 1000, and 2000 μ*M* ATP.
- Quantify incorporation of phosphate signal into substrate at different ATP concentrations, subtracting background values, and calculate mean and SD of triplicate data.
- Calculate K_M[ATP] for WT and DR kinase using nonlinear regression analysis.
- Determination of kinase K_M[ATP] value, and its adoption for kinase assays (see below) imparts an IC_{50} value that is similar to K_i, the inhibition constant (Cheng & Prusoff, 1973).

15. INTACT CELL SYSTEMS FOR ANALYZING DRUG RESISTANCE AND TARGET VALIDATION

Two major "on-target" mechanisms can induce drug resistance in cells in addition to the plethora of signal rewiring that inevitably occurs upon drug exposure. The first involves overexpression and/or an increased activity in the target kinase activity, which enables cells to survive in otherwise lethal concentrations of a drug, exemplified by BCR-ABL overexpression (Gorre et al., 2001) or alternative BRAF splicing (Poulikakos et al., 2011). The second involves direct drug resistance induced by newly acquired or emergent preexisting mutations. Due to the complex spectrum of mutations selected by prolonged drug exposure to a single compound, even *in vitro* (Wacker et al., 2012), we have chosen pursue the specific overexpression of clinically relevant or rationally designed kinase mutants for target validation, which exert resistance effects in a dominant fashion in the presence of endogenous (drug-sensitive) kinases. Moreover, by employing isogenic, inducible stable cell lines individually expressing either WT or DR kinase alleles, effects of overexpression, hyperactivity, and mutations in the resistance tetrad are controlled for simultaneously.

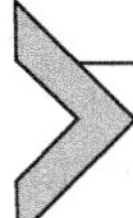

16. GENERATION OF STABLE, ISOGENIC CELL LINES EXPRESSING TETRACYCLINE-INDUCIBLE KINASES

Invitrogen have commercialized plasmid-based vectors (Flp-In™) for creating isogenic stable cell lines suitable for drug-resistance studies in human cells, and are a useful addition to highly mutagenic cell lines such as the colorectal line HCT-116 (Wacker et al., 2012). The single FRT site of integration in parental cells ensures isogenic expression of WT and DR kinase alleles from any Flp-In™ expression vector. Tetracycline (Tet)-inducible lines (Flp-In™ T-REx™) cells are created by stable coexpression of the Tet-repressor plasmid (pCDNA6/TR), in one of four commercially available T-REx™ lines (HEK293, HeLa, CHO, and Jurkat). These are ideal for drug-resistance studies, since expression of the WT or DR allele is induced in parallel by addition of Tet to the culture medium, and direct effects compared in the same stable cells before and after induction of "test" kinases (Fig. 5.2F). A standard protocol for generating isogenic stable cell lines is described below.

16.1. Transfection and selection procedure

- Parental Flp-In™ T-REx™ HEK293 or HeLa cells are cultured in a T75 flask containing DMEM + 10% FCS, 100 μg/ml Zeocin, and 5 μg/ml blasticidin.
- Cells are grown to 70–90% confluence, which is optimal for transfection.
- Add 250 μl of serum-free medium to 0.5 μg WT or DR kinase-encoding plasmid and 4.5 μg pOG44 (encoding Flp recombinase), which mediates recombination at the genomic FRT site.
- Add 5 μl Lipofectamine 2000™ to 250 μl serum-free medium.
- Remove cell medium, wash with PBS, add 2 ml serum-free DMEM, and incubate at 37 °C.
- Mix Lipofectamine and DNA mixtures together and incubate at 20 °C for 5 min.
- Add plasmid/Lipofectamine mixture to plates and rock to mix. Incubate at 37 °C for 5 h.
- Replace transfection medium with 10 ml DMEM + 10% FCS and incubate at 37 °C for 16 h.
- Replace transfection media with selection media (10 ml DMEM + 10% FCS, containing 5 μg/ml Blasticidin and 200 μg/ml Hygromycin B).
- Allow cells to grow until colonies are visible with the naked eye (typically between 7 and 10 days), add new sterile medium every 2–3 days.
- Expand individual colonies, harvest with Trypsin:EDTA solution (pool cloned colonies, which will all contain the gene integrated at the same locus) and transfer to a new T75 flask for experimental validation of expression. We typically employ cell lines for between 20 and 30 passages before discarding, recovering earlier passages from liquid N_2 stocks.
- To test that the kinase of interest has been stably incorporated into the genome, add 1 μg/ml Tet to 70% cell confluent culture dishes. Detach and harvest cells after 24 h using Trypsin:EDTA solution, wash pellet twice with PBS, then resuspend in 800 μl of cell lysis buffer. After 10 min on ice with regular vortexing, centrifuge the lysates at 13,000 × *g* for 10 min at 4 °C. Quantify protein expression by Western blotting. To facilitate rapid detection, an epitope tag (e.g., MYC or FLAG) can be fused in frame with the kinase.

17. ANALYSIS OF KINASE DRUG RESISTANCE TOWARD A CYTOTOXIC INHIBITOR: CELL GROWTH ASSAY BASED ON COLONY FORMATION (FIG. 5.2F)

- Seed cells in a 12-well plate, at between 100 and 1000 cells/well, but typically 500.
- Add 1 μg/ml Tet to experimental wells, or 70% ethanol to controls, and culture for 24 h to induce expression of WT and DR kinase.
- On day 1, add an appropriate dilution series of inhibitor, or DMSO as control, to each well in the presence of 1 μg/ml Tet.
- On day 4, readminister 1 μg/ml Tet and DMSO/inhibitor to each well.
- Leave until desired colony growth is seen in DMSO-treated wells (7–14 days), then fix and stain with 1% (w/v) methylene blue (dissolved in 70% methanol) to visualize colonies.
- Quantify colony formation using spectrophotometric quantification of dye binding or appropriate image software (e.g., Image J). If drug resistance is apparent in Tet-exposed (but not control) DR-expressing cells, then target validation is successful (e.g., Fig. 5.2E). If only some phenotypes are abolished in the presence of the drug, then "off"-target effects are present. New drug-resistance alleles in other potential targets should be evaluated.

18. CONCLUSIONS

The creation and inducible expression of inhibitor-resistant kinase alleles represents a broadly applicable approach to validate inhibitor-related phenotypes and uncover off-target inhibitor effects in cell lines. When exploited in a cellular context, and appropriately controlled for overexpression-induced effects (which may be significant), it provides an ideal physiological environment for analyzing the on- and off-target effects of any inhibitor, and a test bed to evaluate novel compounds that effectively override resistance. Similar approaches using randomly mutated libraries of kinases also have the power to reveal on- and off-target effects of kinase inhibitors. Validated DR mutants can readily be exploited to further investigate phenotypic (e.g., phosphorylated substrates, cell cycle status, and microscopic scrutiny) or cell survival (cloning ability, proliferation, apoptosis, and animal explant studies) involving kinase inhibitors, and are not limited to mammalian systems, being equally applicable in vertebrate and nonvertebrate cell types. By combining bioinformatics, structural kinomics, biochemistry, differential cytotoxicity, and preferably *in vivo* models, we are now in a strong position to utilize drug resistance as an empowering tool for kinome-wide target validation. Indeed, inhibitor-resistant kinase mutants have the unique capacity to untangle complex polypharmacology-driven biology induced by small molecules (Knight, Lin, & Shokat, 2010) and are therefore likely to become critical to empower validation of the biological targets and antitargets of essentially any protein kinase inhibitor (Dar, Das, Shokat, & Cagan, 2012).

REFERENCES

Aliagas-Martin, I., Burdick, D., Corson, L., Dotson, J., Drummond, J., Fields, C., et al. (2009). A class of 2,4-bisanilinopyrimidine aurora A inhibitors with unusually high selectivity against aurora B. *Journal of Medicinal Chemistry*, *52*, 3300–3307.

Anastassiadis, T., Deacon, S. W., Devarajan, K., Ma, H., & Peterson, J. R. (2011). Comprehensive assay of kinase catalytic activity reveals features of kinase inhibitor selectivity. *Nature Biotechnology*, *29*, 1039–1045.

Awad, M. M., Engelman, J. A., & Shaw, A. T. (2013). Acquired resistance to crizotinib from a mutation in CD74-ROS1. *The New England Journal of Medicine*, *369*, 1173.

Azam, M. (2012). An in vitro screening to identify drug-resistant mutations for target-directed chemotherapeutic agents. *Methods in Molecular Biology*, *928*, 175–184.

Azam, M., Latek, R. R., & Daley, G. Q. (2003). Mechanisms of autoinhibition and STI-571/imatinib resistance revealed by mutagenesis of BCR-ABL. *Cell*, *112*, 831–843.

Azam, M., Seeliger, M. A., Gray, N. S., Kuriyan, J., & Daley, G. Q. (2008). Activation of tyrosine kinases by mutation of the gatekeeper threonine. *Nature Structural and Molecular Biology*, *15*, 1109–1118.

Bain, J., Mclauchlan, H., Elliott, M., & Cohen, P. (2003). The specificities of protein kinase inhibitors: An update. *The Biochemical Journal*, *371*, 199–204.

Bain, J., Plater, L., Elliott, M., Shpiro, N., Hastie, C. J., Mclauchlan, H., et al. (2007). The selectivity of protein kinase inhibitors: A further update. *The Biochemical Journal*, *408*, 297–315.

Balla, A., Tuymetova, G., Toth, B., Szentpetery, Z., Zhao, X., Knight, Z. A., et al. (2008). Design of drug-resistant alleles of type-III phosphatidylinositol 4-kinases using mutagenesis and molecular modeling. *Biochemistry*, *47*, 1599–1607.

Balzano, D., Santaguida, S., Musacchio, A., & Villa, F. (2011). A general framework for inhibitor resistance in protein kinases. *Chemistry and Biology*, *18*, 966–975.

Bamborough, P., Drewry, D., Harper, G., Smith, G. K., & Schneider, K. (2008). Assessment of chemical coverage of kinome space and its implications for kinase drug discovery. *Journal of Medicinal Chemistry*, *51*, 7898–7914.

Bantscheff, M., Eberhard, D., Abraham, Y., Bastuck, S., Boesche, M., Hobson, S., et al. (2007). Quantitative chemical proteomics reveals mechanisms of action of clinical ABL kinase inhibitors. *Nature Biotechnology*, *25*, 1035–1044.

Barouch-Bentov, R., & Sauer, K. (2011). Mechanisms of drug resistance in kinases. *Expert Opinion on Investigational Drugs*, *20*, 153–208.

Bikker, J. A., Brooijmans, N., Wissner, A., & Mansour, T. S. (2009). Kinase domain mutations in cancer: Implications for small molecule drug design strategies. *Journal of Medicinal Chemistry, 52*, 1493–1509.

Bishop, A. C., Ubersax, J. A., Petsch, D. T., Matheos, D. P., Gray, N. S., Blethrow, J., et al. (2000). A chemical switch for inhibitor-sensitive alleles of any protein kinase. *Nature, 407*, 395–401.

Blencke, S., Zech, B., Engkvist, O., Greff, Z., Orfi, L., Horvath, Z., et al. (2004). Characterization of a conserved structural determinant controlling protein kinase sensitivity to selective inhibitors. *Chemistry and Biology, 11*, 691–701.

Bradeen, H. A., Eide, C. A., O'Hare, T., Johnson, K. J., Willis, S. G., Lee, F. Y., et al. (2006). Comparison of imatinib mesylate, dasatinib (BMS-354825), and nilotinib (AMN107) in an *N*-ethyl-*N*-nitrosourea (ENU)-based mutagenesis screen: High efficacy of drug combinations. *Blood, 108*, 2332–2338.

Burkard, M. E., & Jallepalli, P. V. (2010). Validating cancer drug targets through chemical genetics. *Biochimica et Biophysica Acta, 1806*, 251–257.

Burkard, M. E., Santamaria, A., & Jallepalli, P. V. (2012). Enabling and disabling polo-like kinase 1 inhibition through chemical genetics. *ACS Chemical Biology, 7*, 978–981.

Carter, T. A., Wodicka, L. M., Shah, N. P., Velasco, A. M., Fabian, M. A., Treiber, D. K., et al. (2005). Inhibition of drug-resistant mutants of ABL, KIT, and EGF receptor kinases. *Proceedings of the National Academy of Sciences of the United States of America, 102*, 11011–11016.

Chen, C., Ha, B. H., Thevenin, A. F., Lou, H. J., Zhang, R., Yip, K. Y., et al. (2014). Identification of a major determinant for serine-threonine kinase phosphoacceptor specificity. *Molecular Cell, 53*, 140–147.

Chen, H., Ma, J., Li, W., Eliseenkova, A. V., Xu, C., Neubert, T. A., et al. (2007). A molecular brake in the kinase hinge region regulates the activity of receptor tyrosine kinases. *Molecular Cell, 27*, 717–730.

Cheng, Y., & Prusoff, W. H. (1973). Relationship between the inhibition constant (K1) and the concentration of inhibitor which causes 50 per cent inhibition (I50) of an enzymatic reaction. *Biochemical Pharmacology, 22*, 3099–3108.

Clark, K., Mackenzie, K. F., Petkevicius, K., Kristariyanto, Y., Zhang, J., Choi, H. G., et al. (2012). Phosphorylation of CRTC3 by the salt-inducible kinases controls the interconversion of classically activated and regulatory macrophages. *Proceedings of the National Academy of Sciences of the United States of America, 109*, 16986–16991.

Cohen, P. (2002). Protein kinases—The major drug targets of the twenty-first century? *Nature Reviews. Drug Discovery, 1*, 309–315.

Cohen, P. (2010). Guidelines for the effective use of chemical inhibitors of protein function to understand their roles in cell regulation. *The Biochemical Journal, 425*, 53–54.

Cohen, P., & Alessi, D. R. (2013). Kinase drug discovery—What's next in the field? *ACS Chemical Biology, 8*, 96–104.

Cools, J., Mentens, N., Furet, P., Fabbro, D., Clark, J. J., Griffin, J. D., et al. (2004). Prediction of resistance to small molecule FLT3 inhibitors: Implications for molecularly targeted therapy of acute leukemia. *Cancer Research, 64*, 6385–6389.

Cooper, M. J., Cox, N. J., Zimmerman, E. I., Dewar, B. J., Duncan, J. S., Whittle, M. C., et al. (2013). Application of multiplexed kinase inhibitor beads to study kinome adaptations in drug-resistant leukemia. *PLoS One, 8*, e66755.

Cuenda, A., Rouse, J., Doza, Y. N., Meier, R., Cohen, P., Gallagher, T. F., et al. (1995). SB 203580 is a specific inhibitor of a MAP kinase homologue which is stimulated by cellular stresses and interleukin-1. *FEBS Letters, 364*, 229–233.

Dar, A. C., Das, T. K., Shokat, K. M., & Cagan, R. L. (2012). Chemical genetic discovery of targets and anti-targets for cancer polypharmacology. *Nature, 486*, 80–84.

Dar, A. C., & Shokat, K. M. (2011). The evolution of protein kinase inhibitors from antagonists to agonists of cellular signaling. *Annual Review of Biochemistry, 80*, 769–795.

Davies, S. P., Reddy, H., Caivano, M., & Cohen, P. (2000). Specificity and mechanism of action of some commonly used protein kinase inhibitors. *The Biochemical Journal, 351*, 95–105.

Davis, M. I., Hunt, J. P., Herrgard, S., Ciceri, P., Wodicka, L. M., Pallares, G., et al. (2011). Comprehensive analysis of kinase inhibitor selectivity. *Nature Biotechnology, 29*, 1046–1051.

Deng, X., Dzamko, N., Prescott, A., Davies, P., Liu, Q., Yang, Q., et al. (2011). Characterization of a selective inhibitor of the Parkinson's disease kinase LRRK2. *Nature Chemical Biology, 7*, 203–205.

Deshpande, A., Reddy, M. M., Schade, G. O., Ray, A., Chowdary, T. K., Griffin, J. D., et al. (2012). Kinase domain mutations confer resistance to novel inhibitors targeting JAK2V617F in myeloproliferative neoplasms. *Leukemia, 26*, 708–715.

Dodson, C. A., Kosmopoulou, M., Richards, M. W., Atrash, B., Bavetsias, V., Blagg, J., et al. (2010). Crystal structure of an aurora-A mutant that mimics aurora-B bound to MLN8054: Insights into selectivity and drug design. *The Biochemical Journal, 427*, 19–28.

Duncan, J. S., Whittle, M. C., Nakamura, K., Abell, A. N., Midland, A. A., Zawistowski, J. S., et al. (2012). Dynamic reprogramming of the kinome in response to targeted MEK inhibition in triple-negative breast cancer. *Cell, 149*, 307–321.

Echalier, A., Cot, E., Camasses, A., Hodimont, E., Hoh, F., Jay, P., et al. (2012). An integrated chemical biology approach provides insight into cdk2 functional redundancy and inhibitor sensitivity. *Chemistry and Biology, 19*, 1028–1040.

Emery, C. M., Vijayendran, K. G., Zipser, M. C., Sawyer, A. M., Niu, L., Kim, J. J., et al. (2009). MEK1 mutations confer resistance to MEK and B-RAF inhibition. *Proceedings of the National Academy of Sciences of the United States of America, 106*, 20411–20416.

Endicott, J. A., Noble, M. E., & Johnson, L. N. (2012). The structural basis for control of eukaryotic protein kinases. *Annual Review of Biochemistry, 81*, 587–613.

Eyers, P. A., Churchill, M. E., & Maller, J. L. (2005). The aurora A and aurora B protein kinases: A single amino acid difference controls intrinsic activity and activation by TPX2. *Cell Cycle, 4*, 784–789.

Eyers, P. A., Craxton, M., Morrice, N., Cohen, P., & Goedert, M. (1998). Conversion of SB 203580-insensitive MAP kinase family members to drug-sensitive forms by a single amino-acid substitution. *Chemistry and Biology, 5*, 321–328.

Eyers, P. A., Van Den, I. P., Quinlan, R. A., Goedert, M., & Cohen, P. (1999). Use of a drug-resistant mutant of stress-activated protein kinase 2a/p38 to validate the in vivo specificity of SB 203580. *FEBS Letters, 451*, 191–196.

Fabian, M. A., Biggs, W. H., 3rd., Treiber, D. K., Atteridge, C. E., Azimioara, M. D., Benedetti, M. G., et al. (2005). A small molecule-kinase interaction map for clinical kinase inhibitors. *Nature Biotechnology, 23*, 329–336.

Fedorov, O., Marsden, B., Pogacic, V., Rellos, P., Muller, S., Bullock, A. N., et al. (2007). A systematic interaction map of validated kinase inhibitors with Ser/Thr kinases. *Proceedings of the National Academy of Sciences of the United States of America, 104*, 20523–20528.

Fitzgerald, C. E., Patel, S. B., Becker, J. W., Cameron, P. M., Zaller, D., Pikounis, V. B., et al. (2003). Structural basis for p38alpha MAP kinase quinazolinone and pyridol-pyrimidine inhibitor specificity. *Nature Structural Biology, 10*, 764–769.

Girdler, F., Gascoigne, K. E., Eyers, P. A., Hartmuth, S., Crafter, C., Foote, K. M., et al. (2006). Validating aurora B as an anti-cancer drug target. *Journal of Cell Science, 119*, 3664–3675.

Girdler, F., Sessa, F., Patercoli, S., Villa, F., Musacchio, A., & Taylor, S. (2008). Molecular basis of drug resistance in aurora kinases. *Chemistry and Biology, 15*, 552–562.

Godl, K., Wissing, J., Kurtenbach, A., Habenberger, P., Blencke, S., Gutbrod, H., et al. (2003). An efficient proteomics method to identify the cellular targets of protein kinase inhibitors. *Proceedings of the National Academy of Sciences of the United States of America, 100*, 15434–15439.

Gorre, M. E., Mohammed, M., Ellwood, K., Hsu, N., Paquette, R., Rao, P. N., et al. (2001). Clinical resistance to STI-571 cancer therapy caused by BCR-ABL gene mutation or amplification. *Science, 293*, 876–880.

Green, C. J., Goransson, O., Kular, G. S., Leslie, N. R., Gray, A., Alessi, D. R., et al. (2008). Use of Akt inhibitor and a drug-resistant mutant validates a critical role for protein kinase B/Akt in the insulin-dependent regulation of glucose and system a amino acid uptake. *The Journal of Biological Chemistry, 283*, 27653–27667.

Guo, X., Gerl, R. E., & Schrader, J. W. (2003). Defining the involvement of p38alpha MAPK in the production of anti- and proinflammatory cytokines using an SB 203580-resistant form of the kinase. *The Journal of Biological Chemistry, 278*, 22237–22242.

Hall-Jackson, C. A., Eyers, P. A., Cohen, P., Goedert, M., Boyle, F. T., Hewitt, N., et al. (1999). Paradoxical activation of Raf by a novel Raf inhibitor. *Chemistry and Biology, 6*, 559–568.

Hall-Jackson, C. A., Goedert, M., Hedge, P., & Cohen, P. (1999). Effect of SB 203580 on the activity of c-Raf in vitro and in vivo. *Oncogene, 18*, 2047–2054.

Hanke, J. H., Gardner, J. P., Dow, R. L., Changelian, P. S., Brissette, W. H., Weringer, E. J., et al. (1996). Discovery of a novel, potent, and Src family-selective tyrosine kinase inhibitor. Study of Lck- and FynT-dependent T cell activation. *The Journal of Biological Chemistry, 271*, 695–701.

Hanks, S. K., & Hunter, T. (1995). Protein kinases 6. The eukaryotic protein kinase superfamily: Kinase (catalytic) domain structure and classification. *FASEB Journal, 9*, 576–596.

Haystead, T. A. (2006). The purinome, a complex mix of drug and toxicity targets. *Current Topics in Medicinal Chemistry, 6*, 1117–1127.

Hegarat, N., Smith, E., Nayak, G., Takeda, S., Eyers, P. A., & Hochegger, H. (2011). Aurora a and aurora B jointly coordinate chromosome segregation and anaphase microtubule dynamics. *Journal of Cell Biology, 195*, 1103–1113.

Heuckmann, J. M., Holzel, M., Sos, M. L., Heynck, S., Balke-Want, H., Koker, M., et al. (2011). ALK mutations conferring differential resistance to structurally diverse ALK inhibitors. *Clinical Cancer Research, 17*, 7394–7401.

Huang, D., Zhou, T., Lafleur, K., Nevado, C., & Caflisch, A. (2010). Kinase selectivity potential for inhibitors targeting the ATP binding site: A network analysis. *Bioinformatics, 26*, 198–204.

Jemaa, M., Vitale, I., Kepp, O., Berardinelli, F., Galluzzi, L., Senovilla, L., et al. (2012). Selective killing of p53-deficient cancer cells by SP600125. *EMBO Molecular Medicine, 4*, 500–514.

Katayama, R., Khan, T. M., Benes, C., Lifshits, E., Ebi, H., Rivera, V. M., et al. (2011). Therapeutic strategies to overcome crizotinib resistance in non-small cell lung cancers harboring the fusion oncogene EML4-ALK. *Proceedings of the National Academy of Sciences of the United States of America, 108*, 7535–7540.

Knight, Z. A., Lin, H., & Shokat, K. M. (2010). Targeting the cancer kinome through polypharmacology. *Nature Reviews. Cancer, 10*, 130–137.

Knight, Z. A., & Shokat, K. M. (2005). Features of selective kinase inhibitors. *Chemistry and Biology, 12*, 621–637.

Kothe, M., Kohls, D., Low, S., Coli, R., Rennie, G. R., Feru, F., et al. (2007). Selectivity-determining residues in Plk1. *Chemical Biology and Drug Design, 70*, 540–546.

Kovarik, P., Stoiber, D., Eyers, P. A., Menghini, R., Neininger, A., Gaestel, M., et al. (1999). Stress-induced phosphorylation of STAT1 at Ser727 requires p38 mitogen-activated protein kinase whereas IFN-gamma uses a different signaling pathway. *Proceedings of the National Academy of Sciences of the United States of America*, *96*, 13956–13961.

Krishnamurty, R., & Maly, D. J. (2010). Biochemical mechanisms of resistance to small-molecule protein kinase inhibitors. *ACS Chemical Biology*, *5*, 121–138.

Kwiatkowski, N., Jelluma, N., Filippakopoulos, P., Soundararajan, M., Manak, M. S., Kwon, M., et al. (2010). Small-molecule kinase inhibitors provide insight into Mps1 cell cycle function. *Nature Chemical Biology*, *6*, 359–368.

Lin, X., Huang, X. P., Chen, G., Whaley, R., Peng, S., Wang, Y., et al. (2012). Life beyond kinases: Structure-based discovery of sorafenib as nanomolar antagonist of 5-HT receptors. *Journal of Medicinal Chemistry*, *55*, 5749–5759.

Littlefield, P., Moasser, M. M., & Jura, N. (2014). An ATP-competitive inhibitor modulates the allosteric function of the HER3 pseudokinase. *Chemistry and Biology*, *21*, 453–458.

Liu, Y., Bishop, A., Witucki, L., Kraybill, B., Shimizu, E., Tsien, J., et al. (1999). Structural basis for selective inhibition of Src family kinases by PP1. *Chemistry and Biology*, *6*, 671–678.

Liu, Y., & Gray, N. S. (2006). Rational design of inhibitors that bind to inactive kinase conformations. *Nature Chemical Biology*, *2*, 358–364.

Lopez, M. S., Choy, J. W., Peters, U., Sos, M. L., Morgan, D. O., & Shokat, K. M. (2013). Staurosporine-derived inhibitors broaden the scope of analog-sensitive kinase technology. *Journal of the American Chemical Society*, *135*, 18153–18159.

Manning, G., Whyte, D. B., Martinez, R., Hunter, T., & Sudarsanam, S. (2002). The protein kinase complement of the human genome. *Science*, *298*, 1912–1934.

Marit, M. R., Chohan, M., Matthew, N., Huang, K., Kuntz, D. A., Rose, D. R., et al. (2012). Random mutagenesis reveals residues of JAK2 critical in evading inhibition by a tyrosine kinase inhibitor. *PLoS One*, 7, e43437.

Melo, J. V., & Chuah, C. (2007). Resistance to imatinib mesylate in chronic myeloid leukaemia. *Cancer Letters*, *249*, 121–132.

Miduturu, C. V., Deng, X., Kwiatkowski, N., Yang, W., Brault, L., Filippakopoulos, P., et al. (2011). High-throughput kinase profiling: A more efficient approach toward the discovery of new kinase inhibitors. *Chemistry and Biology*, *18*, 868–879.

Nichols, R. J., Dzamko, N., Hutti, J. E., Cantley, L. C., Deak, M., Moran, J., et al. (2009). Substrate specificity and inhibitors of LRRK2, a protein kinase mutated in Parkinson's disease. *The Biochemical Journal*, *424*, 47–60.

Ohren, J. F., Chen, H., Pavlovsky, A., Whitehead, C., Zhang, E., Kuffa, P., et al. (2004). Structures of human MAP kinase kinase 1 (MEK1) and MEK2 describe novel non-competitive kinase inhibition. *Nature Structural and Molecular Biology*, *11*, 1192–1197.

O'Keefe, S. J., Mudgett, J. S., Cupo, S., Parsons, J. N., Chartrain, N. A., Fitzgerald, C., et al. (2007). Chemical genetics define the roles of p38alpha and p38beta in acute and chronic inflammation. *The Journal of Biological Chemistry*, *282*, 34663–34671.

Patricelli, M. P., Nomanbhoy, T. K., Wu, J., Brown, H., Zhou, D., Zhang, J., et al. (2011). In situ kinase profiling reveals functionally relevant properties of native kinases. *Chemistry and Biology*, *18*, 699–710.

Patricelli, M. P., Szardenings, A. K., Liyanage, M., Nomanbhoy, T. K., Wu, M., Weissig, H., et al. (2007). Functional interrogation of the kinome using nucleotide acyl phosphates. *Biochemistry*, *46*, 350–358.

Poulikakos, P. I., Persaud, Y., Janakiraman, M., Kong, X., Ng, C., Moriceau, G., et al. (2011). RAF inhibitor resistance is mediated by dimerization of aberrantly spliced BRAF(V600E). *Nature*, *480*, 387–390.

Poulikakos, P. I., Zhang, C., Bollag, G., Shokat, K. M., & Rosen, N. (2010). RAF inhibitors transactivate RAF dimers and ERK signalling in cells with wild-type BRAF. *Nature*, *464*, 427–430.

Ray, A., Cowan-Jacob, S. W., Manley, P. W., Mestan, J., & Griffin, J. D. (2007). Identification of BCR-ABL point mutations conferring resistance to the Abl kinase inhibitor AMN107 (nilotinib) by a random mutagenesis study. *Blood*, *109*, 5011–5015.

Roumiantsev, S., Shah, N. P., Gorre, M. E., Nicoll, J., Brasher, B. B., Sawyers, C. L., et al. (2002). Clinical resistance to the kinase inhibitor STI-571 in chronic myeloid leukemia by mutation of Tyr-253 in the Abl kinase domain P-loop. *Proceedings of the National Academy of Sciences of the United States of America*, *99*, 10700–10705.

Schindler, T., Bornmann, W., Pellicena, P., Miller, W. T., Clarkson, B., & Kuriyan, J. (2000). Structural mechanism for STI-571 inhibition of Abelson tyrosine kinase. *Science*, *289*, 1938–1942.

Schmidt, M., Budirahardja, Y., Klompmaker, R., & Medema, R. H. (2005). Ablation of the spindle assembly checkpoint by a compound targeting Mps1. *EMBO Reports*, *6*, 866–872.

Scutt, P. J., Chu, M. L., Sloane, D. A., Cherry, M., Bignell, C. R., Williams, D. H., et al. (2009). Discovery and exploitation of inhibitor-resistant aurora and polo kinase mutants for the analysis of mitotic networks. *The Journal of Biological Chemistry*, *284*, 15880–15893.

Shah, K., Liu, Y., Deirmengian, C., & Shokat, K. M. (1997). Engineering unnatural nucleotide specificity for Rous sarcoma virus tyrosine kinase to uniquely label its direct substrates. *Proceedings of the National Academy of Sciences of the United States of America*, *94*, 3565–3570.

Shah, N. P., Nicoll, J. M., Nagar, B., Gorre, M. E., Paquette, R. L., Kuriyan, J., et al. (2002). Multiple BCR-ABL kinase domain mutations confer polyclonal resistance to the tyrosine kinase inhibitor imatinib (STI571) in chronic phase and blast crisis chronic myeloid leukemia. *Cancer Cell*, *2*, 117–125.

Shah, N. P., Tran, C., Lee, F. Y., Chen, P., Norris, D., & Sawyers, C. L. (2004). Overriding imatinib resistance with a novel ABL kinase inhibitor. *Science*, *305*, 399–401.

Sheinerman, F. B., Giraud, E., & Laoui, A. (2005). High affinity targets of protein kinase inhibitors have similar residues at the positions energetically important for binding. *Journal of Molecular Biology*, *352*, 1134–1156.

Skaggs, B. J., Gorre, M. E., Ryvkin, A., Burgess, M. R., Xie, Y., Han, Y., et al. (2006). Phosphorylation of the ATP-binding loop directs oncogenicity of drug-resistant BCR-ABL mutants. *Proceedings of the National Academy of Sciences of the United States of America*, *103*, 19466–19471.

Sloane, D. A., Trikic, M. Z., Chu, M. L., Lamers, M. B., Mason, C. S., Mueller, I., et al. (2010). Drug-resistant aurora A mutants for cellular target validation of the small molecule kinase inhibitors MLN8054 and MLN8237. *ACS Chemical Biology*, *5*, 563–576.

Tokarski, J. S., Newitt, J. A., Chang, C. Y., Cheng, J. D., Wittekind, M., Kiefer, S. E., et al. (2006). The structure of dasatinib (BMS-354825) bound to activated ABL kinase domain elucidates its inhibitory activity against imatinib-resistant ABL mutants. *Cancer Research*, *66*, 5790–5797.

Tyler, R. K., Shpiro, N., Marquez, R., & Eyers, P. A. (2007). VX-680 inhibits aurora A and aurora B kinase activity in human cells. *Cell Cycle*, *6*, 2846–2854.

Vieth, M., Sutherland, J. J., Robertson, D. H., & Campbell, R. M. (2005). Kinomics: Characterizing the therapeutically validated kinase space. *Drug Discovery Today*, *10*, 839–846.

Wacker, S. A., Houghtaling, B. R., Elemento, O., & Kapoor, T. M. (2012). Using transcriptome sequencing to identify mechanisms of drug action and resistance. *Nature Chemical Biology*, *8*, 235–237.

Ward, R. A., Anderton, M. J., Ashton, S., Bethel, P. A., Box, M., Butterworth, S., et al. (2013). Structure- and reactivity-based development of covalent inhibitors of the activating and gatekeeper mutant forms of the epidermal growth factor receptor (EGFR). *Journal of Medicinal Chemistry*, *56*, 7025–7048.

Whittaker, S., Kirk, R., Hayward, R., Zambon, A., Viros, A., Cantarino, N., et al. (2010). Gatekeeper mutations mediate resistance to BRAF-targeted therapies. *Science Translational Medicine, 2*, 35ra41.

Wodicka, L. M., Ciceri, P., Davis, M. I., Hunt, J. P., Floyd, M., Salerno, S., et al. (2010). Activation state-dependent binding of small molecule kinase inhibitors: Structural insights from biochemistry. *Chemistry and Biology, 17*, 1241–1249.

Workman, P., & Collins, I. (2010). Probing the probes: Fitness factors for small molecule tools. *Chemistry and Biology, 17*, 561–577.

Young, M. A., Shah, N. P., Chao, L. H., Seeliger, M., Milanov, Z. V., Biggs, W. H., 3rd., et al. (2006). Structure of the kinase domain of an imatinib-resistant Abl mutant in complex with the aurora kinase inhibitor VX-680. *Cancer Research, 66*, 1007–1014.

Yun, C. H., Mengwasser, K. E., Toms, A. V., Woo, M. S., Greulich, H., Wong, K. K., et al. (2008). The T790M mutation in EGFR kinase causes drug resistance by increasing the affinity for ATP. *Proceedings of the National Academy of Sciences of the United States of America, 105*, 2070–2075.

Zhang, S., Wang, F., Keats, J., Zhu, X., Ning, Y., Wardwell, S. D., et al. (2011). Crizotinib-resistant mutants of EML4-ALK identified through an accelerated mutagenesis screen. *Chemical Biology and Drug Design, 78*, 999–1005.

Zhou, T., Commodore, L., Huang, W. S., Wang, Y., Thomas, M., Keats, J., et al. (2011). Structural mechanism of the Pan-BCR-ABL inhibitor ponatinib (AP24534): Lessons for overcoming kinase inhibitor resistance. *Chemical Biology and Drug Design*, 77, 1–11.

Zhou, W., Ercan, D., Chen, L., Yun, C. H., Li, D., Capelletti, M., et al. (2009). Novel mutant-selective EGFR kinase inhibitors against EGFR T790M. *Nature, 462*, 1070–1074.

Zuccotto, F., Ardini, E., Casale, E., & Angiolini, M. (2010). Through the "gatekeeper door": Exploiting the active kinase conformation. *Journal of Medicinal Chemistry, 53*, 2681–2694.

Zunder, E. R., Knight, Z. A., Houseman, B. T., Apsel, B., & Shokat, K. M. (2008). Discovery of drug-resistant and drug-sensitizing mutations in the oncogenic PI3K isoform p110 alpha. *Cancer Cell, 14*, 180–192.

CHAPTER SIX

FLiK: A Direct-Binding Assay for the Identification and Kinetic Characterization of Stabilizers of Inactive Kinase Conformations

Jeffrey R. Simard[*,1,2], **Daniel Rauh**[*,†,2]
*Chemical Genomics Centre of the Max Planck Society, Dortmund, Germany
†Fakultät Chemie, Chemische Biologie, Technische Universität Dortmund, Dortmund, Germany
[1]Current address: Amgen Inc., 360 Binney Street, Cambridge, Massachusetts, USA
[2]Corresponding author: e-mail address: simard@amgen.com; daniel.rauh@tu-dortmund.de

Contents

Abstract

Despite the hundreds of kinase inhibitors currently in discovery and preclinical phases, the number of FDA-approved kinase inhibitors remains very low by comparison, a discrepancy which reflects the challenges which accompanies kinase inhibitor development. Targeting protein kinases with ATP-competitive inhibitors has been the classical approach to inhibit kinase activity, but the highly conserved nature of the ATP-binding site often contributes to the poor inhibitor selectivity. To address this problem, we developed a high-throughput screening technology that can discriminate for inhibitors, which stabilize inactive kinase conformations by binding within allosteric

ISSN 0076-6879
http://dx.doi.org/10.1016/B978-0-12-397918-6.00006-9

pockets in the kinase domain. Here, we describe how to use the Fluorescence Labels in Kinases approach to measure the K_d of ligands as well as how to kinetically characterize the binding and dissociation of ligands to the kinase. We also describe how this technology can be used to rapidly screen small molecule libraries in high throughput.

1. INTRODUCTION

A major roadblock in protein kinase inhibitor research and development is the challenge of poor selectivity and the likelihood of unwanted off-target inhibition, which are largely a consequence of the highly conserved ATP-binding site shared by all protein kinases. Although drugs, which specifically target and inhibit the activity of a single kinase, are highly desirable, evidence is gathering, which suggests that polypharmacology may be beneficial in the fight against complex diseases such as cancer (Knight, Lin, & Shokat, 2010; Morphy, 2010). Polypharmacology strategies focus on the concept of "controlled unselectivity" to target a specific cluster of kinases within one or more specific pathways to achieve the desired therapeutic outcome. An analogous approach is combinatorial-personalized therapy using multiple kinase inhibitors to achieve the same goal.

Regardless of the strategy employed, it is becoming more evident that examining drug–target residence times will provide a more complete context for fully understanding kinase inhibitor selectivity *in vivo* (Copeland, 2011; Copeland, Pompliano, & Meek, 2006). Ideally, inhibitors should have high rates of association (k_{on}) and slow rates of dissociation (k_{off}) to maximize residence time with the target enzyme. In the case of kinases, an inhibitor, which appears to be relatively unselective *in vitro*, can be rendered more selective if it dissociates most slowly from the kinase of interest. Thus, medicinal chemistry efforts should not judge selectivity-based solely on inhibitor affinities (K_d) and potencies (IC_{50}) obtained from *in vitro* profiling of compounds against the entire kinome (Anastassiadis, Deacon, Devarajan, Ma, & Peterson, 2011; Copeland, 2011; Karaman et al., 2008). Although these types of studies undoubtedly contain valuable information and provide a solid groundwork for further inhibitor development, it is possible for a wide range of k_{on} and k_{off} values to result in the same overall affinity ($K_d = k_{off}/k_{on}$; Markgren et al., 2002). Thus, in addition to affinity, lead optimization strategies should consider the kinetic components contributing to affinity. By identifying aspects of ligand structure, which prolong k_{off} relative to k_{on}, medicinal chemistry efforts can be directed to facilitate the design of molecules with improved residence times with the desired kinase.

This approach would likely minimize the unwanted consequences of high-affinity off-target binding *in vivo*.

Emerging data suggest that the issue of kinase inhibitor selectivity can be addressed by moving away from classical ATP-competitive (Type I) inhibitors and targeting the DFG-out pocket with Type II and Type III inhibitors (Davis et al., 2011). The DFG-out pocket is adjacent to the ATP-binding site and is frequently referred to an allosteric pocket or kinase-switch pocket. In comparison to the ATP-binding site, the allosteric pocket is a more restrictive binding cavity and is only accessible upon a change in conformation. Therefore, this pocket tends to be less accessible to small molecules. However, the amino acids lining this pocket are much less conserved across the kinome, providing opportunities for additional H bonding and hydrophobic interactions between the kinase and ligands, which can bind within this pocket (Liu & Gray, 2006). Once bound, these factors tend to reduce the k_{off} of Type II and Type III inhibitors relative to k_{on} (Copeland, 2011). Thus, a logical methodology for improving kinase inhibitor selectivity by prolonging drug–target residence times should focus on the identification and kinetic optimization of ligands, which can bind preferentially to the DFG-out conformation.

The availability of the DFG-out pocket requires the kinase-activation loop to adopt a catalytically deficient conformation in which the ATP-binding site becomes partially occluded by the Phe side chain of the DFG motif (Backes, Zech, Felber, Klebl, & Müller, 2008a, 2008b; Rabiller et al., 2010). While the DFG-out conformation is more favorable in the unphosphorylated kinase, phosphorylation of the activation loop shifts conformational equilibrium to the more active DFG-in conformation, increases kinase activity, and often reduces the affinity of Type II and Type III inhibitors. Although the search for chemical scaffolds, which have affinity for the DFG-out pocket, is moving to the forefront of kinase inhibitor research, efforts have been constrained by the lack of high-throughput assay technologies, which can identify and discriminate for ligands which bind to and stabilize enzymatically inactive kinase conformations.

We have developed FLiK (Fluorescence Labels in Kinases) as a widely applicable assay system for both identifying and characterizing DFG-out-binding ligands. Kinases are site specifically labeled with an environmentally sensitive fluorophore, which reports on conformational changes induced by the binding of specific types of ligands (Fig. 6.1A; Simard, Getlik, et al., 2009; Simard, Kluter, et al., 2009). Changes in kinase conformation alter the charged microenvironment and solvation of the fluorophore, resulting

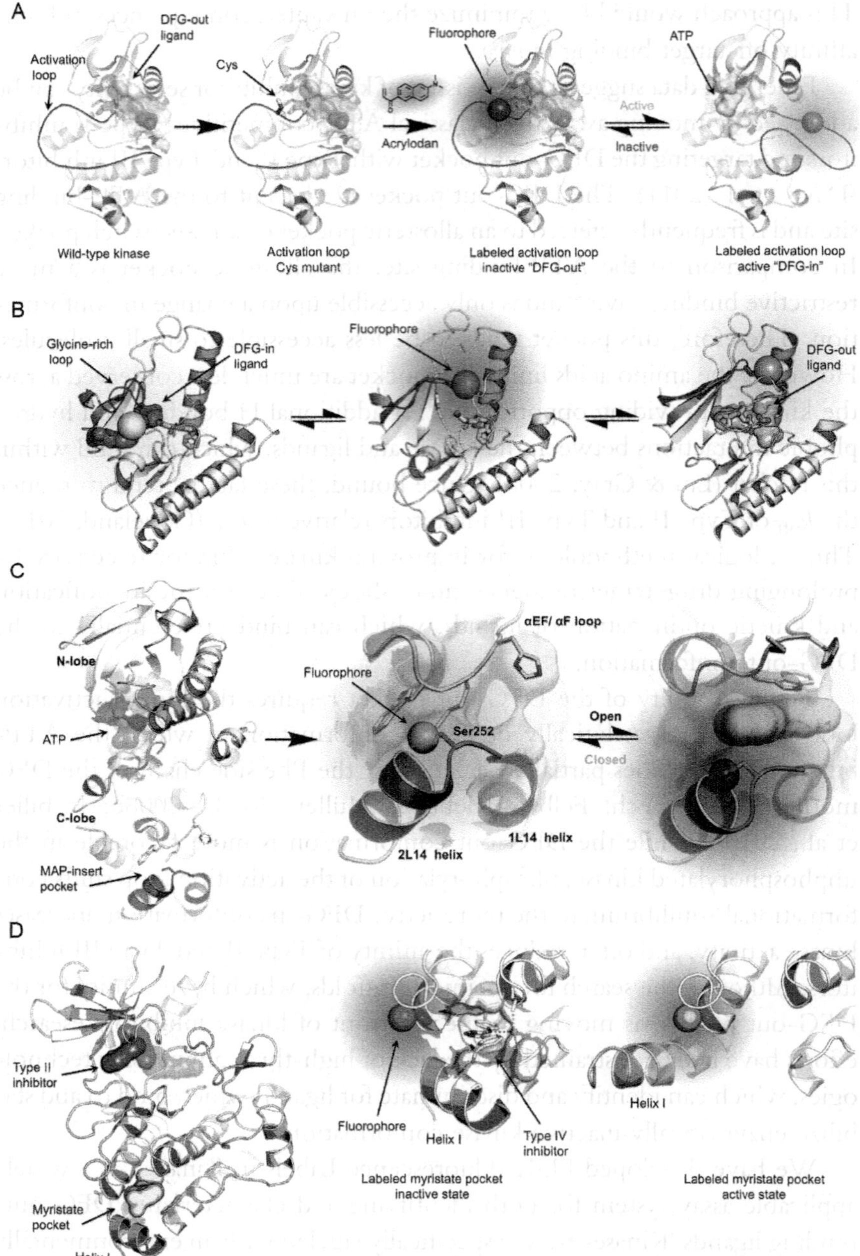

Figure 6.1 Overview of various FLiK assays for kinase-binding sites. (A) Labeling the activation loop. Kinases are regulated by an activation loop, which can adopt active (DFG-in) and inactive (DFG-out) conformations. As a general approach, a Cys residue is mutated into the desired labeling position and is used for the attachment of an

in distinct and quantifiable changes in its emission spectrum, which, in turn, provide a straightforward binding assay methodology for determining the K_d of the ligand. The assay also allows for follow-up characterization of identified compounds by permitting the determination of k_{on} and k_{off} to better understand the kinetic factors, which contribute to the measured K_d. A key advantage of this approach is that enzyme activity is not required. Using unphosphorylated kinase, the sensitivity for ligands, which bind preferentially to the DFG-out conformation, is enhanced when compared to measuring IC_{50} values using traditional activity-based assays, which rely on the use of phosphorylated active kinase.

The FLiK approach has been used to successfully monitor conformational changes in the activation loop of both Ser/Thr and Tyr kinases associated with the slow binding of DFG-out inhibitors (Grutter et al., 2012; Simard, Kluter, et al., 2009; Simard, Getlik, et al., 2009; Simard, Grutter, et al., 2009). To date, we have applied the FLiK approach to other kinase structural elements (Fig. 6.1B), including labeling of the P loop (glycine-rich loop) to identify more selective Type I ligands, which engage the flexible P loop in certain kinases (Simard et al., 2010). Additionally, we have recently reported a labeling strategies aimed at remote-binding sites outside of the ATP-binding cleft as a method for identifying more selective allosteric (Type IV) ligands, including assays for the MAP-insert pocket of p38α (Fig. 6.1C) as well as the myristate pocket of Abl kinase (Fig. 6.1D;

environmentally sensitive fluorophore (colored sphere). This FLiK assay enables specific detection of ligands, which bind to and stabilize the DFG-out conformation, such as Type II and Type III kinase inhibitors. (B) Labeling of the P loop. The selectivity of some ATP-competitive inhibitors can be improved by introducing chemical moieties, which are capable of engaging the P loop of certain kinases. (C) Labeling of the MAP insert. The MAP insert is a structural feature found in MAP kinases, cyclin-dependent kinases, and glycogen synthase kinase (Akella, Moon, & Goldsmith, 2008). This FLiK assay enables the specific detection of ligands, which occupy the MAP-insert pocket of a selected group of kinases. Although it is not yet clear if binding to this site effects catalytic activity of the kinase, it may alter signaling cascades by modulating the docking and/or scaffolding interactions of the MAP insert with other proteins. (D) Labeling the Abl myristate pocket. Stable binding to this site requires the bending of helix I of Abl. This FLiK assay enables the specific detection of allosteric Abl inhibitors. *Panel (A): Reproduced from Simard, Kluter, et al. (2009) with permission from Nature Publishing Group. Panel (B): This panel has been reprinted with permission from Simard et al. (2010). Copyright 2010 American Chemical Society. Panel (C): This panel has been reprinted with permission from Getlik et al. (2012). Copyright 2012 American Chemical Society. Panel (D): This panel has been reprinted with permission from Schneider et al. (2012). Copyright 2012 American Chemical Society.*

Getlik et al., 2012; Schneider et al., 2012). However, these strategies are beyond the scope of this chapter. To highlight the potential of FLiK as an high-throughput screening (HTS) method for rapidly identifying slow-binding DFG-out ligands, we will describe its application to the activation loop of p38α in more detail and provide a step-by-step protocol for researchers who are interested in evaluating their own kinase targets.

2. DESIGN AND PREPARATION OF KINASES FOR FLiK

The FLiK approach requires the removal of solvent-exposed cysteines and the insertion of a cysteine into a desired position in the kinase, which will serve as the attachment point for a thiol-reactive fluorophore. The kinase mutant is then expressed and purified, labeled, and characterized using standard biochemical and biophysical methods. However, it is necessary to have a strategy in place when designing the protein construct. Identifying optimal fluorophore-labeling positions is critical to enabling a high-throughput assay that reports specific ligand-induced conformational changes.

2.1. Selection of the labeling position on the activation loop

Fluorophores such as acrylodan are commonly employed for generating fluorescent protein conjugates which report on conformational changes (Copeland, 2011; de Lorimier et al., 2002; Richieri, Ogata, & Kleinfeld, 1999). Ideally, fluorophores should be highly sensitive to polarity and/or the charged microenvironment that is characteristic of nearby amino acid chains in the protein. It is also advantageous to choose fluorophores, which are thiol reactive. In contrast to amine-reactive probes, labeling by thiol-reactive fluorophores is typically complete and more specific due to the lower abundance of free thiols in proteins. However, to ensure the specificity of labeling for the FLiK assay, naturally occurring Cys residues, which are solvent exposed, should be mutated away, where possible, leaving the inserted Cys as the only solvent-exposed anchor point for the fluorophore.

For the FLiK assay, it is critical to insert a Cys residue into an amino acid position on the kinase that exhibits significant movement upon ligand binding and is somewhat solvent-exposed to enable the covalent attachment of the added fluorophore. To date, we have reported on several labeling strategies, which we have successfully used to develop FLiK assays for various kinases. These assays enable rapid and specific detection of inhibitors and ligands with unique binding modes (Getlik et al., 2012; Grutter et al.,

2012; Schneider et al., 2012; Simard, Grutter, et al., 2009; Simard et al., 2010; Simard, Getlik, et al., 2009; Simard, Kluter, et al., 2009). Although the specific labeling site may vary for each kinase or binding site, a general approach can be applied in the design of kinase constructs compatible with the FLiK approach. For any structural feature, which is known to exhibit flexibility or undergo conformational changes upon ligand binding, a kinase construct, which is compatible with the FLiK assay, can be designed as follows:

(1) Perform a BLAST search using the full amino acid sequence of the kinase of interest as a method of finding other kinases with the highest percentage of sequence identity.

(2) Using PyMol (DeLano, 2002), analyze protein crystal structures of the kinase of interest, if available. If no published structures are available, several online-modeling tools such as Swiss Modeller (Arnold, Bordoli, Kopp, & Schwede, 2006) or ESyPred3D (Lambert, Leonard, De Bolle, & Depiereux, 2002) can be used to generate 3D structural models based on available structural templates in the Protein Data Bank (http://www.pdb.org). The kinases identified in the BLAST search may serve as convenient starting points.

(3) Using online tools such as Clustal W, perform amino acid sequence alignments of the kinase of interest with a number of other highly homologous kinases. This method may help identify positions in the sequence, which are compatible with the FLiK approach. Sequence alignments may help identify highly conserved regions, common phosphorylation sites, or positions known to be involved in key structural interactions. Additionally, sequence alignments may reveal certain positions, which have a naturally occurring Cys. Such positions may tolerate mutations in which a Cys is introduced for specific labeling with the desired fluorophore.

2.2. Preparation of p38α MAP kinase construct for FLiK

The FLiK assay was initially developed as an assay, which discriminates for Type II/Type III ligands, which bind to and stabilize the inactive DFG-out conformation of the kinase-activation loop (Simard, Getlik, et al., 2009, Simard, Kluter, et al., 2009). To date, the FLiK assay has also been utilized to monitor several types of ligand-induced conformational changes and different binding sites in p38α as well as other kinases (Getlik et al., 2012; Grutter et al., 2012; Schneider et al., 2012; Simard, Getlik, et al., 2009;

Simard, Kluter, et al., 2009; Simard et al., 2010). The expression and purification of a p38α MAP kinase FLiK construct is described later as an example.

2.2.1 Expression of p38α MAP kinase

For compatibility with the FLiK assay, the gene sequence for full-length human p38α was mutated by site-directed mutagenesis to introduce the desired mutations for specific labeling of the activation loop (C119S/C162S/A172C) (Simard, Getlik, et al., 2009). For assay development with other kinases, we have also successfully utilized synthetic genes generated with the appropriate mutations already incorporated into the DNA sequence. The mutant p38α gene was then cloned into an expression vector, transformed into *Escherichia coli* and expressed as described previously (Simard, Getlik, et al., 2009). The expression is accomplished as follows:

1. Clone the desired gene into a pOPIN-F expression vector.
2. Transform into BL21(DE3) *E. coli*.
3. Grow bacterial culture at 37 °C in LB media until an OD_{600} of 0.6 is reached.
4. Cool to RT in 30 min using a chilled water bath if necessary and induce protein expression with 1 m*M* IPTG.
5. Incubate the cultures overnight (~20 h) at 18 °C while shaking at 160 rpm.

2.2.2 Purification of p38α MAP kinase

His-tagged p38α was purified from *E. coli* as described previously (Simard, Getlik, et al., 2009). During purification, PreScission Protease was used to cleave the N-terminal His tag from the construct. Tagged kinase constructs may perform differently in the FLiK assay when compared to their cleaved counterparts. Full details of the purification are as follows:

1. Pellet the bacteria by centrifugation and then resuspend in Buffer A (50 m*M* Tris, pH 8.0, 500 m*M* NaCl + 5% glycerol + 25 m*M* imidazole).
2. Load sample onto a 30-mL Ni column, wash with at least three column volumes (CV) of Buffer A, then elute with a 0–50% linear gradient of Buffer B (Buffer A + 500 m*M* imidazole) over two CV.
3. Cleave the N-terminal tag by pooling the protein fractions and incubating with PreScission Protease (50 μg/mL final concentration) in a 12–30 mL capacity 10K-MWCO dialysis cassette overnight at 4 °C in

Dialysis Buffer (50 m*M* Tris, pH 7.5, 5% glycerol, 150 m*M* NaCl, 1 m*M* EDTA, 1 m*M* DTT).

4. Centrifuge the protein for 15 min at ~13,000 rpm to remove any precipitate that may have formed during the cleavage step.
5. Collect and dilute the supernatant fourfold in Buffer C (50 m*M* Tris, pH 7.4, 5% glycerol, 50 m*M* NaCl, 1 m*M* DTT).
6. Load the sample onto a 1 mL Sepharose Q FF column and wash with 10 CV of Buffer C. Elute with a 0–100% linear gradient of Buffer D (Buffer C + 600 m*M* NaCl) over 20 CV.
7. Pool and concentrate the protein down to a volume of 2–4 mL using a 10-MWCO centricon.
8. Load the sample onto a Sephadex HiLoad 26/60 Superdex 75 column equilibrated with Buffer E (20 m*M* Tris, pH 7.4, 5% glycerol, 200 m*M* NaCl, 1 m*M* DTT) and elute at a rate of 2 mL/min.
9. Concentrate the eluted fractions of pure protein to ~10 mg/mL, then aliquot and store at −80 °C.

3. LABELING OF p38α MAP KINASE WITH ACRYLODAN

A wide selection of thiol- or amine-reactive fluorescent probes is commercially available (http://www.invitrogen.com) and can be used for the labeling of FLiK kinases. Given the lower abundance of cysteine in proteins, the use of thiol-reactive fluorophores is more straightforward and significantly reduces the number of sites available for nonspecific labeling. When choosing a fluorophore for establishing a reliable kinase biosensor for the FLiK assay, special consideration should be given to molecules, which are highly sensitive to changes in the charged microenvironment. These fluorophores will be the most sensitive detectors of conformational changes (de Lorimier et al., 2002). For example, acrylodan responds with a change in overall fluorescence intensity as well as a >40 nm shift in the emission maximum, thus enabling the use of ratiometric fluorescence readouts. Fluorescence ratios are widely used in assay development since they correct for subtle variations in sample volume across several samples. Further details regarding the fluorescence properties of acrylodan in the FLiK assay are provided elsewhere (Simard, Getlik, et al., 2009). A general procedure for labeling FLiK kinase mutants is described as follows:

1. Measure the molar concentration of the protein stock using traditional methods. Calculate the volume of acrylodan stock (prepared in DMSO) needed to achieve a 1.5:1 (mol:mol) ratio of acrylodan:protein.

2. Dilute the acrylodan stock in buffer until the DMSO concentration is <1% (v/v). Add the concentrated protein to the prediluted acrylodan. For thiol-reactive fluorophores, such as acrylodan, the chosen buffer should have pH < 8 and should not contain reducing agents such as DTT (*Note*: TCEP is compatible with the labeling of cysteine thiols). Refer to the manufacturer's instructions for ideal labeling conditions for the selected fluorophore.
3. Incubate sample on ice or refrigerate (4–8 °C) overnight. The labeling procedure may be optimized for each protein/fluorophore combination by altering the temperature and length of incubation. For example, the progress of the acrylodan-labeling reaction can be followed over time by monitoring increasing emission intensity between 460 and 510 nm (acrylodan excitation wavelength = 386 nm) under various conditions.
4. Remove the remaining unreacted fluorophore through repeated buffer exchange cycles using a 10-MWCO centricon. Concentrate, wash, and dilute the protein at least three times when using this method. For large quantities of protein, unreacted acrylodan can also be removed by passing the protein over an appropriate size exclusion column.
5. Concentrate the final washed protein sample and store at −20 to −80°C.
6. Subsequent analysis of the extent and specificity of labeling should be done using mass spectroscopy methods as described elsewhere (Simard, Kluter, et al., 2009).

4. ASSAY CHARACTERIZATION AND VALIDATION

Before using the newly labeled kinase in a costly HTS campaign to assay large compound libraries, the following list of tests should be performed to characterize the fluorescence response induced by ligand binding. Ideally, a reference ligand should be chosen, which is known to induce the desired conformational change. In the case of p38α labeled at the activation loop, Type II or III inhibitors were available as positive controls. Type I inhibitors were also useful as negative controls. These ligands can be synthesized based on the available literature procedures, purchased from commercial sources, or may be included in proprietary compound libraries.

Several quantifiable variables (ΔR_{max} and ΔI_{std}) can be determined to assess the quality of a fluorescent kinase as a biosensor (de Lorimier et al., 2002). However, the Z' factor is the gold-standard statistical measure of assay quality and is used to predict if assays will be useful for HTS. The Z' factor is calculated using the equation:

$$Z' \text{ factor} = 1 - \frac{3 \times (\sigma_p + \sigma_n)}{|\mu_p - \mu_n|} \tag{6.1}$$

where both the mean (μ) and the standard deviation (σ) of both the positive (p) and the negative (n) controls (μ_p, σ_p and μ_n, σ_n, respectively) are taken into account. It is important to examine these variables for every FLiK construct, since shifting the labeling position on a protein can have mild or drastic effects on the ligand-induced fluorescence response. The following types of measurements are recommended for thorough assay characterization and validation.

4.1. Measure emission spectra for each kinase conformation

Using reference inhibitors, which are known to induce specific conformational changes, the first test should focus on the emission spectrum of the FLiK kinase and how it changes in response to ligands, which induce the desired conformational change (Fig. 6.2A). It is also necessary to check various other compounds to determine whether the assay is capable of specifically detecting the desired subset of compounds with a certain binding mode. This can be done as follows:

1. Prepare a cuvette containing a mini stir bar and a suspension of 50–100 n*M* FLiK kinase in Assay Buffer and place the cuvette into the sample holder of the fluorescence spectrometer. If possible, the instrument should have magnetic plate under the cuvette to enable rapid sample mixing.
2. Using the appropriate excitation wavelength for the chosen fluorophore, scan fluorescence emission over a wide range of wavelengths to capture a full spectrum.
3. *Positive control*: Add an excess (10–20 μ*M*) of a reference ligand (any compound which is known to stabilize or bind to the desired kinase conformation) and stir the sample briefly to ensure full dispersion of the compound and DMSO (vehicle). Measure the emission spectrum repeatedly over a period of time until no further changes are observed in the measured spectrum.
4. Overlay the spectra series using graphing software to visualize the changes in emission spectra that occur with ligand binding over time.
5. Identify the emission maxima of the FLiK kinase in the absence of ligand and compare that to the spectrum obtained in the presence of a saturating amount of ligand. For each kinase conformation, use the emission

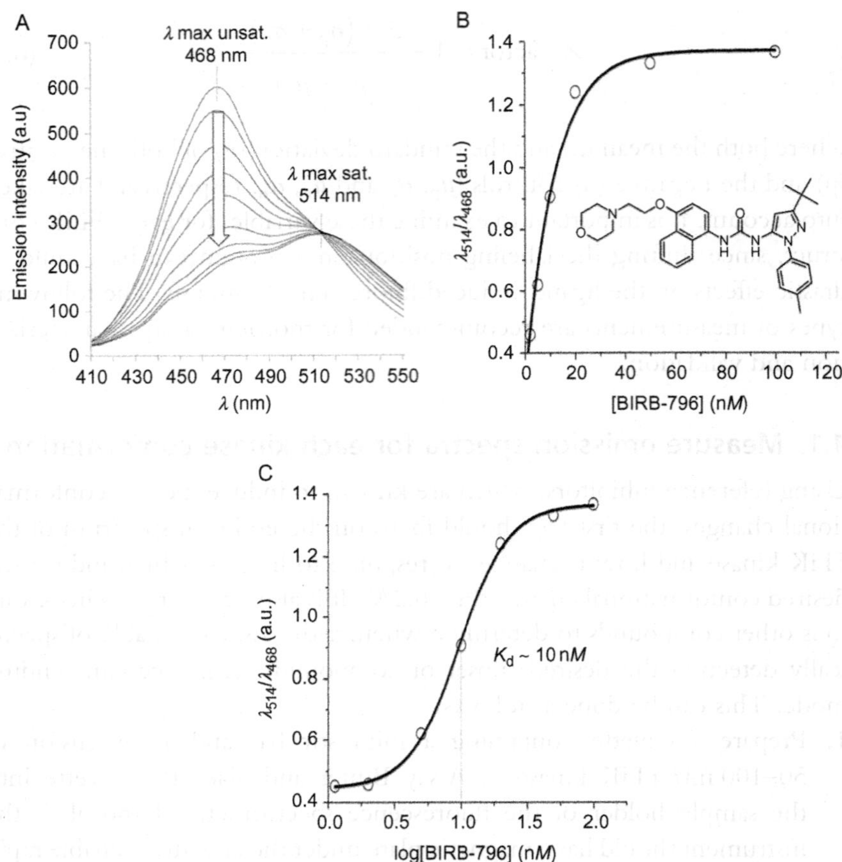

Figure 6.2 K_d determination for a DFG-out-binding ligand. This figure presents a typical set of endpoint measurement data obtained from the described K_d determination experiments. Shown data were obtained using acrylodan-labeled p38α and the Type II inhibitor BIRB-796. (A) The ratiometric fluorescence values are obtained from the emission spectra obtained in the presence of increasing inhibitor concentration. (B) These data can be plotted to observe the saturation of the kinase in a particular conformation or (C) these data can be used to directly obtain the K_d for the ligand. *Panels have been reprinted with permission from Simard, Getlik, et al. (2009). Copyright 2009 American Chemical Society.*

intensities at the two chosen wavelengths to determine the ratiometric output for the bound and unbound kinases. These values will give an indication of the lowest and the highest possible ratio values, or the "assay window." For ratiometric readouts, it is preferable to use fluorophores with at least two emission maxima or a single maximum, which shifts significantly (>10 nm) upon ligand binding. In the case of acrylodan, one maximum does not change while the other is very sensitive to conformational changes in the protein.

6. Repeat the measurement several times to obtain multiple values for the unbound and bound states of the kinase. Use these values to calculate the Z' factor for the assay as shown in Eq. (6.1) earlier. The Z' factor should ideally be >0.5 for HTS.
7. *Buffer optimization*: If needed, adjust various buffer conditions to make the fluorophore more sensitive to conformation changes. Since environmentally sensitive fluorophores respond to local changes in charge and polarity, the magnitude and sensitivity of fluorescence change can sometimes be enhanced by adjusting ionic strength (salt concentration), pH, or by adding a detergent at an optimal concentration. Buffer optimization can be time consuming and should only be necessary if the initial response of the construct was found to be not robust enough for HTS.
8. *Negative control*: Repeat steps 1–5 above using any ligand, which is a known inhibitor of the kinase, but is known to bind in an alternate binding mode or binding site and does not induce the desired conformation change. Ideally, the FLiK construct should not respond to such ligands.

4.2. K_d determination

In the FLiK assay, K_d values are typically obtained from endpoint measurements in which ratiometric values of emission intensities at two different wavelengths are plotted as a function of inhibitor concentration (Fig. 6.2B and C). These experiments can be carried out using one of two possible methods. The most favorable method should be considered based on the knowledge of protein availability, required binding times for ligands, and stability of the kinase in buffer over time. In the first method, ligand is titrated into a single cuvette containing the labeled protein, which is rapidly stirred, and spectra are measured after each addition of ligand. The advantage to this approach is that it requires less protein to obtain a single K_d curve. However, the disadvantage is that this approach may be time consuming when a very slow-binding ligand is titrated with the kinase. For slow-binding ligands, a more preferable method for determining the K_d involves addition of a single dose of inhibitor added to a series of cuvettes at increasing concentrations. Following a long incubation time to ensure complete binding, each cuvette is measured to examine changes in the emission spectrum associated with the binding of a particular concentration of inhibitor. The advantage to this approach is that the sample does not require constant rigorous stirring over a significant time period. This method is especially advantageous when analyzing the very slow binding of Type II/III inhibitors to kinases (Pargellis et al., 2002). A typical K_d determination is outlined below for each method.

4.2.1 Titration of ligand with the FLiK kinase (for rapidly binding ligands)

1. Prepare a cuvette containing a mini stir bar and 50–100 n*M* suspension of the labeled kinase. Place it into the fluorescence spectrometer and then measure the emission spectra for the labeled protein before adding any ligand.
2. Calculate and record the ratio of intensities at the two chosen emission wavelengths.
3. Using inhibitor stocks (prepared in DMSO), inject increasing amounts of compound into the stirring protein suspension. If possible, use concentrated stocks to minimize the required volume needed for each compound addition. Ideally, the total (%v/v) DMSO at the end of the titration should remain <1% to avoid undesirable effects on protein stability over time.
4. After each addition, allow sufficient time for the binding to reach completion (stable spectrum). For ligands which bind rapidly, stirring for 30 s should be sufficient to ensure that equilibrium has been reached. Measure the emission spectrum following each addition of ligand, calculate and record the emission ratio as in step no. 2.
5. Plot the data using desired graphing software. A plot of inhibitor concentration as a function of emission ratio will generate a saturable binding curve. Using a logarithmic plot of inhibitor concentration as a function of emission ratio values will generate a sigmoidal binding curve. The K_d value is obtained from the midpoint of the curve, which is the concentration of ligand required to occupy 50% of the FLiK kinase. The %binding to the kinase is a way to quantify the percent of acrylodan-labeled kinase in the well which has adopted the DFG-out conformation in response to the binding of a compound and provides a straightforward comparison to the positive control inhibitor. The %binding is calculated from the fluorescence ratio (R) using the following equation:

$$\% \text{ Binding} = \left(\left(R_{\text{hit}} - R_{\text{neg}}\right) / \left(R_{\text{pos}} - R_{\text{neg}}\right)\right) \times 100. \qquad (6.2)$$

4.2.2 Titration of ligand with the FLiK kinase (for slow-binding ligands)

1. Prepare a series of cuvettes (one cuvette per inhibitor concentration) containing a mini stir bar and a 50–100 n*M* suspension of the labeled kinase.
2. Starting at low concentrations, add the appropriate amount of inhibitor to each cuvette in the series using the prepared stocks. Briefly mix each sample by placing the cuvette on a magnetic stir plate for 10 s to disperse the inhibitor and DMSO throughout the sample.
3. Place the cuvettes into the fluorescence spectrometer and measure the emission spectra for each sample. Measuring the cuvettes repeatedly over time may provide valuable information about the slow time scale of binding for these ligands and reveal the time dependence of K_d values. For a slow-binding DFG-out-binding ligand, K_d curves typically shift leftward over time. Additionally, these measurements allow an easy assessment of the effect that incubation time has on the assay window. When the assay window starts to decrease, this may be a sign that the protein is no longer stable.
4. Calculate and record the ratio of intensities for all samples as described in Section 4.2.1.
5. Plot the data as described in step no. 5 of Section 4.2.1 to obtain time-dependent K_d values.

4.3. Kinetic measurements

In addition to measuring K_d values, the FLiK assay also allows for the easy determination of kinetic rate constants for binding (k_{on}) and dissociation (k_{off}). Kinetic rate constants such as k_{on} and k_{off} provide information and insight into the affinity of compounds since they are related to the K_d ($K_d = k_{off}/k_{on}$). Typically, these parameters are determined using surface plasmon resonance to guide further medicinal chemistry optimization of compounds (de Kloe et al., 2010; Markgren et al., 2002), but in this case, the FLiK assay can also be used to characterize the kinetic components and contributions to measured affinities. These parameters can be obtained by monitoring changes in fluorescence in real-time at a single wavelength (Fig. 6.3A). The association rate constant, k_{on}, is the fastest observed binding rate (k_{obs}) for a ligand. The k_{on} can be derived from the linear fit of a plot of k_{obs} values obtained at several inhibitor concentrations. The linear range of this plot will result in a straight line with slope equal to k_{on} (Fig. 6.3B). Similar plots and experiments can be used to experimentally

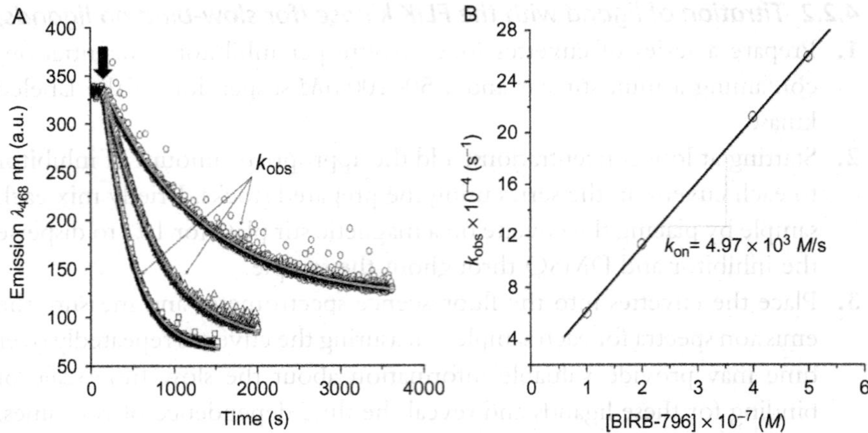

Figure 6.3 Determination of k_{on} for a DFG-out-binding ligand. This figure presents a typical set of kinetic-binding measurements for determining the k_{on} of BIRB-796 using acrylodan-labeled p38α. (A) The fluorescence emission is monitored at a single wavelength upon the addition of increasing concentrations of inhibitor. (B) These values can be plotted as a function of inhibitor concentration to obtain a linear curve with slope = k_{on}. *Panels have been reprinted with permission from Simard, Getlik, et al. (2009). Copyright 2009 American Chemical Society.*

derive the maximum k_{off} value for a ligand by adding increasing amounts of unlabeled kinase to a suspension of the fluorescent FLiK kinase prebound with different concentrations of inhibitor, thereby inducing dissociation of the ligand (Fig. 6.4). For slower binding ligands (>5 s), a standard fluorescence spectrometer is sufficient for resolving kinetic events. However, for rapid binders (<5 s), a stopped-flow apparatus might improve the time resolution of the measurement and enable accurate curve fitting. A typical k_{on} and k_{off} determination is outlined below for each method:

4.3.1 Determination of k_{on}

1. Prepare a cuvette containing a mini stir bar and a suspension of 50–100 n*M* kinase. Place it into the sample holder of the fluorescence spectrometer and set the excitation wavelength according to the fluorophore used to label the kinase.
2. Turn on the magnetic stirrer to begin mixing the sample. If temperatures above or below room temperature are required, the instrument should be equipped with temperature controls. Measured rate constants can be affected by sample temperature.

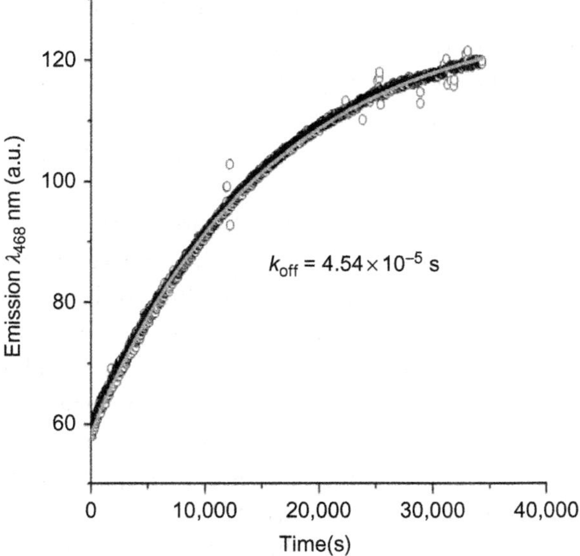

Figure 6.4 Determination of k_{off} for a DFG-out-binding ligand. This figure presents a typical kinetic measurement for determining the k_{off} of BIRB-796 using acrylodan-labeled p38α. The fluorescence emission is monitored at a single wavelength upon the addition of an excess of unlabeled kinase to scavenge the prebound ligand from the acrylodan-labeled kinase. *Panel has been reprinted with permission from Simard, Getlik, et al. (2009). Copyright 2009 American Chemical Society.*

3. Monitor the emission intensity at a wavelength, which is the most sensitive to ligand binding. This wavelength can be determined by performing the experiments described in Section 4.2.1.
4. Measure baseline fluorescence signal for ~30 s to ensure that the baseline is relatively flat and linear. Minor downward drift in the fluorescence signal over time can occur with certain fluorophores. However, the magnitude of this downward drift is often insignificant compared to the fluorescence changes induced by ligand binding.
5. Using the K_d curve, choose ligand concentrations which are in the linear range of the curve. Use these concentrations for examining the kinetic parameters.
6. Using a 1–10-μL pipettor, add the inhibitor to the stirring sample. For the best results, make sure the pipette tips are long enough to reach far into the cuvette to ensure that inhibitor is rapidly dispersed while mixing. To prevent outside light from entering the measurement chamber of the spectrometer, instruments which have an injection port

positioned above the cuvette are recommended. This allows ligand to be added while measuring fluorescence signal simultaneously.

7. Measure the change in emission over time; allow the fluorescence change to proceed until it linearizes. Repeat for at least four different ligand concentrations, using a new sample of protein for each measurement.
8. Plot the data using desired graphing software. Fit the kinetic trace using a first order fitting function to determine the observed rate constant (k_{obs}) for each dose of added ligand. If the graphing software provides the fit result as a half-time of fluorescence decay ($t_{1/2}$), the following equation should be applied to determine k_{obs}:$k_{obs} = \ln 2/t_{1/2}$. The units of $t_{1/2}$ should be converted to seconds prior to the conversion.
9. Plot k_{obs} (units of s) as a function of inhibitor concentration (units of M), fit the data using a linear fitting function. The slope of the resulting line is the true k_{on} (units of M/s) of the ligand. The k_{off} for the ligand can also be derived from the y-intercept of this plot (y-intercept $= k_{off}$).

4.3.2 Determination of k_{off}

1. Prepare a cuvette containing the FLiK kinase as described in step nos. 1 and 2 of Section 4.3.1.
2. Before measuring, add and preincubate the FLiK kinase with an inhibitor known to bind to a desired binding site and conformation of the kinase. Choose a concentration of ligand, which is in the linear range of K_d curve and allow the ligand to reach binding equilibrium.
3. Monitor fluorescence emission over time as described earlier in step no. 4 of Section 4.3.1.
4. Use a 1–10-μL pipettor to add a concentrated stock of unlabeled kinase to the sample, while measuring the change in emission over time until it becomes linear. Repeat by adding unlabeled kinase to a suspension of FLiK kinase preincubated with several concentrations of ligand.
5. Plot as the data using desired graphing software. Plot the kinetic trace and fit the curve as described above in step nos. 8 and 9 of Section 4.3.1 to determine the k_{off} for the ligand.

5. HTS WITH FLiK

5.1. Adaptation to HTS formats

The FLiK assay technology can easily be adapted to high-throughput assay plates (96-, 384-, and 1536-well formats). Adaptation of the assay to small volume microtiter plates dramatically reduces the amount kinase required while increasing throughput to enable rapid screening of large compound libraries. The following steps should be taken for adapting the FLiK assay to HTS formats:

1. Prepare a suspension of 50 n*M* FLiK kinase and place equal volumes of sample into several wells of a black assay plate. Place the HTS plate into a fluorescence plate reader and set the excitation wavelength according to the fluorophore used to label the kinase.
2. Measure the emission spectrum of the kinase in the presence and absence of the positive control inhibitor to verify that the desired maxima have not shifted in the HTS format when compared to the cuvette format. For some kinases, we have observed as much as a 10 nm shift in one or both of the maxima when measured in HTS plates. If changes are observed, adjust the wavelengths used to calculate the ratio in HTS plates.
3. Calculate the Z' factor of the HTS assay using several wells containing positive and negative controls.
4. Assess protein stability and estimate the required incubation time for ligand binding to reach equilibrium. Prepare a row of wells containing a dilution series of the compound and a fixed concentration of the FLiK kinase. Measure the same plates over a period of time and monitor the leftward shift of the K_d-binding curve. Additionally, recalculate the Z' factor at each time point. Choose an incubation time, which produces both a stable K_d value and maximal assay window. A drop in the assay window or a decrease in data point reproducibility often indicates that the protein is no longer stable, resulting in lower calculated Z'-factor values. To minimize these effects during long incubation times, keep the plate sealed (in a humidity chamber or with adhesive seals) and protected from light.
5. If the K_d values and Z' factor need to be improved, several assay variables can be tested to examine their effects on the quality and reproducibility of the assay. These optimizations should be carried out for each FLiK assay. DMSO concentration (added with the compound), detergent type, detergent concentration, kinase concentration, volume in each well, plate type, incubation time, and incubation temperature are several examples. We have found that DMSO is critical to protein stability while detergent type and concentration can shift measured K_d values, depending on the assay. Detergent concentrations $<0.01\%$ (v/v) are recommended.

6. Using the optimized assay conditions, run a mock HTS scenario using liquid-handling robots to dispense kinase into two assay plates. Add compound to one plate (positive control) while adding only vehicle (buffer with DMSO) to the other plate. Incubate the plates and measure. Assess the full assay plates for good signal uniformity and calculate the Z' factor for the final optimized HTS assay.

5.2. HTS of compound libraries

We have performed several HTS using various FLiK kinases and found that 50–100 n*M* kinase is sufficient for avoiding many fluorescence artifacts and ensures that the signal from the fluorophore-labeled kinase is large enough to overcome the intrinsic fluorescence of many compounds when screened at a maximum concentration of 10–20 μ*M*. A typical primary screen involves assay plates containing one compound per well at a maximum concentration of 10–20 μ*M*. To assess background fluorescence of the compounds alone, each compound plate can be screened twice, once with only buffer and once with the FLiK kinase in the same buffer to enable background subtraction. Typically, compounds which bind at least 50% compared to the positive control at 10–20 μ*M* are chosen for follow-up screening. After a primary screen, hits are selected and rescreened in a dose–response format to confirm binding to the kinase and to determine the K_d value. A general guideline for an HTS is provided below.

1. Map out the planned arrangement of a 384-well screening plate with the location of the compounds which will be added to the screening plate. Be sure to include several wells for both negative (DMSO) and positive (DFG-out inhibitor) controls.
2. Prepare all compound stocks (single point or dose–response) in 100% DMSO. Transfer compounds to a deep well assay plate and predilute in Assay Buffer. Place positive and negative controls into the appropriate wells of the predilution plate. This plate is the inhibitor stock plate for the assay. Ideally, the concentration of DMSO in the inhibitor stock plate should be 5% (v/v) or lower.
3. Use multichannel pipettors or liquid-handling devices to transfer compounds from the prediluted inhibitor stock plate to an empty small volume (20 μL), black, 384-well HTS plate.
4. Use multichannel pipettors or liquid-handling devices to transfer FLiK kinase in buffer to the assay plate. Addition of a larger volume of kinase solution should be sufficient to mix the contents of each well. The final

mixture of kinase and compound should be at 1 × in the assay plate and the final concentration of DMSO should ideally be 1% (v/v) or lower.

5. Seal and store the plates in darkness as described in step no. 4 of Section 5.1 using the previously determined incubation time and temperature.
6. After incubation, remove the seal from each plate and place the plate inside a microtiter plate reader to measure the fluorescence emission intensity at both desired wavelengths. Depending on the software which accompanies the instrument, the readout will either be the calculated ratio value or the two sets of individually measured emission intensities at each wavelength.
7. Identify compounds which trigger the expected fluorescence response and rescreen these compounds in a dose–response format using a similar procedure as described earlier for the primary screen.

5.3. Data analysis, fluorescence artifacts, and pitfalls

As is the case with any fluoresce-based assay, the identification of false positives can be tricky and sometimes difficult. Especially in the case of the FLiK assay, identifying such compounds can be very time consuming, especially since it is difficult to run all the necessary controls needed to identify all the possible types of fluorescence artifacts. The method described earlier, in which a background plate is used, will quickly identify intrinsic compound fluorescence which can be easily subtracted to reveal the emission signal contribution of the fluorophore label used in the FLiK kinase. In many cases, subtraction of the intrinsic signal from the compound reveals that the calculated ratio of emission intensities of the fluorophore is changing in a dose-dependent manner. To date, we have identified and confirmed the binding of several compounds, which also possess intrinsic background fluorescence.

The use of background plates will not account for other types of fluorescence artifacts such as quenching or synergizing interactions between the fluorophore and certain compounds. The artifacts are more difficult to identify and may not necessarily be observed if the compounds are placed into suspension with free, unreacted fluorophore since the fluorescence characteristics and behavior of reactive fluorophores often change upon conjugation to proteins. Once conjugated, the fluorescence emission of environmentally sensitive fluorophores depends on the local environment provided by the kinase conformation. By binding to the kinase, the ligands come into close proximity of the fluorophore and this close distance may enable the observed quenching or synergizing interactions which would otherwise not occur with free fluorophore in solution. However, we should note that we have identified and confirmed the binding of some compounds, which also quench or augment the emission spectrum of the FLiK kinase (Simard, Grutter, et al., 2009). We have found that these types of compounds tend to change the overall intensity of the entire spectrum but also induce the expected change in the calculated ratio value in a dose-dependent manner. In other words, although the absolute intensity of the spectrum has been changed, the behavior of the fluorophore in response to binding has not been affected.

It is important to note that hit rates in the FLiK assay tend to be low due to the ability of the assay to report on ligands which only induce certain conformational changes. Although many more compounds may bind, only those inducing the desired conformational change will be selected for follow-up, making it possible to advance smaller collections of hits to subsequent screening in alternative assays. Rather than developing unique FLiK assays aimed at each type of fluorescence artifact, we instead use the multipronged approach described below to weed out and remove ligands remaining in this small collection of hits that may not truly bind or induce the desired conformational change.

1. Perform a primary screen as described in Section 5.2 and identify hits, which change the ratio of the FLiK kinase.
2. Prepare hits as a dilution series and retest them in a dose–response format in the FLiK assay using a background plate in addition to an assay plate containing the FLiK kinase. Incubate and then measure each plate (kinase plate and buffer plate) at the two chosen wavelengths. Calculate the ratio values for each compound concentration using values for each wavelength. Calculate the ratio for both the background corrected and uncorrected data. Compounds, which are not fluorescent, produce equivalent results regardless of whether background fluorescence is factored into the calculation. In contrast, fluorescent compounds produce very different results. These compounds should warrant closer analysis of the raw data before deciding if the compound is a valid hit. Typically, valid hits will produce a binding curve when screened in this format. For fluorescent compounds, the binding curve may only be visible following background subtraction at each concentration. For compounds, which are suspected of synergizing or quenching of the fluorophore, validation of binding can be more difficult. Such problematic compounds may be deemed less desirable for follow-up studies and removed from the

screen. However, a more conservative approach would include these compounds in follow-up studies with alternative assays.

3. Examine the binding kinetics of each hit in the FLiK assay. Fluorescence artifacts are less likely to show time dependence. This is especially useful when examining potential slow-binding hits identified with the DFG-out FLiK assay screen in which the kinase is labeled with a fluorophore on the activation loop.
4. Choose a hit list for subsequent testing in alternative binding assays for the kinase of interest, if available. Fluorescence polarization assays are an example. Displacement of a fluorescent prebound probe will change the polarization of the fluorophore and confirm that the hits are binding to the kinase.
5. Test the compounds in activity-based assays, if available, to confirm inhibition of catalytic activity. Since the phosphorylated kinase is often required for these types of assays, shifts in ligand potency may be observed if testing hits identified in a DFG-out FLiK assay in which the unphosphorylated kinase was used to enhance the likelihood of achieving the DFG-out conformation.
6. Attempt to crystallize hits, which show a positive response in all of the above assays. A crystal structure will reveal the details of the binding mode and validate the FLiK assay used to initially identify the compound.

6. SUMMARY

The poor selectivity of kinase inhibitors can be addressed by moving away from classical ATP-competitive inhibitors and targeting alternative kinase conformations and allosteric-binding sites. Profiling compounds against the entire kinome provides valuable information about the affinity and selectivity of compounds *in vitro*. However, optimizing compounds to lengthen drug–target residence times will ultimately provide a more complete context for fully understanding kinase inhibitor selectivity *in vivo*. By considering the kinetic components (k_{on} and k_{off}) of affinity, lead optimization strategies may improve *in vivo* selectivity and efficacy by directing medicinal chemistry efforts around improving the residence time of the ligand for the desired kinase.

We have developed FLiK as a HTS technology, which enables the rapid and robust identification of ligands, which bind to and stabilize specific kinase conformations. The FLiK approach allows straightforward determination of ligand affinity (K_d values) and kinetic characterization (k_{on} and k_{off}) to better understand the kinetic factors, which contribute to the measured affinity. Moreover, FLiK does not require kinase activity or prior knowledge of the substrate, which may be advantageous when studying novel or less-characterized kinases. To date, the FLiK approach has been used to successfully monitor conformational changes in the activation loop of both Ser/Thr and Tyr kinases associated with the slow binding of DFG-out inhibitors. It has also been adapted to detect ATP-competitive inhibitors, which engage the glycine-rich loop (P loop) as well as Type IV ligands, which bind to remote allosteric-binding sites outside of the ATP-binding cleft.

ACKNOWLEDGMENTS

The German Federal Ministry for Education and Research supported this work through the German National Genome Research Network-Plus (NGFNPlus) (Grant No. BMBF 01GS08104). MSD, Bayer HealthCare, Merck-Serono, and Bayer CropScience are thanked for financial support.

REFERENCES

Akella, R., Moon, T. M., & Goldsmith, E. J. (2008). Unique MAP kinase binding sites. *Biochimica et Biophysica Acta*, *1784*, 48–55.

Anastassiadis, T., Deacon, S. W., Devarajan, K., Ma, H., & Peterson, J. R. (2011). Comprehensive assay of kinase catalytic activity reveals features of kinase inhibitor selectivity. *Nature Biotechnology*, *29*, 1039–1045.

Arnold, K., Bordoli, L., Kopp, J., & Schwede, T. (2006). The SWISS-MODEL workspace: A web-based environment for protein structure homology modelling. *Bioinformatics*, *22*, 195–201.

Backes, A. C., Zech, B., Felber, B., Klebl, B., & Müller, G. (2008a). Small-molecule inhibitors binding to protein kinases. Part I: Exceptions from the traditional pharmacophore approach of type I inhibition. *Expert Opinion on Drug Discovery*, *3*, 1409–1425.

Backes, A. C., Zech, B., Felber, B., Klebl, B., & Müller, G. (2008b). Small-molecule inhibitors binding to protein kinase. Part II: The novel pharmacophore approach of type II and type III inhibition. *Expert Opinion on Drug Discovery*, *3*, 1427–1449.

Copeland, R. A. (2011). Conformational adaptation in drug–target interactions and residence time. *Future Medicinal Chemistry*, *3*, 1491–1501.

Copeland, R. A., Pompliano, D. L., & Meek, T. D. (2006). Drug–target residence time and its implications for lead optimization. *Nature Reviews. Drug Discovery*, *5*, 730–739.

Davis, M. I., Hunt, J. P., Herrgard, S., Ciceri, P., Wodicka, L. M., Pallares, G., et al. (2011). Comprehensive analysis of kinase inhibitor selectivity. *Nature Biotechnology*, *29*, 1046–1051.

de Kloe, G. E., Retra, K., Geitmann, M., Kallblad, P., Nahar, T., van Elk, R., et al. (2010). Surface plasmon resonance biosensor based fragment screening using acetylcholine binding protein identifies ligand efficiency hot spots (LE hot spots) by deconstruction of nicotinic acetylcholine receptor alpha7 ligands. *Journal of Medicinal Chemistry*, *53*, 7192–7201.

de Lorimier, R. M., Smith, J. J., Dwyer, M. A., Looger, L. L., Sali, K. M., Paavola, C. D., et al. (2002). Construction of a fluorescent biosensor family. *Protein Science*, *11*, 2655–2675.

DeLano, W. L. (2002). *The PyMOL molecular graphics system. http://www.pymol.org.*
Getlik, M., Simard, J. R., Termathe, M., Grutter, C., Rabiller, M., van Otterlo, W. A., et al. (2012). Fluorophore labeled kinase detects ligands that bind within the MAPK insert of p38alpha kinase. *PLoS One*, 7, e39713.
Grutter, C., Simard, J. R., Mayer-Wrangowski, S. C., Schreier, P. H., Perez-Martin, J., Richters, A., et al. (2012). Targeting GSK3 from Ustilago maydis: Type-II kinase inhibitors as potential antifungals. *ACS Chemical Biology*, 7, 1257–1267.
Karaman, M. W., Herrgard, S., Treiber, D. K., Gallant, P., Atteridge, C. E., Campbell, B. T., et al. (2008). A quantitative analysis of kinase inhibitor selectivity. *Nature Biotechnology*, *26*, 127–132.
Knight, Z. A., Lin, H., & Shokat, K. M. (2010). Targeting the cancer kinome through polypharmacology. *Nature Reviews. Cancer*, *10*, 130–137.
Lambert, C., Leonard, N., De Bolle, X., & Depiereux, E. (2002). ESyPred3D: Prediction of proteins 3D structures. *Bioinformatics*, *18*, 1250–1256.
Liu, Y., & Gray, N. S. (2006). Rational design of inhibitors that bind to inactive kinase conformations. *Nature Chemical Biology*, *2*, 358–364.
Markgren, P. O., Schaal, W., Hamalainen, M., Karlen, A., Hallberg, A., Samuelsson, B., et al. (2002). Relationships between structure and interaction kinetics for HIV-1 protease inhibitors. *Journal of Medicinal Chemistry*, *45*, 5430–5439.
Morphy, R. (2010). Selectively nonselective kinase inhibition: Striking the right balance. *Journal of Medicinal Chemistry*, *53*, 1413–1437.
Pargellis, C., Tong, L., Churchill, L., Cirillo, P. F., Gilmore, T., Graham, A. G., et al. (2002). Inhibition of p38 MAP kinase by utilizing a novel allosteric binding site. *Nature Structural Biology*, *9*, 268–272.
Rabiller, M., Getlik, M., Kluter, S., Richters, A., Tuckmantel, S., Simard, J. R., et al. (2010). Proteus in the world of proteins: Conformational changes in protein kinases. *Archiv der Pharmazie (Weinheim)*, *343*, 193–206.
Richieri, G. V., Ogata, R. T., & Kleinfeld, A. M. (1999). The measurement of free fatty acid concentration with the fluorescent probe ADIFAB: A practical guide for the use of the ADIFAB probe. *Molecular and Cellular Biochemistry*, *192*, 87–94.
Schneider, R., Becker, C., Simard, J. R., Getlik, M., Bohlke, N., Janning, P., et al. (2012). Direct binding assay for the detection of type IV allosteric inhibitors of Abl. *Journal of the American Chemical Society*, *134*, 9138–9141.
Simard, J. R., Getlik, M., Grutter, C., Pawar, V., Wulfert, S., Rabiller, M., et al. (2009). Development of a fluorescent-tagged kinase assay system for the detection and characterization of allosteric kinase inhibitors. *Journal of the American Chemical Society*, *131*, 13286–13296.
Simard, J. R., Getlik, M., Grutter, C., Schneider, R., Wulfert, S., & Rauh, D. (2010). Fluorophore labeling of the glycine-rich loop as a method of identifying inhibitors that bind to active and inactive kinase conformations. *Journal of the American Chemical Society*, *132*, 4152–4160.
Simard, J. R., Grutter, C., Pawar, V., Aust, B., Wolf, A., Rabiller, M., et al. (2009). High-throughput screening to identify inhibitors which stabilize inactive kinase conformations in p38alpha. *Journal of the American Chemical Society*, *131*, 18478–18488.
Simard, J. R., Kluter, S., Grutter, C., Getlik, M., Rabiller, M., Rode, H. B., et al. (2009). A new screening assay for allosteric inhibitors of cSrc. *Nature Chemical Biology*, *5*, 394–396.

CHAPTER SEVEN

Discovery of Allosteric Bcr–Abl Inhibitors from Phenotypic Screen to Clinical Candidate

Nathanael S. Gray[*,1], **Doriano Fabbro**[†]
[*]Department of Biological Chemistry and Molecular Pharmacology, Harvard Medical School, Department of Cancer Biology, Dana-Farber Cancer Institute, Boston MA, USA
[†]PIQUR Therapeutics AG, Hochbergerstrasse 60C, Basel, Switzerland
[1]Corresponding author: e-mail address: nsgray01@gmail.com

Contents

Abstract

The development of imatinib, an ATP-competitive inhibitor of the BCR–ABL oncoprotein, has revolutionized the treatment of chronic myelogenous leukemia (CML). Unfortunately, the leukemia eventually becomes resistant imatinib as a result of emergence of cells expressing drug insensitive BCR–ABL mutant proteins. This has motivated the development of several next-generation ATP-competitive drugs. This chapter describes the discovery and development of a complementary strategy involving inhibiting BCR–ABL by targeting an allosteric binding site. Compounds that bind to the myristate-binding pocket of BCR–ABL are able to induce formation of an "inactive" state and are able to overcome resistance mutations located in the ATP-binding pocket including the recalcitrant T315I "gatekeeper" mutation. Myristate-pocket inhibitors are

Methods in Enzymology, Volume 548
ISSN 0076-6879
http://dx.doi.org/10.1016/B978-0-12-397918-6.00007-0

also able to function synergistically with ATP-competitive inhibitors in cellular and murine models of CML and this dual inhibitory strategy is currently being investigated in the clinic.

1. DEVELOPMENT OF ATP-SITE-DIRECTED INHIBITORS OF BCR–ABL FOR THE TREATMENT OF CML

The development of imatinib for the treatment of chronic myelogenous leukemia (CML) provided the key proof-of-concept that selective small molecules that specifically target the underlying oncogenic drivers of cancer can provide effective therapy. The molecular hallmark of CML is a reciprocal chromosomal translocation t(9;22)(q34;q11) that generates an aberrant fusion protein, BCR–ABL, which results in constitutive activation of the ABL tyrosine kinase (TK). Three clinically important variants are the p190, p210, and p230 isoforms (Advani & Pendergast, 2002; Melo, 1996). p190 is generally associated with acute lymphoblastic leukemia (ALL), while p210 is generally not only associated with chronic myeloid leukemia but can also be associated with ALL. p230 is usually associated with chronic neutrophilic leukemia (Pakakasama et al., 2008).

Imatinib (Glivec, Gleevev, and STI-571) is a small molecule developed in the late 1990s that inhibits BCR–ABL kinase activity by binding to its

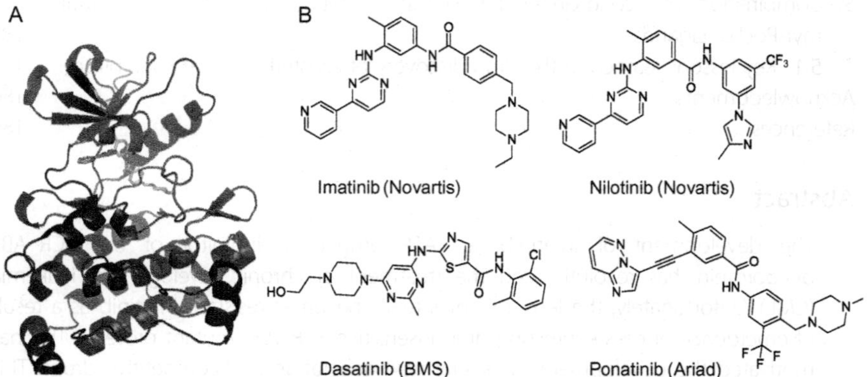

Figure 7.1 Bcr–Abl kinase inhibitors. (A) Crystal structure of the kinase domain of Abl (blue ribbons) with imatinib (green sticks) bound in the ATP pocket. The glycine-rich loop (purple ribbons) which forms the "roof" of the ATP-binding site and the activation loop (red ribbons), both of which undergo conformational rearrangements upon imatinib binding are indicated. (B) Chemical structures of first- (imatinib), second- (nilotinib and dasatinib), and third-generation (ponatinib) FDA approved Bcr–Abl inhibitors. (See the color plate.)

ATP site, thereby leading to apoptosis of BCR–ABL transformed cells (Fig. 7.1). Imatinib traps the inactive conformation of the ABL kinase domain by binding to both the nucleotide pocket and an adjacent hydrophobic cleft. Imatinib has revolutionized the treatment of CML especially for patients with the chronic phase which precedes the accelerated phase and the blast crisis phase of CML. Patients with more advanced disease (late-stage chronic phase, accelerated phase, or blast crisis phase of CML) become resistant to imatinib. The most common form of resistance is due to mutations in BCR–ABL located predominantly in the ABL kinase domain that are less effectively inhibited by imatinib. The first discovered and most difficult to inhibit mutation occurs at the so-called "gatekeeper" residue of the ABL kinase domain, T315I, which further activates the kinase activity and introduces a steric impediment to imatinib binding (Fig. 7.2). Second-generation efforts to develop compounds that could overcome imatinib resistant BCR–ABL mutants resulted in the development of nilotinib (Weisberg et al., 2005), dasatinib (Lombardo et al., 2004), and bosutinib (Cortes et al., 2011, 2012). Nilotinib binds to BCR–ABL in a similar fashion to imatinib but possess significantly improved molecular recognition of BCR–ABL that results in improved cellular potency against both wild-type

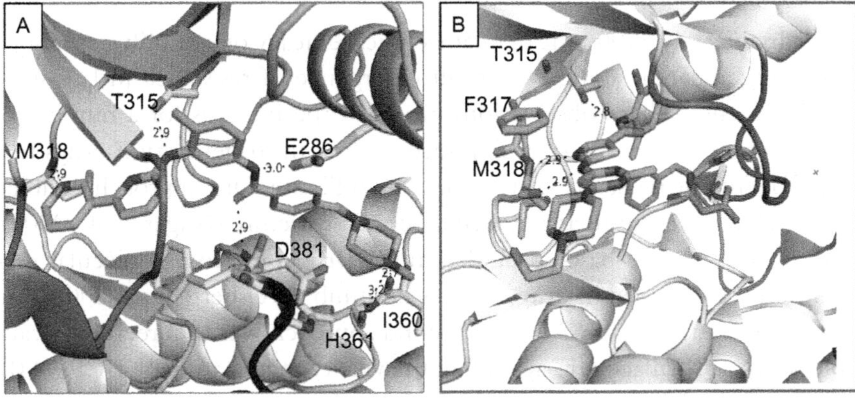

Figure 7.2 Crystal structures of imatinib and dasatinib with Abl highlighting the protein–ligand interactions. (A) Imatinib (green sticks, carbons), nitrogen (blue), oxygen (red) forms five hydrogen-bond interactions with the ATP-binding pocket of Abl as indicated by dotted lines with distances shown in angstroms. The benzamide portion of the inhibitor is situated in a pocket made accessible by the flip of the DFG-motif (yellow sticks). (B) Similar figure for dasatinib. The "DFG-motif" is in not flipped in this structure, thereby closing the pocket accessed by the benzamide group of imatinib. Note both inhibitors make key hydrogen bonds to the "gatekeeper" residue T315 and traverse close to this position suggesting that they require the side-chain of the gatekeeper to be small in order to bind well. (See the color plate.)

and mutant forms of BCR–ABL (Fig. 7.2). Dasatinib, originally designed as an inhibitor of LCK as a potential immunosuppressant, was later repurposed as a highly potent inhibitor BCR–ABL which binds to the active conformation of BCR–ABL. Bosutinib was originally developed as a "dual" SRC–ABL kinase inhibitor also binds to active conformation of ABL (Puttini et al., 2006). However, neither nilotinib, bosutinib nor dasatinib is capable of inhibiting T315I at clinically relevant doses which has spurred the development of the so-called "third-generation" ATP-site-directed TKIs such as ponatinib (Huang et al., 2010) and numerous other preclinical stage compounds (O'Hare et al., 2005). While ponatinib effectively inhibits T315I in patients, it displays promiscuous inhibition of a large number of kinases and exhibits poor tolerability in humans (Mayer, Gielen, Willinek, Muller, & Wolf, 2014). The story of the development of ATP-site-directed BCR–ABL inhibitors has been well reviewed elsewhere (Buchdunger, Matter, & Druker, 2001) and our intention in this chapter is instead to describe how a new mechanistic class of allosteric BCR–ABL inhibitors was discovered and optimized to the clinical development stage.

2. DISCOVERY AND CHARACTERIZATION OF NON-ATP-SITE-DIRECTED BCR–ABL INHIBITORS

In the early 2000s, the dramatic clinical efficacy of imatinib was rapidly becoming apparent along with swift emergence of resistance in late-stage CML patients. These observations provided the impetus for a number of efforts whose goal was to develop second- and third-generation drugs that could overcome resistance by targeting the ATP site. One approach, enabled by the elucidation of imatinib-ABL cocrystal structure (Nagar et al., 2002), was to use structure-based drug design (SDBB) to improve molecular recognition that resulted in drugs such as nilotinib (Weisberg et al., 2005) and ponatinib (Huang et al., 2010). A second approach was to screen the available kinase pharmacopeia against both wild-type and mutant BCR–ABL which resulted in the repurposing of dasatinib (Lombardo et al., 2004) and bosutinib (Golas et al., 2003) as broad-spectrum Bcr–Abl inhibitors. We decided to use phenotypic screening as a third approach using a BCR–ABL transformed 32D cell line to screen for compounds that could selectively inhibit its proliferation relative to their isogenic parental controls (Adrian et al., 2006). The 32D cells are a murine pre-B-cell that require the cytokine IL-3 for growth and survival. The IL-3 requirement of 32D cells can be bypassed by oncogenes, including BCR–ABL, thereby allowing the cells to grow in the absence of IL-3 (Warmuth, Kim, Gu, Xia, & Adrian, 2007). Inhibiting an oncogenic kinase in transformed 32D results in their apoptotic death which can be rescued by IL-3. Thus, 32D or Ba/F3 cells transformed with oncogenic TKs have become a powerful tool for the discovery and optimization of TK inhibitors because a simple proliferation assays in conjunction with the autophosphorylation status (by ELISA) of the oncogenic TK provides accurate structure–activity relationships that faithfully report on the relevant intracellular kinase conformation (Warmuth et al., 2007). On-target inhibition of BCR–ABL can be distinguished from nonspecific cytotoxicity by the IL-3 cytokine rescue or by counter-screening the inhibitors against wild-type untransformed Ba/F3 cells grown in the presence of IL-3. Our goal was to identify novel BCR–ABL inhibitors or to identify compounds targeting the oncogenic BCR–ABL signaling in 32D cells. The 32D BCR–ABL cell-based screen was performed against large combinatorial heterocyclic library constructed using solid-phase chemistry encoded by nanokan™-directed sorting (Ding, Gray, Wu, Ding, & Schultz, 2002). After confirming that "hit" compounds exhibited selectivity for BCR–ABL transformed cells relative to parental untransformed Ba/F3 cells, resynthesized compounds were tested for their ability to inhibit the activity of recombinant Abl kinase domain (SH1) and for their ability to inhibit BCR–ABL autophosphorylation and phosphorylation of canonical BCR–ABL effectors such as the transcription factor STAT5. As expected, we identified a number of ATP-competitive BCR–ABL inhibitors from structurally well recognizable classes such as phenylaminopyrimidines and pyridopyrimidinones as well compounds that were later shown to be nucleotide-dependent inhibitors of HSP90. Identification of these screening "hits" validated the ability of the screen to identify both the expected "on-target" BCR–ABL inhibitors as well as "collaborating" proteins such as HSP90 which are known to be essential to maintain BCR–ABL protein homeostasis. However, we were most intrigued by the discovery of a series of exceedingly simple 4-trifluoromethoxyaniline-substituted pyrimidines such as compound **1** which validated as being differentially cytotoxic and inhibited BCR–ABL auto- and substrate phosphorylation but surprisingly were unable to inhibit the enzymatic ABL kinase activity. A series of analogs were prepared to understand the structure–activity relationships required to provide differential cytotoxicity which results in the discovery of GNF-2 and -5 (Fig. 7.3). One interesting feature gleaned from this SAR series was the requirement for the pyrimidine substitution pattern to be 4,6 and not 2,4 which is

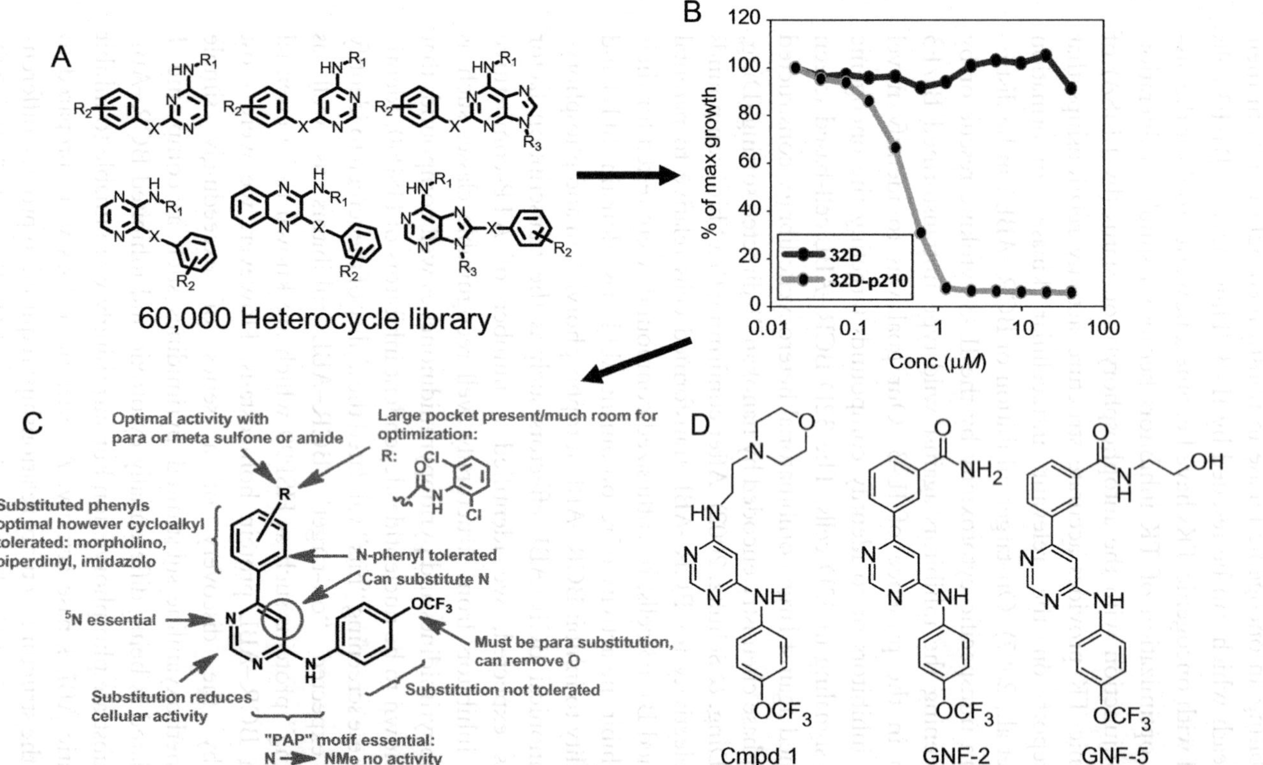

Figure 7.3 Process that lead to the discovery of GNF-2 and -5. (A) Representative heterocyclic structures that were synthesized using solid-phase combinatorial chemistry encoded using directed sorting. (B) The library was screen for differential cytotoxicity between cells addicted to Bcr–Abl kinase activity for survival and proliferation versus isogenic "parental" cells grown under conditions not requiring Bcr–Abl kinase activity. (C) Summary of the salient structure–activity relationships developed by follow-up chemistry performed on the screening "hits." (D) Chemical structures of compound **1** (actual compound identified from the combinatorial library) and of GNF-2 and -5. (See the color plate.)

typically preferred based on the 2,4-pyrimidines being better binders of the kinase "hinge" segment that serves as the critical connection between the N- and C-terminal kinase domains (Deng et al., 2010). The SAR studies also enabled the design of a linker modified GNF-2 which was used to demonstrate binding to SH1 and to BCR–ABL in cell lysates. Competition experiments with ATP or imatinib demonstrated that GNF-2 was most likely not binding to the ATP pocket. At this point, the project hit a bottleneck as we were convinced that GNF-2 was mechanistically interesting but we were unable to cocrystallize the compound with ABL had no idea how it was inhibiting cellular BCR–ABL kinase activity.

3. CHARACTERIZATION OF THE BINDING OF THE NON-ATP-SITE-DIRECTED BCR–ABL INHIBITOR GNF-2

The first X-ray structure of ABL together with the regulatory SH2 and SH3 domains lacking the last exon (SH3SH2SH1) in conjunction with detailed biochemical and cellular analysis provided key insights into how c-ABL is normally regulated (Nagar et al., 2002, 2003). In this structure, the N-terminus of ABL is myristoylated and the C14 hydrocarbon lipid of myristate binds into a cylindrical pocket (myr-pocket) located at the C-terminus of the kinase domain (Fig. 7.4). This interaction was demonstrated to serve as a molecular "lasso" that clamps the regulatory SH2 and SH3 onto the kinase domain thereby enforcing an assembled autoinhibited conformation. This suggested that unlike the structurally related SRC kinase where the SH2 domain binds to a phosphotyrosine (Y527) located C-terminal to the kinase domain, ABL instead exploits this myristoyl lasso interaction to achieve an autoinhibited conformation. When BCR becomes fused to ABL, the N-terminus of ABL along with the myristoyl group is lost providing one of several reasons that BCR–ABL possesses constitutive kinase activity. For example, mutation of the N-terminal glycine to alanine (G2A) of c-ABL, which results in the production of a nonmyristoylated kinase, is sufficient to render ABL constitutively active. As GNF-2 is not competitive with ATP, we hypothesized that GNF-2, using the trans-conformation enabled by the 4,6-pyrimidine substitution pattern, could exploit this myr-pocket to awake an endogenous regulatory mechanism to reset the full-length ABL (SH3SH2SH1) into the inactive, assembled state. This hypothesis was validated first by demonstrating the introduction of point mutations into the myristate-binding pocket that disrupted GNF-2 from binding also conferred resistance to the ability of GNF-2 to inhibit

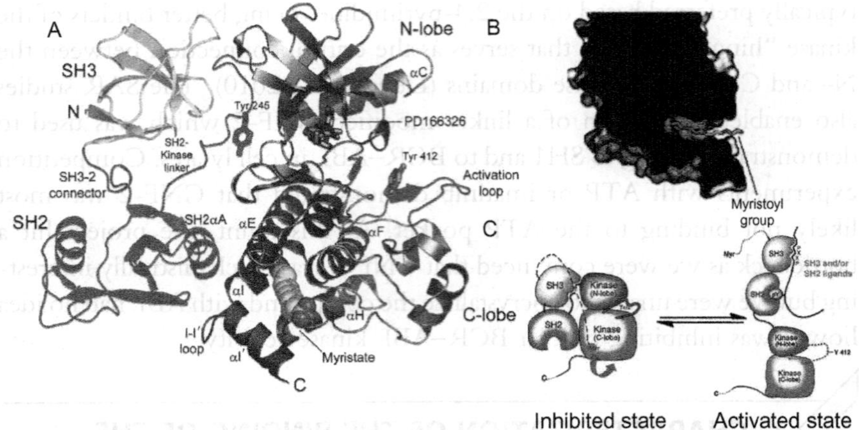

Figure 7.4 Depiction of the proposed "latching" mechanism of Abl enabled by N-terminal myristoylation. (A) Ribbon structure of Abl containing the kinase domain (blue ribbons), the phosphotyrosine-binding SH2 domain (green ribbons), the linker-binding SH3 domain (yellow ribbons), and well as several other key regions as indicated. The myristate lipid is shown as a space-filling model (yellow balls) nestled in a hydrophobic cavity located at the C-terminus of the kinase domain. (B) Proposed model of how N-terminal myristoylation of Abl enforces the autoinhibited conformation. The inhibited state is characterized by binding of the N-terminal myristate into a pocket at the kinase C-terminus which reinforces the binding of the SH2 and SH3 domains onto the kinase domain and linker connecting the kinase N-terminus and the SH2 domain. Activation of Abl is achieved by engagement of the SH2 and SH3 domains with ligands, phosphorylation of the activation loop, and disengagement of the myristate and likely numerous other mechanisms that have not been elucidated. (See the color plate.)

cellular BCR–ABL activity and second by solving a GNF-2 ABL costructure which demonstrated unequivocable binding of GNF-2 to the myristate pocket (Adrian et al., 2006; Zhang et al., 2010). Secondly, it was shown by an elegant NMR reporter assay that myristate and GNF-2 were docking to the myr-pocket in the SH1 without affecting its activity. We next focused on investigating how the binding of GNF-2 to the myr-pocket resulted in the inhibition of BCR–ABL kinase activity.

A possible answer to this question was provided by structural analysis enabled by both NMR and crystallography together with the use of H/D-exchange mass spectrometry (Iacob et al., 2009; Iacob, Zhang, Gray, & Engen, 2011; Jahnke et al., 2010). Structural and NMR analyses of the SH1 demonstrated that GNF-2 was indeed binding in the myr-pocket (Iacob et al., 2009, 2011; Jahnke et al., 2010). More importantly, these studies demonstrated that upon GNF-2 binding to the myristate pocket, a kinked conformation of the αI helix is induced which is conducive to docking of SH2 domain to the kinase C-terminus (Fabbro et al., 2010; Skora, Mestan, Fabbro, Jahnke, & Grzesiek, 2013; Figs. 7.5 and 7.6), which led to further studies on the role of the SH2 domain in the regulation of c-Abl and other kinases (Filippakopoulos, Muller, & Knapp, 2009; Grebien et al., 2011). Based upon these mechanistic studies, resonances in the HSQC spectra of GNF-2 with SH1 were identified that were diagnostic of the kinked αI helix conformation. This information was used to design a fragment-based screen to find additional scaffolds capable of inducing an autoinhibited conformation. Several fragments were crystallized with ABL and subsequent SDBB informed by biochemical and cellular BCR–ABL assays arrived at a series of highly potent inhibitors. Further confirmation of the functional importance of kinking the α-I helix upon binding of GNF-2 was provided by the discovery of allosteric activators of ABL (Jahnke et al., 2010; Yang et al., 2011). These activator compounds were discovered serendipitously from a screening campaign intended to discover biochemical ABL kinase inhibitors. Cocrystallography with these activator compounds, which are structurally distinct from GNF-2/-5, demonstrated that they also bind to the myristate pocket without bending the α-I helix, but rather by enforcing a straight confirmation of the α-I helix (Fig. 7.5 & Fig 7.6). In addition, H/D-exchange mass spectrometry demonstrated that GNF-2 ligation stabilized peptides both in the myr-pocket site and also in the kinase ATP pocket located more than 30 Å away. This result not only confirmed that GNF-2 bound to the myr-pocket but also revealed an

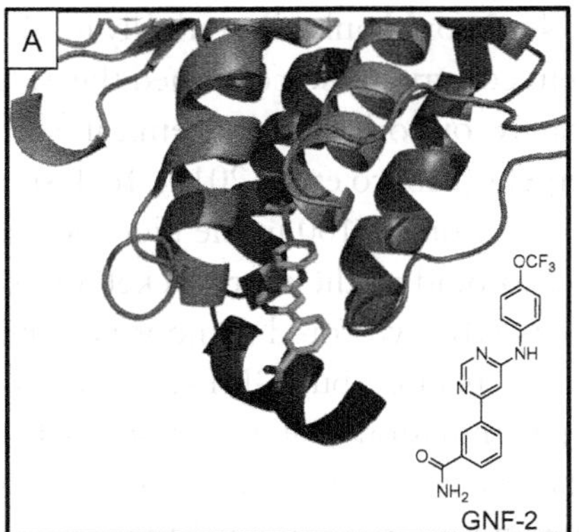

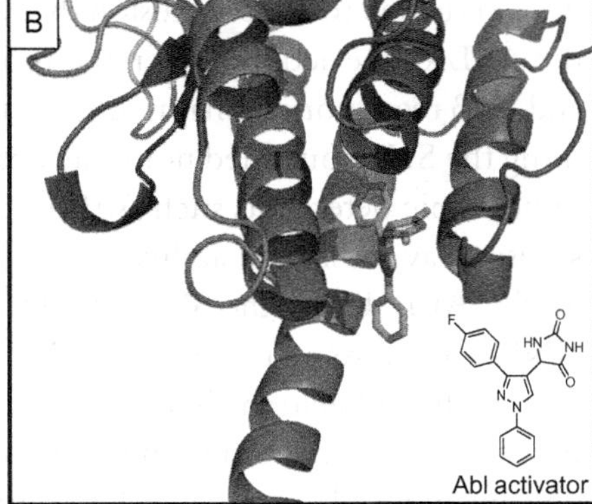

Figure 7.5 Crystal structures of two myristate-site ligands bound to Abl. (A) GNF-5 and (B) Abl-activator compound (carbon atoms of ligands depicted with light blue sticks). The chemical structures of each are shown in the insets. Note the kinking of the red helix (labeled αI and αI′ in Fig. 7.3A) in the GNF-5 costructure which is in sharp contrast to the linear conformation in the Abl-activator costructure. (See the color plate.)

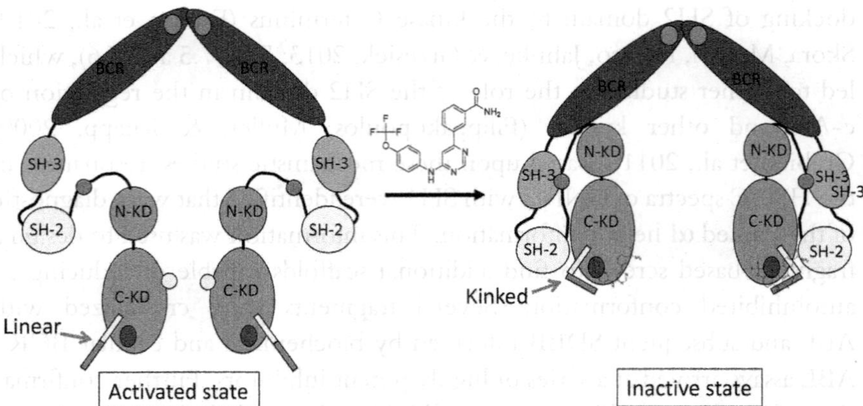

Figure 7.6 Cartoon depictions for how ATP and myristate ligands inhibit Abl kinase. Binding of an ATP site ligand such as imatinib blocks the ATP-binding site, thereby preventing ATP binding but still allows for motion of the SH2 and SH3 domain relative to the kinase domain. Binding of a ligand such as GNF-5 to the myristate pocket induces conformational changes, most notably the dramatic kinking of the αI and αI′ helix (see Fig. 7.7), that favors binding SH2 and SH3 domains. Note that the Bcr-domain, in addition to its established function to oligomerize Abl, is also fused in such a way as to remove the myristate ligand present at the Abl C-terminus in the normal c-Abl. Thus, GNF-5 type compounds exploit a dormant endogenous mechanism of Abl regulation. (See the color plate.)

allosteric effect that is transmitted to the remote ATP-binding pocket. As is true for most allosteric inhibitors, it currently remains unclear exactly what is the molecular "route" by which this information is transmitted. All these discoveries led to the conclusion that GNF-2 could only allosterically inhibit the ABL kinase activity when the protein construct also contained the SH2 and SH3 regulatory domains. Unfortunately, our original biochemical assays using the SH1 contained nonionic detergent (Fabbro et al., 2010). Inclusion of nonionic detergents such as Brij-35 or Triton-X-100 in the biochemical kinase assay masks the ability of GNF-2 to bind to the myr-pocket and to inhibit ABL. Using the SH3SH2SH1-wt-ABL, we could demonstrate that GNF-2 is more active than the ABL myr-peptide but GNF-2 could not inhibit the recalcitrant T315I gatekeeper mutation (Fabbro et al., 2010). In conclusion, choosing the right constructs for the biochemical assay upfront would have saved a lot of time and effort to unravel the binding mode of GNF-2. On the other hand, these biochemical and structural investigations helped to establish the myr-pocket as a *bona fide* allosteric regulatory site capable of mediating the pharmacological action of both agonists and antagonists—the first such site characterized for any kinase.

4. THERAPEUTIC POTENTIAL OF FIRST-GENERATION myr-POCKET BINDERS

4.1. Single-agent activity

Further optimization of GNF-2 to improve its pharmacokinetic properties resulted in the development of GNF-5 (Deng et al., 2010; Zhang et al., 2010; Fig. 7.3). GNF-5 demonstrated single-agent activity in xenografts employing BCR–ABL-dependent cells and in bone-marrow transplantation models. The degree of efficacy was the modest which may be a consequence of the modest cellular potency of GNF-5 and disease progression was observed while the animals were being treated. This prompted us to investigate how resistance to GNF-5 emerges in cell culture. By selecting for resistance to GNF-5 in BCR–ABL-transformed Ba/F3 cells, we isolated many BCR–ABL alleles harboring mutations in and around the myristate pocket. As expected, these mutations blocked binding of GNF-5 to the myr-pocket as confirmed using the GNF-5 affinity matrix (Fig. 7.7). Unexpectedly, we also recovered a number of resistant clones that expressed the gatekeeper T315I mutation which also confers resistance to approved inhibitors including imatinib, nilotinib, and dasatinib (O'Hare et al., 2005). The T315I mutation

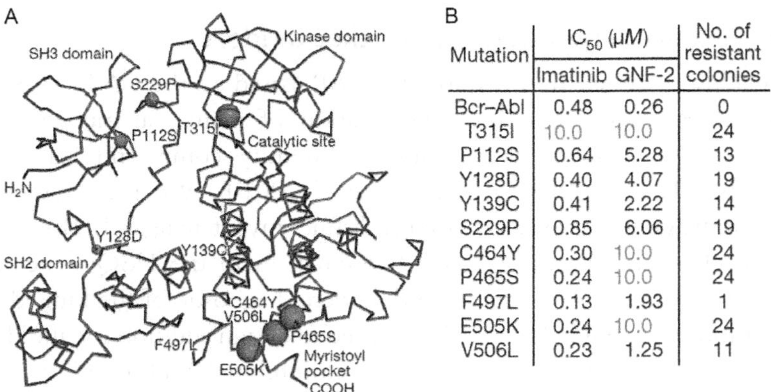

Mutation	IC_{50} (μM)		No. of resistant colonies
	Imatinib	GNF-2	
Bcr–Abl	0.48	0.26	0
T315I	10.0	10.0	24
P112S	0.64	5.28	13
Y128D	0.40	4.07	19
Y139C	0.41	2.22	14
S229P	0.85	6.06	19
C464Y	0.30	10.0	24
P465S	0.24	10.0	24
F497L	0.13	1.93	1
E505K	0.24	10.0	24
V506L	0.23	1.25	11

Figure 7.7 Screen and characterization of GNF-2 resistance mutations in Bcr–Abl. (A) Ribbon diagram of SH1, SH2, and SH3 domains showing location of GNF-5 resistance mutations (red balls, sized in proportion to the number of resistant clones obtained from the screen). The majority of the mutations are clustered in the myristate-binding pocket as expected but also at the gatekeeper residue (T315I). (B) Cell proliferation IC50s in micromolar for imatinib and GNF-2 for Ba/F3 cells engineered to express the Bcr–Abl point mutations recovered from the screen. Also indicated is the number of clones expressing a particular mutation recovered from the screen. (See the color plate.)

did not substantially change the affinity of GNF-5 for binding to ABL but completely blocked the ability of GNF-5 to downregulate BCR–ABL kinase activity. Indeed, H/D-exchange mass spectrometry demonstrated that T315I prevented the allosteric communication from the myristate to the ATP-binding site (Iacob et al., 2011). Thus, T315I is an allosteric resistance mutation to the GNF-5 class of myristate ligands (Fig. 7.7).

4.2. Combinations with ATP-competitive ligands

The observation of allosteric communication between the myristate and ATP-binding sites provided a natural impetus for exploring the efficacy of combined inhibition with ligands targeting both sites. Indeed, in both biochemical and cellular assays, GNF-5 displayed at least additive inhibitory activity with both nilotinib and dasatinib. The combination of GNF-5 with nilotinib significantly increased the survival of mice in bone-marrow transplantation CML models relative to treatment with either agent alone (Zhang et al., 2010). These studies helped to provide the motivation for developing further optimized myristate ligands that would exhibit superior-binding potency and "drug-like" properties with the hope of being able to overcome allosteric resistance incurred by mutations in the ATP site such as the gatekeeper T315I.

4.3. Second-generation myr-pocket binders

An extensive campaign to explore the SAR based on the GNF-2/-5 "lead compounds" resulted in the identification of many variants that were approximately equipotent but did not identify compounds capable of inhibiting T315I BCR–ABL (Deng et al., 2010). We next launched a new campaign to uncover structurally distinct myristate ligands. Based upon our mechanistic studies of GNF-2, resonances in the HSQC spectra of GNF-2 with ABL were identified that were diagnostic of the kinked conformation of the α-I helix. SDBB was used to design additional scaffolds that were capable of potently inducing the autoinhibited conformation of both wild-type and the recalcitrant T315I-ABL mutant (Fabbro et al., 2010). Some details of this medicinal chemistry effort have been published (Fabbro et al., 2010) and some are in recent published patent applications (Dodd et al., 2013; Furet et al., 2013), suggesting that these second-generation myr-pocket binders engage in additional interactions in and around the myristate site of Abl relative to GNF-2 and -5 resulting in potent activity toward the full-length (SH3SH2SH1)-wt-ABL and T315I-ABL in both biochemical- and cell-based assays (Fabbro et al., 2010). These second-generation myr-pocket inhibitors are among the most potent and specific kinase inhibitors known to date.

5. COMBINATIONS OF SECOND-GENERATION ATP-SITE INHIBITORS WITH SECOND-GENERATION myr-POCKET LIGANDS

The improved potency of second-generation myr-pocket binders against wt-Abl and T315-Abl also translated into synergy in Ba/F3 cells transformed with Bcr–ABL-T315I when combined ATP-site-directed inhibitors such as nilotinib or dasatinib as noted in previous studies (Zhang et al., 2010). A dynamic equilibrium between several c-Abl conformations that are modulated by both myr-pocket binders and ATP-site-directed inhibitors has been demonstrated using a combination of solution NMR and SAXS. These studies revealed the existence of an open state of the c-Abl in the absence of inhibitors. The addition of a catalytic site inhibitor such as imatinib induces a large structural rearrangement characterized by the detachment of the SH3–SH2 domains from the kinase domain and the formation of an "open" inactive state. Further addition of the myr-pocket binder induces conversion to a "closed" inactive state (Skora et al., 2013). These data may explain the unusual dose–response in cellular inhibition assays and may provide insight into the domain motions of the "full-length ABL" by the combined effect of the two inhibitor types, which appears to be able to overcome drug resistance. These findings on the allosteric actions of the two classes of inhibitors reveal molecular details of their recently reported synergy to overcome drug resistance and may help to devise new strategies for drug development.

5.1. Key lessons learned in the drug discovery of allosteric BCR–ABL inhibitors

1. We were initially misled by our biochemical assays that suggested that GNF-2 could not inhibit the enzymatic activity of SH1. An important lesson from this experience is that the biochemical analysis of allosteric inhibitors is the best done with as native and "full-length" an enzyme as possible in order to have access to diverse regulatory interactions.
2. This program initiated in 2001 and phase I studies with ABL001 are set to begin in 2014, thus 13 years elapsed in this drug discovery campaign. Had this project been considered as an ordinary pipeline drug discovery project

it would have been terminated numerous times due to the slow progress. However, given the importance of BCR–ABL as a target to the organization combined with the intellectual challenge that the research inspired among the key scientific drivers provided the environment where this project could be sustained. Elucidating the molecular mechanism of this class of ligand was required to effectively advance the project.

3. The mechanistic work in this project required a serious commitment of intellectual and labor in diverse disciplines including medicinal chemistry, pharmacology, biochemistry, biophysics, cell, and structural biology. Key to success was that we had key contributors in all these areas that provided leadership and insight required to advance the project.
4. Allosteric kinase inhibitors that perform their job far removed from the ATP site are highly selective.

ACKNOWLEDGMENTS

This was only possible with the diligent efforts of a highly interdisciplinary research team. Special thanks to Francisco Adrián, Jianming Zhang, Qiang Ding, Markus Warmuth, Taebo Sim, Anastasia Velentza, Christine Sloan, Yi Liu, Guobao Zhang, Wooyoung Hur, Sheng Ding, Paul Manley, Stephanie Dodd, Pascal Furet, Robert Grotzfeld, Darryl Jones, Andreas Marzinzik, Xavier Pelle, Bahaa Salem, Joseph Schoepfer, Jürgen Mestan, Doriano Fabbro, Wolfgang Jahnke, Sandra Cowan-Jacob, Allen Li, Roxana Iacob, John Powers, Christine Dierks, Fangxian Sun, Gui-Rong Guo, Barun Okram, Yongmun Choi, Amy Wojciechowski, Xianming Deng, Guoxun Liu, Gabriele Fendrich, Andre Strauss, Navratna Vajpai, Stephan Grzesiek, Tove Tuntland, Badry Bursulaya, Mohammad Azam, John Engen, Ellen Weisberg, and George Daley.

REFERENCES

Adrian, F. J., Ding, Q., Sim, T., Velentza, A., Sloan, C., Liu, Y., et al. (2006). Allosteric inhibitors of Bcr-abl-dependent cell proliferation. *Nature Chemical Biology*, *2*, 95–102.

Advani, A. S., & Pendergast, A. M. (2002). Bcr-Abl variants: Biological and clinical aspects. *Leukemia Research*, *26*, 713–720.

Buchdunger, E., Matter, A., & Druker, B. J. (2001). Bcr-Abl inhibition as a modality of CML therapeutics. *Biochimica et Biophysica Acta*, *1551*, M11–M18.

Cortes, J. E., Kantarjian, H. M., Brummendorf, T. H., Kim, D. W., Turkina, A. G., Shen, Z. X., et al. (2011). Safety and efficacy of bosutinib (SKI-606) in chronic phase Philadelphia chromosome-positive chronic myeloid leukemia patients with resistance or intolerance to imatinib. *Blood*, *118*, 4567–4576.

Cortes, J. E., Kim, D. W., Kantarjian, H. M., Brummendorf, T. H., Dyagil, I., Griskevicius, L., et al. (2012). Bosutinib versus imatinib in newly diagnosed chronic-phase chronic myeloid leukemia: Results from the BELA trial. *Journal of Clinical Oncology*, *30*, 3486–3492.

Deng, X., Okram, B., Ding, Q., Zhang, J., Choi, Y., Adrian, F. J., et al. (2010). Expanding the diversity of allosteric bcr-abl inhibitors. *Journal of Medicinal Chemistry*, *53*, 6934–6946.

Ding, S., Gray, N. S., Wu, X., Ding, Q., & Schultz, P. G. (2002). A combinatorial scaffold approach toward kinase-directed heterocycle libraries. *Journal of the American Chemical Society*, *124*, 1594–1596.

Dodd, S. K., Furet, P., Grotzfeld, R. M., Jones, D. B., Manley, P., Marzinzik, A., et al. (2013). *Benzamide derivatives for inhibiting the activity of ABL1, ABL2 and BCR-ABL1.* WO 2013/171639 A1.

Fabbro, D., Manley, P. W., Jahnke, W., Liebetanz, J., Szyttenholm, A., Fendrich, G., et al. (2010). Inhibitors of the Abl kinase directed at either the ATP- or myristate-binding site. *Biochimica et Biophysica Acta*, *1804*, 454–462.

Filippakopoulos, P., Muller, S., & Knapp, S. (2009). SH2 domains: Modulators of nonreceptor tyrosine kinase activity. *Current Opinion in Structural Biology*, *19*, 643–649.

Furet, P., Grotzfeld, R. M., Jones, D. B., Manley, P., Marzinzik, A., Moussaoui, S., et al. (2013). *Compounds and compositions for inhibiting the activity of ABL1, ABL2 and BCR-ABL1.* WO 2013/171641 A1.

Golas, J. M., Arndt, K., Etienne, C., Lucas, J., Nardin, D., Gibbons, J., et al. (2003). SKI-606, a 4-anilino-3-quinolinecarbonitrile dual inhibitor of Src and Abl kinases, is a potent antiproliferative agent against chronic myelogenous leukemia cells in culture and causes regression of K562 xenografts in nude mice. *Cancer Research*, *63*, 375–381.

Grebien, F., Hantschel, O., Wojcik, J., Kaupe, I., Kovacic, B., Wyrzucki, A. M., et al. (2011). Targeting the SH2-kinase interface in Bcr-Abl inhibits leukemogenesis. *Cell*, *147*, 306–319.

Huang, W. S., Metcalf, C. A., Sundaramoorthi, R., Wang, Y., Zou, D., Thomas, R. M., et al. (2010). Discovery of 3-[2-(imidazo[1,2-b]pyridazin-3-yl)ethynyl]-4-methyl-N-{4-[(4-methylpiperazin-1-y l)methyl]-3-(trifluoromethyl)phenyl}benzamide (AP24534), a potent, orally active pan-inhibitor of breakpoint cluster region-abelson (BCR-ABL) kinase including the T315I gatekeeper mutant. *Journal of Medicinal Chemistry*, *53*, 4701–4719.

Iacob, R. E., Pene-Dumitrescu, T., Zhang, J., Gray, N. S., Smithgall, T. E., & Engen, J. R. (2009). Conformational disturbance in Abl kinase upon mutation and deregulation. *Proceedings of the National Academy of Sciences of the United States of America*, *106*, 1386–1391.

Iacob, R. E., Zhang, J., Gray, N. S., & Engen, J. R. (2011). Allosteric interactions between the myristate- and ATP-site of the Abl kinase. *PLoS One*, *6*, e15929.

Jahnke, W., Grotzfeld, R. M., Pelle, X., Strauss, A., Fendrich, G., Cowan-Jacob, S. W., et al. (2010). Binding or bending: Distinction of allosteric Abl kinase agonists from antagonists by an NMR-based conformational assay. *Journal of the American Chemical Society*, *132*, 7043–7048.

Lombardo, L. J., Lee, F. Y., Chen, P., Norris, D., Barrish, J. C., Behnia, K., et al. (2004). Discovery of N-(2-chloro-6-methyl-phenyl)-2-(6-(4-(2-hydroxyethyl)-piperazin-1-yl)-2-methylpyrimidin-4-ylamino)thiazole-5-carboxamide (BMS-354825), a dual Src/Abl kinase inhibitor with potent antitumor activity in preclinical assays. *Journal of Medicinal Chemistry*, *47*, 6658–6661.

Mayer, K., Gielen, G. H., Willinek, W., Muller, M. C., & Wolf, D. (2014). Fatal progressive cerebral ischemia in CML under third-line treatment with ponatinib. *Leukemia*, *28*, 976–977.

Melo, J. V. (1996). The molecular biology of chronic myeloid leukaemia. *Leukemia*, *10*, 751–756.

Nagar, B., Bornmann, W. G., Pellicena, P., Schindler, T., Veach, D. R., Miller, W. T., et al. (2002). Crystal structures of the kinase domain of c-Abl in complex with the small molecule inhibitors PD173955 and imatinib (STI-571). *Cancer Research*, *62*, 4236–4243.

Nagar, B., Hantschel, O., Young, M. A., Scheffzek, K., Veach, D., Bornmann, W., et al. (2003). Structural basis for the autoinhibition of c-Abl tyrosine kinase. *Cell*, *112*, 859–871.

O'Hare, T., Walters, D. K., Stoffregen, E. P., Jia, T., Manley, P. W., Mestan, J., et al. (2005). In vitro activity of Bcr-Abl inhibitors AMN107 and BMS-354825 against clinically relevant imatinib-resistant Abl kinase domain mutants. *Cancer Research, 65*, 4500–4505.

Pakakasama, S., Kajanachumpol, S., Kanjanapongkul, S., Sirachainan, N., Meekaewkunchorn, A., Ningsanond, V., et al. (2008). Simple multiplex RT-PCR for identifying common fusion transcripts in childhood acute leukemia. *International Journal of Laboratory Hematology, 30*, 286–291.

Puttini, M., Coluccia, A. M., Boschelli, F., Cleris, L., Marchesi, E., Donella-Deana, A., et al. (2006). In vitro and in vivo activity of SKI-606, a novel Src-Abl inhibitor, against imatinib-resistant Bcr-Abl\+ neoplastic cells. *Cancer Research, 66*(23), 11314–11322. Epub 2006 Nov 17. PMID:17114238.

Skora, L., Mestan, J., Fabbro, D., Jahnke, W., & Grzesiek, S. (2013). NMR reveals the allosteric opening and closing of Abelson tyrosine kinase by ATP-site and myristoyl pocket inhibitors. *Proceedings of the National Academy of Sciences of the United States of America, 110*, E4437–E4445.

Warmuth, M., Kim, S., Gu, X. J., Xia, G., & Adrian, F. (2007). Ba/F3 cells and their use in kinase drug discovery. *Current Opinion in Oncology, 19*, 55–60.

Weisberg, E., Manley, P. W., Breitenstein, W., Bruggen, J., Cowan-Jacob, S. W., Ray, A., et al. (2005). Characterization of AMN107, a selective inhibitor of native and mutant Bcr-Abl. *Cancer Cell, 7*, 129–141.

Yang, J., Campobasso, N., Biju, M. P., Fisher, K., Pan, X. Q., Cottom, J., et al. (2011). Discovery and characterization of a cell-permeable, small-molecule c-Abl kinase activator that binds to the myristoyl binding site. *Chemical Biology, 18*, 177–186.

Zhang, J., Adrian, F. J., Jahnke, W., Cowan-Jacob, S. W., Li, A. G., Iacob, R. E., et al. (2010). Targeting Bcr-Abl by combining allosteric with ATP-binding-site inhibitors. *Nature, 463*, 501–506.

CHAPTER EIGHT

The Logic and Design of Analog-Sensitive Kinases and Their Small Molecule Inhibitors

Michael S. Lopez, Joseph I. Kliegman, Kevan M. Shokat
Howard Hughes Medical Institute and Department of Cellular & Molecular Pharmacology, University of California, San Francisco, California, USA

Contents

Abstract

Analog-sensitive AS Kinase technology allows for rapid, reversible, and highly specific inhibition of individual engineered kinases in cells and in mouse models of human diseases. The technique consists of two parts: a kinase containing a space-creating mutation in the ATP-binding pocket and a bulky ATP-competitive small molecule inhibitor that

Methods in Enzymology, Volume 548
ISSN 0076-6879
http://dx.doi.org/10.1016/B978-0-12-397918-6.00008-2

complements the shape of the mutant ATP pocket. This strategy enables dissection of phospho-signaling pathways, elucidation of the physiological function of individual kinases, and characterization of the pharmacology of clinical-kinase inhibitors. Here, we present an overview of AS technology and describe a stepwise approach for generating AS Kinase mutants and identifying appropriate small molecule inhibitors. We also describe commonly encountered technical obstacles and provide strategies to overcome them.

1. INTRODUCTION

1.1. Overview of analog-sensitive-kinase technology

Analog-sensitive (AS) Kinase technology is a chemical-genetic technique that enables systematic generation of highly specific inhibitors for individual kinases (Shokat et al., 2000). This approach combines the advantages of a small molecule inhibitor (temporal resolution, reversibility, etc.) with the specificity of a genetic perturbation. When a residue at a structurally conserved position in the kinase active site, termed the gatekeeper (Liu et al., 1999), is mutated from the natural amino acid (methionine, leucine, phenylalanine, threonine, etc.) to a residue bearing a smaller side chain (glycine or alanine), a novel pocket not found in wild-type (WT) kinases is created within the ATP-binding site. The engineered kinase is termed AS because it can be potently and specifically inhibited with ATP analogs containing a bulky substituent that complements the enlarged ATP-binding pocket (Bishop et al., 1999). AS Kinases are powerful tools for deciphering phospho-signaling networks, elucidating the cellular function of individual kinases, and characterizing the pharmacology of clinical therapeutics.

Since nearly all kinases contain a bulky gatekeeper residue, extending AS Kinase technology to the entire kinome is possible in principle. Toward this end, we developed a systematic approach for engineering functional AS Kinases and identifying suitable inhibitors from a small panel of rationally designed small molecules (Fig. 8.1A). This strategy has enabled the successful application of the AS technique to more than 80 kinases reported to date. Here, we describe the stepwise protocol for the identification of the gatekeeper residue in diverse kinases, the proper choice for gatekeeper substitution, and several strategies to rescue the activity of compromised AS Kinases (Fig. 8.1A–D). We also explain the logic of AS Kinase inhibitor design, describe the strengths and limitations of common AS Kinase inhibitors, and detail protocols for their synthesis and characterization. Throughout the text, we also review key applications of this method in order to highlight experiments in which this technique may prove most useful.

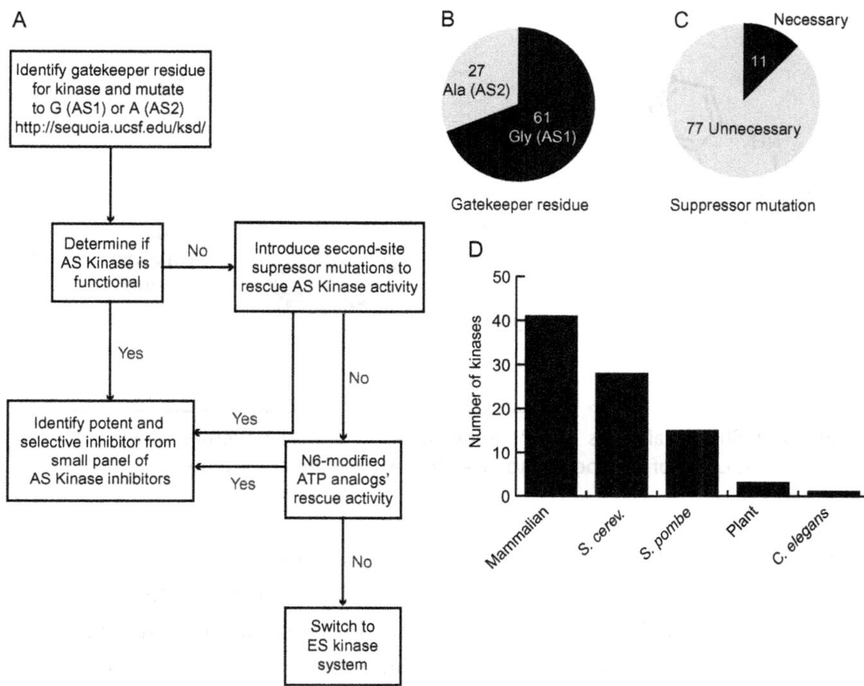

Figure 8.1 Overview of AS Kinase technology. (A) Flowchart outlines systematic approach for identifying functional AS Kinases. (B) Distribution of Ala and Gly gatekeeper residues used in reported AS Kinases. (C) Fraction of reported AS Kinases requiring second-site suppressor mutations. (D) Distribution of reported AS Kinases according to species. (See the color plate.)

1.2. The gatekeeper governs access to the ATP-binding pocket of protein kinases

The amino acid position analogous to c-Src T338 is called the gatekeeper because the size of this residue's side chain governs the volume and shape of the ATP-binding site and thereby determines the size of ATP-competitive ligands that can be accommodated. Kinases with small gatekeeper residues allow binding of molecules with large bulky groups such as the C3-tolyl ring of PP1 that cannot bind to kinases with larger gatekeepers such as methionine or phenylalanine. This phenomenon was first observed with PP1 (Fig. 8.2A), a potent ATP-competitive inhibitor of Src-family kinases (Hanke et al., 1996). Liu, Shah, Yang, Witucki, and Shokat (1998) used bioinformatics and site-directed mutagenesis to demonstrate that PP1 is capable of binding any kinase with a small amino acid residue (Thr, Val, Ala, Gly) at the position analogous to c-Src T338 (Liu et al.,

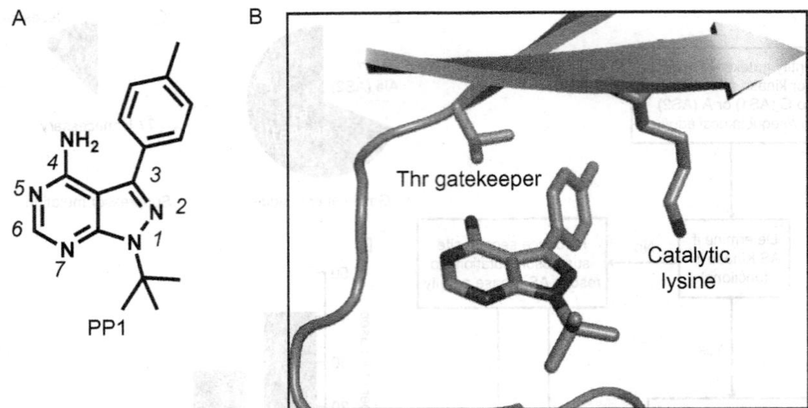

Figure 8.2 Structural basis of PP1 activity. (A) Chemical structure of PP1. (B) X-ray cocrystal structure of PP1 bound to Hck kinase. (See the color plate.)

1998). Importantly, Liu and Shokat recognized the potential for any kinase to be made sensitive to inhibitors such as PP1 by exchanging the native gatekeeper with a smaller amino acid by site-directed mutagenesis. They also noted that kinases sensitive to PP1 become resistant upon mutation of the gatekeeper residue to isoleucine, thereby predicting resistance mutations at the gatekeeper position that would later emerge in the clinic (Branford, Rudzki, Walsh, & Grigg, 2002).

Soon after the discovery of the gatekeeper and the characterization of PP1 selectivity by Liu and Shokat, the structure of PP1 bound to the Src-family kinase Hck showed the pyrazolopyrimidine core of PP1 mimics the adenine ring of ATP in its binding to the nucleobase pocket while the *p*-tolyl group at the C3 position projects into a deep hydrophobic cleft situated between the gatekeeper T338 and the catalytic lysine residue K295 (Fig. 8.2B; Schindler et al., 1999). Based on these structural features, it was predicted that PP1 analogs with an enlarged C3 substituent would suffer a steric clash with the WT gatekeeper residue while mutation of the gatekeeper to glycine or alanine, which are exceedingly rare at this position, would create space to accommodate the enlarged C3 substituent. Thus, the same phenomenon that results in the selectivity of PP1 for a subset of WT kinases can be used to design inhibitors with selectivity for kinases engineered to contain glycine or alanine gatekeepers. It has been demonstrated that potent and specific inhibitors can be readily identified for various AS Kinases by screening a small panel of PP derivatives with enlarged C3 substituents (Shokat et al., 2000).

2. CONSTRUCTING AS KINASES

The challenge of constructing a novel AS Kinase varies; however, the technique has become a robust and rapid method for kinase characterization and the steps to engineering an AS Kinase is now routine. The following steps outline a straightforward strategy to identify the gatekeeper residue by primary or structure-based sequence alignment, mutate the gatekeeper, and find a selective inhibitor analog from a small panel of molecules developed in the Shokat Lab. In cases where the AS Kinase mutant suffers compromised activity, there are standard second-site suppressor mutations that may restore kinase activity to the AS Kinase. Additionally, bulky ATP analogs are more efficiently used by AS Kinases and can be employed to restore the activity of hypomorphic AS Kinase mutants in cells. Finally, a complementary technique that relies on covalent targeting of a cysteine gatekeeper kinase (termed an electrophile-sensitive (ES) kinase) offers the unique experimental advantages of irreversible inhibition.

2.1. Identifying the gatekeeper residue

As illustrated in Fig. 8.1A, there is a systematic approach for engineering functional AS Kinases. The first step is to correctly identify the gatekeeper residue of the kinase of interest (KOI). This can be done by submitting the amino acid sequence of the KOI-kinase domain to the Kinase Database at http://sequoia.ucsf.edu/ksd/. Alternatively, the amino acid sequence of the kinase domain can be aligned with c-Src and other protein kinases to identify the position equivalent to c-Src T338 (Fig. 8.3). Typically, the gatekeeper residue is preceded by two hydrophobic amino acids (I336 and V337 in c-Src) and followed by an acidic residue (E339 in c-Src) and another hydrophobic amino acid (Y340 in c-Src). While atypical kinases may require additional structural information to accurately identify the gatekeeper position, this simple sequence alignment is usually sufficient for canonical tyrosine and serine/threonine protein kinases.

To identify the gatekeeper by structural alignment, you must first download appropriate protein-modeling software such as MODELLER (http://salilab.org/modeller/). Then, identify the structure that will be used as a template to model your KOI. This may be done with an integrated feature of MODELLER that searches the PDB (rcsb.org) for solved structures with sequence homology to your KOI. If this feature is used, the quality of the homology can be evaluated with an integrated script that outputs statistics

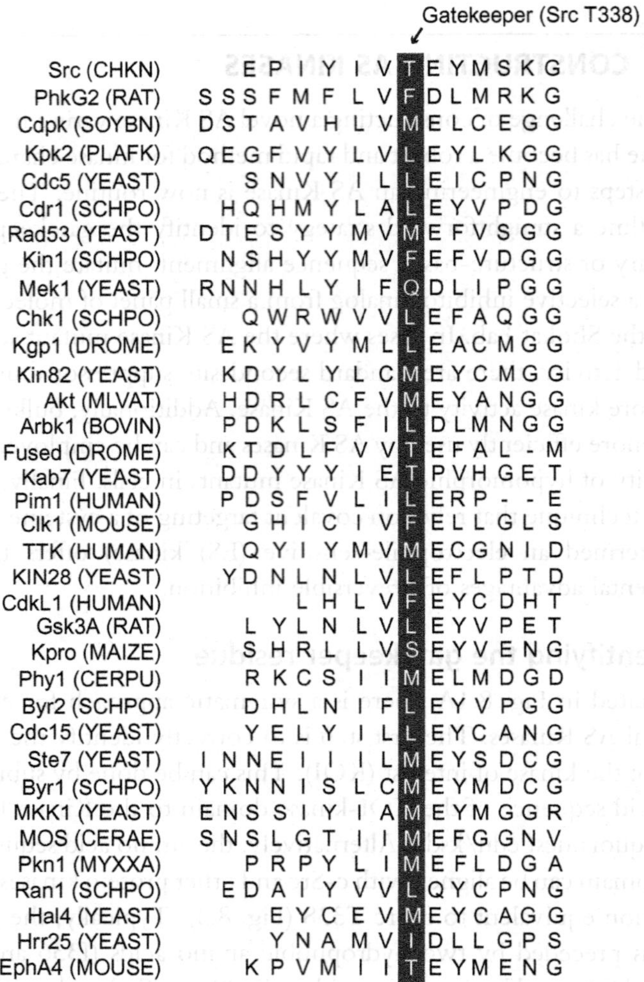

Figure 8.3 Representative alignment of the hinge region of kinase domains from diverse organisms and kinase families. (See the color plate.)

that inform the choice of template. Alternatively, the phylogenetic tree of kinases may be used to determine the most related kinase with a published structure. Once the appropriate template is selected, model building is a simple two-step process of (1) aligning your KOI to the published structure and (2) building the model. The commands for accomplishing these steps are clearly described (http://salilab.org/modeller/tutorial/basic.html). Thus, structural alignment using MODELLER allows for the identification of gatekeeper residues in atypical kinases with low sequence conservation.

2.2. Second-site suppressor mutations

A challenge that is sometimes encountered in the construction of AS Kinases is the loss of activity that results from mutation of the gatekeeper to glycine or alanine. The gatekeeper residue of protein kinases contributes to the hydrophobic spine of the kinase domain and plays an important role in promoting enzymatic activity (Azam, Seeliger, Gray, Kuriyan, & Daley, 2008). Thus, mutation of this position to a smaller amino acid may disrupt this structural feature and lead to a reduction in the catalytic efficiency of the AS Kinase. For example, mutation of the budding yeast kinase Cdc28 from the native phenylalanine gatekeeper to glycine results in a 10-fold reduction in ATP affinity and a sixfold reduction in k_{cat} (Shokat et al., 2000). Fortunately, this reduced activity is often acceptable and does not result in biological defects. The activity of the kinase may be easily evaluated by testing the ability of the AS allele to complement loss of the WT gene. In cases where compromised kinase activity is prohibitive, our lab has developed a series of steps to remedy this limitation.

The glycine gatekeeper mutation (termed AS1) creates the maximal expansion of the ATP-binding pocket and typically maximizes the difference in sensitivity between AS and WT alleles to AS Kinase inhibitors. However, if the AS1 kinase suffers a prohibitive reduction in activity, the alanine gatekeeper (termed AS2) can be employed to increase enzyme activity while retaining sufficient sensitivity to AS Kinase inhibitors (Fig. 8.1B). In cases where the KOI is intolerant of both glycine and alanine gatekeepers second-site suppressor mutations may be introduced to restore AS Kinase activity (Fig. 8.1C). Mutation of residues in the antiparallel β-sheet of the N-terminal lobe of the kinase domain to branched aliphatic residues such as valine or isoleucine can restore activity in the context of the AS gatekeeper (Zhang et al., 2005). For example, Cdc5 and MEKK1 contain cysteine residues in place of the typical valine at the position analogous to Src V284. Mutation of this position to valine restores the activity of both Cdc5-AS2 and MEKK1-AS1 to near-WT levels.

In certain cases, the activity of an impaired AS Kinase can be rescued by rational substitution of residues that are not conserved in closely related but well-behaved AS Kinases. For example, Niswender et al. observed that the homologous kinases PKAα and PKAβ have differential tolerance for the gatekeeper mutation (Niswender et al., 2002). PKAα-AS2 (M120A) retained greater than 75% activity relative to PKAα-WT, while PKAβ-AS1 (M120G) and PKAβ-AS2 (M120A) retained only 10% and 40% of

PKAβ-WT activity, respectively. The authors observed that most amino acid differences between the two kinases were relatively conservative substitutions with the exception of position 46, an isoleucine residue in PKAα and a lysine in PKAβ. Thus, combining the gatekeeper mutation M120A with the K46I substitution yields an AS PKAβ mutant that retains greater than 75% activity. As the number of AS Kinases described in the literature continues to grow, they will increasingly be used as a roadmap for indentifying mutations that rescue the activity of kinases intolerant of alanine and glycine gatekeepers.

2.3. Unnatural ATP analogs rescue enzyme activity

An effect of the smaller gatekeeper residue is a reduction in affinity of the kinase for its substrate ATP. Thus, introduction of an ATP surrogate with enhanced affinity for the expanded ATP-binding pocket rescues the activity of impaired AS Kinases. AS Kinases have enhanced catalytic activity when ATP analogs bearing hydrophobic N6 substituents complementing the gatekeeper pocket are used in the place of ATP (Shah & Shokat, 2002). Recently, Merrick et al. (2011) used this strategy to rescue the hypomorphic Cdk2-AS mutant in human epithelial cells. In this case, the gatekeeper mutation resulted in a structural defect that disrupted the proper binding of Cdk2 to one of its activating subunits cyclin A. The addition of 3MB-PP1 corrected the structural defect as evidenced by proper binding of cyclin A to Cdk2, but inhibited kinase activity as one would expect. In contrast, the addition of N6-benzylaminopurine (6-BAP) rescued cyclin A binding but also retained kinase activity as evidenced by unperturbed cell proliferation. Importantly, the cellular activity of 6-BAP suggests that this nucleobase is converted into the active nucleotide in cells. This hypothesis is supported by work showing conversion of a related N6-modified nucleobase kinetin into kinetin tri-phosphate in HeLa cells (Hertz et al., 2013).

2.4. Cysteine gatekeeper alternative

In extreme cases where glycine and alanine residues are not tolerated at the gatekeeper position, our lab has developed an approach that utilizes a cysteine gatekeeper (Garske, Peters, Cortesi, Perez, & Shokat, 2011). This engineered kinase is termed an ES kinase and relies on inhibitors that target the cysteine gatekeeper through covalent chemistry. The cysteine gatekeeper has a larger, nonpolar side chain that maintains the integrity of the hydrophobic spine and thereby retains greater enzymatic activity. For example, mutation of the threonine gatekeeper of c-Src to glycine reduces the catalytic efficiency from 4.99 to 0.59 min/μM, while mutation to cysteine retains robust activity of 8.34 min/μM (Garske et al., 2011). The expansion of the ATP-binding pocket of ES kinases is minimal; however, the unique reactivity of the cysteine thiol side chain allows selective covalent targeting of the gatekeeper with inhibitors bearing appropriate electrophiles. Like glycine and alanine, cysteine gatekeepers are exceedingly rare in the human kinome thereby reducing the chance of off-target activity of ES-kinase inhibitors.

The irreversible nature of covalent inhibitors also provides unique experimental advantages. For example, an ES kinase might be titrated with inhibitor to identify the precise activity threshold required for a specific function. Blair et al. (2007) used a similar approach to determine the quantitative relationship between EGFR stimulation by its ligand EGF and its downstream-signaling outputs (Blair et al., 2007). Another exciting advantage of this technique is the potential to target two different engineered kinases with orthogonal inhibitors. Taken together, glycine, alanine, and cysteine gatekeeper residues can be used in combination with appropriate second-site suppressor mutations to engineer nearly any kinase into a functional enzyme that is sensitive to specific small molecule inhibition.

3. AS KINASE INHIBITORS

Once a functional AS Kinase is generated, the next step is to identify a suitable inhibitor from a small panel of molecules specifically designed to have broad generality toward diverse AS Kinases and minimal activity toward WT kinases or other cellular proteins. There are two major classes of AS Kinase inhibitors: PPs are derived from the semipromiscuous kinase inhibitor PP1 and staralogs are derivatives of the pan-kinase inhibitor staurosporine. The only (ES-kinase inhibitors reported to date are based on the PP and quinazoline scaffolds) (Garske et al., 2011). A new class of inhibitors based on BEZ235 that target AS mutants of PI3-like kinases in yeast have recently been described (Kliegman et al., 2013). These molecules function based on the same logic as staralogs and PP analogs with a larger hydrophobic moiety creating a steric clash with WT PI3-like kinases, but binding to the engineered kinase with an expanded affinity pocket due to a size reduction of the gatekeeper residue. Here, we describe the limitations and advantages of each class of inhibitors and provide protocols for their synthesis and use.

3.1. Pyrazolo[3,4-*d*]pyrimidine inhibitors

Initially, 1NA-PP1 and 1NM-PP1 (Fig. 8.4A) were identified as the most efficacious AS Kinase inhibitors and are the most commonly used in peer-reviewed publications. These molecules also have the advantage that they are now commercially available from several companies. However, one limitation of 1NA-PP1 is that it inadvertently inhibits a small set of WT human kinases (Zhang et al., 2013). 1NM-PP1 is more selective but lacks potency against some AS Kinase. To develop molecules that overcome these limitations we recently reported the development of 3-substituted benzyl PPs (Fig. 8.4A) with improved selectivity and potency for AS Kinase. These new molecules have found widespread utility and allowed the application of the AS technique to 10 kinases that were previously recalcitrant to inhibition with 1NA-PP1 and 1NM-PP1 (Zhang et al., 2013).

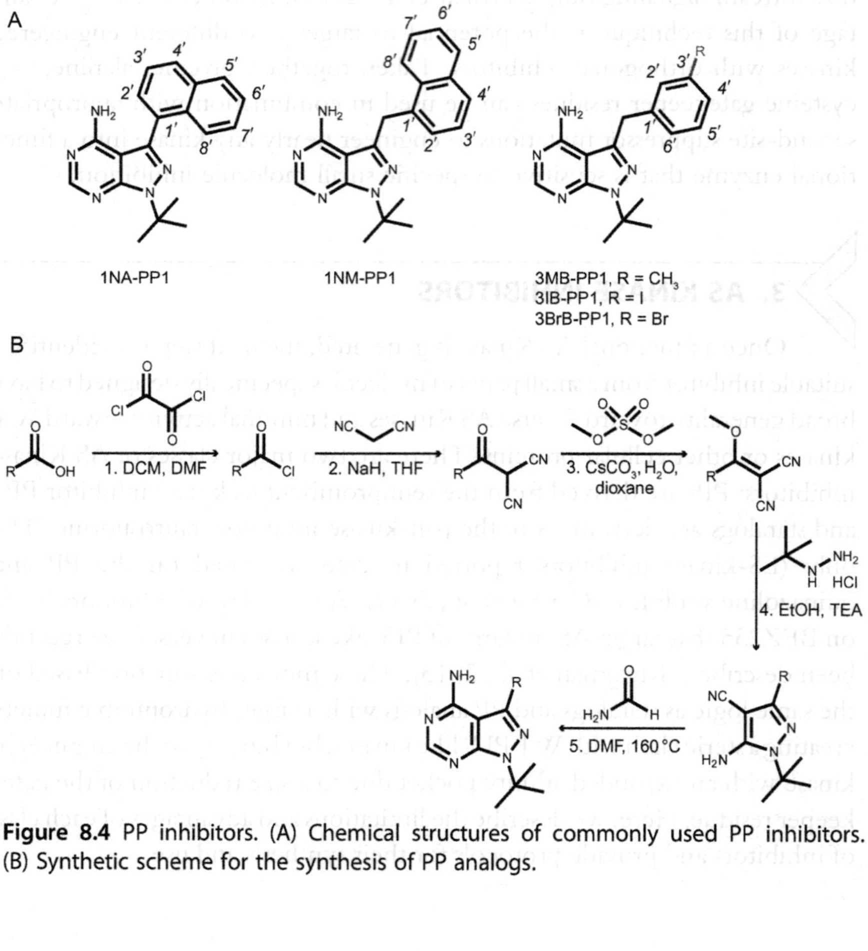

Figure 8.4 PP inhibitors. (A) Chemical structures of commonly used PP inhibitors. (B) Synthetic scheme for the synthesis of PP analogs.

3.2. Synthesis of PP inhibitors

Pyrazolo[3,4-*d*]pyrimidine (PP) inhibitor synthesis protocol (Fig. 8.4B):

(1) Dissolve 1 equiv. of the appropriate carboxylic acid in dry CH_2Cl_2 (0.2% DMF) in a clean dry, round-bottom flask (rbf) under argon (Ar). Slowly add 3 equiv. oxalyl chloride (dissolved 1:4 in CH_2Cl_2) over 15 min via an addition funnel, and allow the reaction to stir for 1 h. Remove the solvent and volatiles *in vacuo* and carry the crude concentrate to the next step.

(2) Dissolve 1 equiv. of malononitrile in THF, cool to 0 °C, and add 2 equiv. of sodium hydride. Next, redissolve the crude product 1 in a minimal volume of THF and add to the stirring reaction. Allow the reaction to warm to RT and proceed for ~30 min. Concentrate the reaction and carry the crude product on to the next step.

(3) Dissolve crude intermediate 2 in 1:1 dioxane:H_2O then add 8 equiv. $CsCO_3$ and 7 equiv. dimethylsulfate. Heat the reaction to 70 °C for 4 h, cool to RT, and allow it to proceed overnight. Dilute the reaction into brine and extract with several portions of EtOAc. Combine the organics, dry with $MgSO_4$, filter, and concentrate. Triturate the yellow concentrate with diethyl ether to yield a solid.

(4) Combine 1 equiv. of intermediate 3, 1 equiv. *t*-butyl hydrazine HCl, and 2 equiv. TEA in EtOH and allow the reaction to proceed for 1 h at RT. Concentrate and purify by SiO_2 chromatography ($CH_2Cl_2 \rightarrow$ 9:1 CH_2Cl_2:MeOH).

(5) Dissolve intermediate 4 in formamide in a clean, dry rbf equipped with water-cooled condenser and purge with Ar gas. The solid may not dissolve completely until heated. Heat the reaction at 165 °C overnight. Allow the reaction to cool to RT then transfer it dropwise to ice-cold water whereupon the product should precipitate. Purify by C18 HPLC (H_2O + 0.1% formic acid $\rightarrow$ 1:4 H_2O:MeCN + 0.1% formic acid).

3.3. Staralog inhibitors

Staralog inhibitors are C7-substituted indolocarbazole kinase inhibitors (Lopez et al., 2013). The primary advantage of these molecules is their high level of selectivity. For example, 1 μ*M* Star 12 retains potency against most AS Kinases but does not inhibit greater than 60% activity of a single WT kinase *in vitro* (308 tested). A similar analog, Star 17, was found to be the ideal inhibitor for targeting EphA4-AS1 in cells; this is because EphA4 is inherently sensitive to PP-based inhibitors thereby preventing complete inhibition of EphA4-AS1 without also inhibiting WT activity. In contrast, 10 μ*M* Star 12 is able to completely inhibit EphA4-AS1 without affecting the activity of EphA4-WT in 293T cells.

3.4. Synthesis of staralog inhibitors

Staralog inhibitor synthesis protocol (Fig. 8.5):

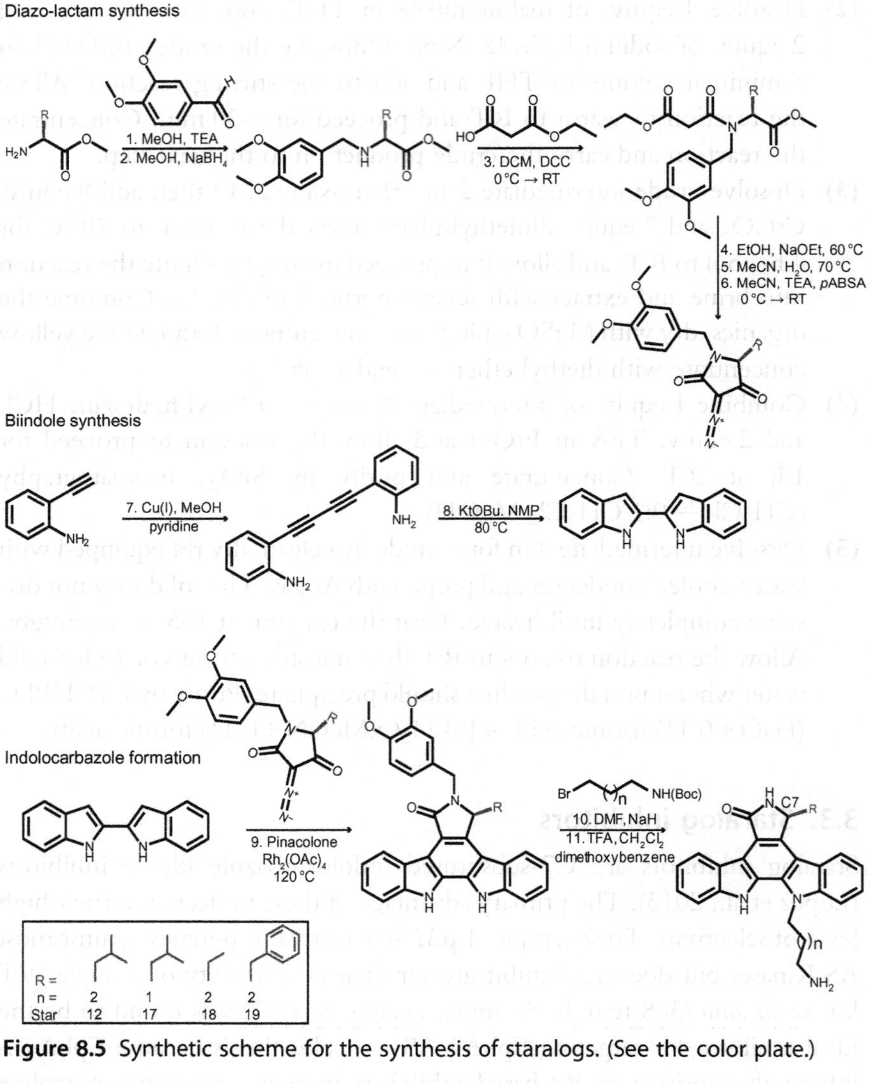

Figure 8.5 Synthetic scheme for the synthesis of staralogs. (See the color plate.)

(1) Place appropriate amino acid methyl ester, dimethoxybenzaldehyde, and dry methanol into a clean, dry rbf-containing Teflon stir. Add dry triethylamine (TEA) and stir at room temperature for 4–12 h.

(2) Add $NaBH_4$ in small portions over a period over 1 h. This reaction is exothermic and tends to boil over if $NaBH_4$ is added too quickly. After $NaBH_4$ addition is complete, stir for 1 h at RT, concentrate, redissolve concentrate in EtOAc, and wash with several portions of aqueous $NaCO_3$ in a separation funnel. Dry the organics with $MgSO_4$, filter, and concentrate. Redissolve concentrate in EtOAc, and purify by SiO_2 chromatography (hexanes → 1:1 hexanes:EtOAc).

(3) In a clean, dry rbf, dissolve ethyl hydrogen malonate and purified product from step 2 in CH_2Cl_2 and cool to 0 °C with stirring. In a separate vial, dissolve DCC in a small volume of CH_2Cl_2 at RT. Add the DCC/CH_2Cl_2 solution to the stirring, cooled reaction dropwise. The DCU byproduct should precipitate with 30 s. Allow the reaction to warm to room temperature and stir for 1 h. Remove DCU by filtration through a plug of celite, concentrate, redissolve in a minimal volume of CH_2Cl_2, and filter once more. Remove solvent *in vacuo* to yield a yellow oil substance. Redissolve oil in EtOAc and purify by SiO_2 chromatography (hexanes → 1:1 hexanes/EtOAc).

(4) Wash ~1.2 equiv. of sodium metal (in mineral oil) with hexanes and place in a preweighed glass vial. Quickly evaporate the residual hexanes with a stream of Ar gas, close vial screwcap, and record mass of sodium. Prepare a fresh solution of sodium ethoxide by dissolving 1.2 equiv. of sodium metal in dry ethanol. This reaction is exothermic and evolves hydrogen gas. In a separate vial, dissolve purified product 3 in a small volume of dry ethanol and add dropwise to the stirring solution of sodium ethoxide. Fit the reaction vessel with a water-cooled condenser and stir in 60 °C oil bath for 10 min, allow reaction to cool to RT, and remove ethanol *in vacuo* to yield a white solid. Suspend the white solid in 1 *M* HCl and extract with three portions of CH_2Cl_2. Combine the organic portions and remove solvent *in vacuo* to yield an orange colored oil. Carry the crude reaction on to the next step.

(5) Dissolve the crude product 4 in 95:5 MeCN:H_2O in a rbf fitted with a water-cooled reflux condenser and heat at 70 °C for 1 h. Allow the reaction to cool to RT then remove solvent *in vacuo*. Carry the crude concentrate on to the next step.

(6) Dissolve the crude product 5 and 1.1 equiv. of *p*ABSA in MeCN and cool to 0 °C. Add 1.1 equiv. of TEA and continue to stir reaction until ice melts and the reaction warms to RT (2–3 h). The solution should change color from pale yellow to deep red-orange upon the addition of TEA. Remove the solvent *in vacuo*, redissolve in 0.1 *M* NaOH, and extract with several portions of EtOAc. Combine the organic portions and concentrate to yield a dark red solid. Redissolve solid in EtOAc and purify by SiO_2 chromatography (hexanes → 3:7 hexanes/EtOAc) to yield an orange, gummy solid.

(7) Dissolve 2-alkynylaniline in 1:1 pyridine/MeOH and add 1 equiv. of $Cu(OAc)_2$. Stir the reaction overnight, remove MeOH *in vacuo*, and partition between H_2O and EtOAc. Remove the aqueous layer and wash with three more portions of 0.1 *M* HCl, followed by three portions of 1 *M* copper sulfate (to remove pyridine) and another three portions of H_2O. Concentrate the organics and carry the crude product on to the next step.

(8) Dissolve the crude product 7 in NMP, add KOtBu, and stir the reaction at 80 °C overnight. Dilute the reaction into EtOAc and wash with 0.1 *M* NaCl, followed by several portions of H_2O. Remove the organic solvent *in vacuo*, purify by recrystallization from EtOAc.

(9) Combine purified products 6 and 8 in a clean, dry-sealed reaction vessel and dissolve in freshly distilled pinacolone. Thoroughly degas this solution by purging under Ar or N_2. Add 0.1 equiv. of $Rh_2(OAc)_4$ and degas once more, then seal reaction vessel and move to 120 °C oil bath for 2–3 h. Allow reaction to cool to RT then remove pinacolone *in vacuo*. Redissolve concentrate in EtOAc and wash with 0.1 *M* NaCl and several portions of H_2O. Retain the organic portion, dry with $MgSO_4$, filter, and concentrate. Redissolve the concentrate in EtOAc and purify by SiO_2 chromatography (0–100% EtOAc in hexanes).

(10) Dissolve purified product 9 in dry DMF and cool to 0 °C. Add 1.2 equiv. of NaH and allow reaction to proceed for ~10 min at 0 °C. Add appropriate alkyl halide and allow the ice to melt and the reaction to warm to RT. If reaction progress stalls after several hours, dilute reaction into water, extract with EtOAc, dry, concentrate, and resubject. Once reaction is complete, dilute in H_2O, extract with several portions of EtOAc, dry with $MgSO_4$, filter, and concentrate. Carry crude concentrate on to the next step.

(11) Dissolve crude product 10 in a solution of 95:5 TFA:H_2O and 100 equiv. of dimethoxybenzene precooled to 0 °C. Allow the reaction to warm to RT then stir for an additional 1 h. Remove the solvent *in vacuo* and purify by C18 High-Performance Liquid Chromatography with a gradient of 20% MeCN → 100% MeCN in H_2O + 0.1% formic acid.

3.5. ES-kinase inhibitors

While AS Kinase inhibitors such as 1NA-PP1 project a bulky nonpolar group that complements the shape of the glycine or alanine gatekeeper pocket, ES-kinase inhibitors project groups that complement the chemical reactivity of the cysteine gatekeeper. To date, the only ES-kinase inhibitors reported are based on the PP and quinazoline scaffolds (Garske et al., 2011). It is important to remember that ES-kinase inhibitors are sensitive to thiol nucleophiles such as β-mercaptoethanol and dithiothreitol and should not be stored with or unnecessarily exposed to high concentrations of these reagents (Fig. 8.6).

3.6. Synthesis of ES-kinase PP inhibitors

The ES-kinase inhibitor 3VS-PP1 is closely related to other PP inhibitors in structure and synthetic design. A procedure similar to that described in Section 3.3 can be used to prepare the 3-nitrophenyl PP intermediate. The following protocol outlines the conversion of this intermediate to the final compound 3VS-PP1.

3VS-PP1 inhibitor synthesis protocol:

(1) Dissolve 1 equiv. of 3-nitrophenyl PP and 30 equiv. of zinc dust (<10 μm) in a clean, dry rbf and purge with Ar gas. Add 20 equiv of glacial acetic acid and stir reaction overnight. Filter the reaction through a plug of celite, dilute into sat. $NaHCO_3$, and extract with several portions of EtOAc. Combine the organics, dry with $MgSO_4$, filter,

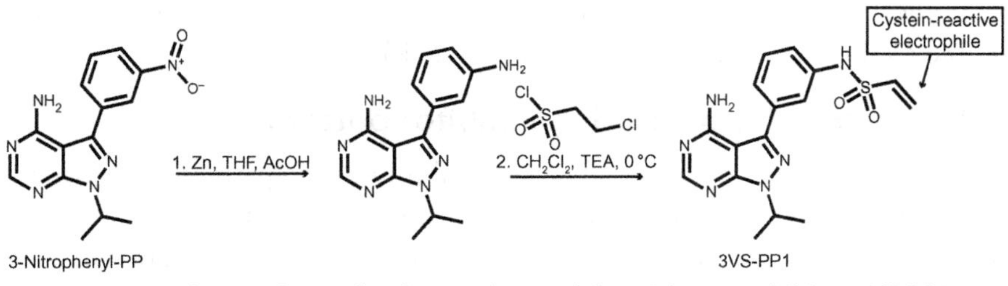

Figure 8.6 Synthetic scheme for the synthesis of the ES-kinase inhibitor 3VS-PP1.

concentrate, and purify by SiO_2 chromatography (100% $CH_2Cl_2 \rightarrow$ 9:1 CH_2Cl_2:MeOH).

(2) Dissolve 1.1 equiv. of intermediate 1 and 1.2 equiv. of TEA in dry CH_2Cl_2 and cool to 0 °C. Add 1 equiv. 2-chloro-1-ethane sulfonylchloride and allow the reaction to proceed for 1 h at RT. Quench reaction by dilution into sat. $NaHCO_3$ and extract with three portions of CH_2Cl_2. Combine the organic portions, dry with $MgSO_4$, filter, and concentrate. Purify by C18 HPLC (H_2O + 0.1% formic acid $\rightarrow$ 1:1 H_2O:MeCN + 0.1% formic acid).

3.7. AS Kinase inhibitors for PI3-like kinases

For PI3-like kinases, structural consensus is emerging that a functionalized quinoline scaffold can act as an effective hinge-binding element in addition to projecting into the pocket toward the gatekeeper residue. Kliegman et al. found that in at least two cases in *Saccharomyces cerevisiae* (TOR2-as2 and MEC1-as2), commercially available BEZ235 was selective for the AS Kinase over the WT enzyme in cells (Kliegman et al., 2013). QL-IX-55 is similar in structure to BEZ235 but has a smaller amino pyridine substituent in place of the quinoline group. This difference in substituent size explains why BEZ235 but not QL-IX-55 forms a steric clash with the yeast TOR2-WT gatekeeper and is selective for the AS kinase. In contrast, QL-IX-55 is a potent inhibitor of TOR1-WT and TOR2-WT (Hsieh et al., 2012) (Figure 8.7). This BEZ-based scaffold represents a new structural class of inhibitors of engineered PI3-like kinases that should be explored further.

In contrast to yeast, PI3 kinase-like kinases in mammals are intolerant of the typical gatekeeper mutation, rendering mTOR catalytically inactive. However, the nontraditional gatekeeper mutant (Y2225A) retains activity and is potently inhibited by typical PP1-based inhibitors. While it was formerly understood that PI3-like kinases were functionally inaccessible using the AS Kinase strategy (Alaimo & Shokat, 2001), recent work has shown that expanding the selection of analogs as well as creative alternatives to the canonical gatekeeper residue can be employed to make these proteins amenable to the technique (Kliegman et al., 2013).

3.8. Protocol for measuring inhibitor potency

In order to identify the most efficacious inhibitor for a particular AS Kinase, a simple assay should be devised to measure the potency and selectivity of a panel of inhibitors. If the purified WT and AS Kinases are available, it may be ideal to carry out this experiment *in vitro* using a peptide or protein substrate and ^{32}P-ATP. The half-maximal inhibitory concentration

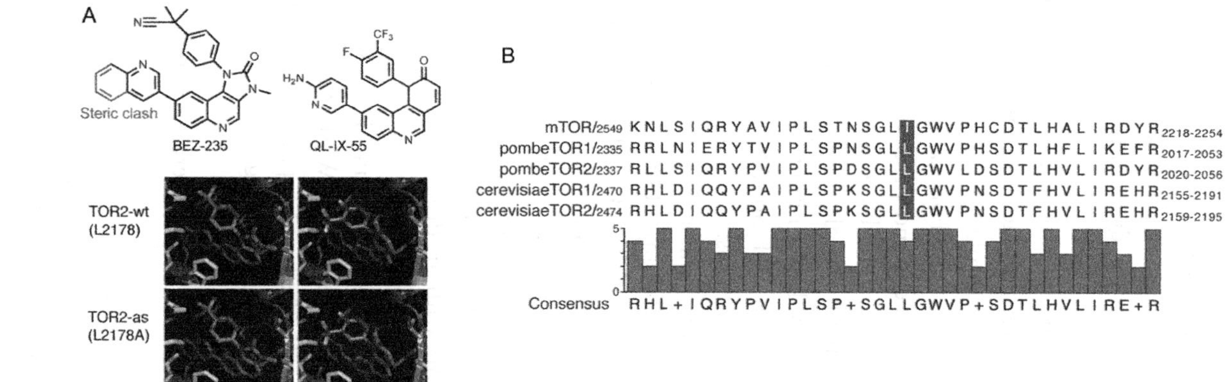

Figure 8.7 (A) Homology model of mTOR based on the structure of PI3Kγ shown with the gatekeeper residue in gray. The known *S. cerevisiae* TOR1/TOR2 inhibitor QL-IX-55 (purple) and BEZ235 (magenta) were oriented based on a typical H-bonding interaction with the backbone carbonyl of valine in the active site at VanDer Waals distances away from other residues that form the ATP-binding pocket. The isoleucine gatekeeper clash with BEZ235 is exacerbated by mutation to leucine and alleviated by mutation to alanine. The smaller QL-IX-55 does not sense this residue. (B) Sequence alignment shows that the gatekeeper residue (in purple) is isoleucine in mTOR and leucine in all other cases. The active site is highly conserved. (See the color plate.)

(IC_{50}) of each molecule can be determined by measuring the inhibitor dose-dependent activity of the kinase. Typically, appropriate AS Kinase inhibitors have *in vitro* IC_{50} values below 100 n*M* for the AS mutant and above 1 μ*M* for the WT kinase. The advantage of this approach is that precise IC_{50} values can be measured for several inhibitors in one experiment. For example, one 96-well plate can accommodate eight concentrations of four inhibitors in triplicate. The major disadvantage is that many kinases are difficult to express and purify.

If the purified proteins are unavailable, an alternative way to measure kinase inhibition is by overexpressing the WT and AS Kinases in cells and analyzing for substrate protein phosphorylation by western blot or measuring an appropriate phenotypic readout. For example, many kinases can be transiently transfected into 293T cells and probed for substrate or autophosphorylation using commercially available antibodies. One disadvantage of this strategy is that it may difficult to scale this experiment to allow measuring dose–response curves for multiple inhibitors.

Protocol for IC_{50} determination with Src-AS1 *in vitro*:

1. Prepare the following three solutions:

   2.5 × Reaction mix (5 n*M* Src kinase, 125 m*M* Tris (pH 8.0), 25 m*M* $MgCl_2$, 0.25 mg/mL BSA, 250 μ*M* substrate peptide (IYGEFKKK)),
   2.5 × ATP mix (250 μ*M* ATP, 0.2 μCi ^{32}P ATP), and
   5 × Inhibitor dilution in 10% DMSO.
2. In a 96-well plate, combine 10 μL of reaction mix with 5 μL drug dilution for each reaction, centrifuge to ensure removal of air bubbles, and incubate at room temperature for 15 min.
3. Combine 10 μL of ATP mix with the drug dilution/reaction mix to initiate the kinase reaction.
4. Quench 3 μL of the reaction by spotting onto P31 ion exchange paper and quickly evaporating under heat lamp. Wash the paper 3 × with 1% phosphoric acid and dry under the heat lamp. Expose dried P31 paper to a phosphorimaging screen and analyze using a Typhoon FLA 9500 (GE Healthcare Life Sciences).
   - It is preferable to quench samples at several time points (15, 30, and 45 min).
5. Quantify the signal of each reaction spot using imaging software such as ImageQuant (GE Healthcare Life Sciences), plot values as a function of log(drug concentration), and fit to a sigmoid to determine the IC_{50} values.

Protocol for testing inhibitors against EphA4-AS1 in 293T cells:

1. Plate 0.3×10^6 293T cells in 2 mL DMEM (10% FBS) into each well of a 6-well tissue culture plate and grow cells 12–18 h at 37 °C in 5% CO_2.
2. Transfect 0.1 µg of pCS2+ plasmid-bearing full-length EphA4-WT or EphA4-AS1 using lipofectamine®, LTX, and PLUS reagent (Life Technologies) and return cells to 37 °C 5% CO_2 incubator for 12–18 h.
3. Remove medium and replace with 1.5 mL of fresh DMEM (10% FBS) containing inhibitor and a final concentration of 1% DMSO. Incubate for 1–2 h at 37 °C 5% CO_2.
   - Typical concentration ranges (10 µ*M* to 39 n*M*, fourfold dilutions).
4. Remove drug medium and wash cells with cold PBS.
5. Add 150 µL cold lysis buffer, incubate on ice for 10 min, and pipette up and down several times to ensure complete lysis.
   - Lysis buffer: 50 m*M* Tris (pH 7.4), 150 m*M* NaCl, 1 m*M* EDTA, 1 m*M* EGTA, 1 m*M* Na_3VO_4, 10 m*M* sodium-β-glycerophosphate, 1% triton, 50 m*M* NaF, 5 m*M* sodium pyrophosphate, 0.27 *M* sucrose, 50 m*M* benzamidine, one complete mini protease inhibitor tablet (Roche), 1 m*M* PMSF, one PhosSTOP (Roche) phosphatase inhibitor tablet, 0.1 mg/mL RNAse A, and 0.1 mg/mL DNAse I.
6. Centrifuge lysates to precipitate insoluble debris then transfer the soluble lysate to fresh tubes.
7. Normalize sample protein concentration, separate lysate proteins by SDS–PAGE, and transfer to a nitrocellulose membrane.
8. Probe cells for phopsho-Eph and total EphA4 using appropriate primary antibodies and IRDye® 800 secondary antibodies (LiCOR). Image and quantify results using a LI-COR Odyssey Quantitative Imaging System and plot the ratio of pEph/EphA4 as a function of drug concentration. EC_{50} values can be determined by plotting the pEph/EphA4 ratio as a function of the log(drug concentration) and fitting the data to a sigmoid function using graphing software.

4. AS KINASES IN CELLS

AS technology is most commonly used to study kinase-signaling pathways in single-cell eukaryotes, such as yeast, or in cell lines derived from multicellular organisms such as mice and humans. This allows ease of genetic manipulations and avoids the complication of pharmacokinetics (PK) and pharmacodynamics (PD) that are encountered when using small molecules in intact multicellular organisms. The advantages of the approach are most apparent *in vivo* since the technique permits rapid and tunable interrogation of catalytic function that is not possible using genetic manipulations alone. Many processes in biology (signaling events in particular) occur on timescales that genetic perturbations (which are subject to adaptation at the transcriptional and translational timescales) are unable to capture. Chemical perturbation offers precise experimental control over cellular events such as cell division that occur within minutes.

4.1. AS Kinases in yeast

The genetic tractability of yeast and their ability to exist in a stable haploid state has lead to widespread use of AS technology in both *S. cerevisiae* and *Schizosaccharomyces pombe*. For example, there are 113 conventional protein-kinase genes in budding yeast (Hunter & Plowman, 1997) and the AS technique has been applied to 26 individual kinases thereby covering 23% of the yeast kinome. More recently, Gregan and coworkers developed functional AS alleles and identified suitable inhibitors for 13 of the 17 essential kinases in *S. pombe* (Cipak et al., 2011). Some of the key discoveries made with various AS Kinases in yeast include the large-scale identification of Cdc28 substrates (Holt et al., 2009) and the discovery of allosteric activation of the RNAse activity of Ire1 upon ligand binding to the ATP-binding site within its kinase domain (Papa, Zhang, Shokat, & Walter, 2003).

Recently, Kliegman and coworkers combined AS technology with high-throughput yeast genetics in order to measure dose-dependent genetic interactions between the catalytic function of a kinase and each deleted gene. This technique offers a new systematic approach that addresses the functional connectivity of individual enzymes without the destruction of protein complexes that is typically queried by traditional gene deletion studies. The ease of using the AS approach in yeast chemical-genetic applications offers a systematic way to use this technology to elucidate eukaryotic kinase-signaling pathways and generate testable hypotheses for kinase–substrate relationships.

4.2. AS Kinases in mammalian cells

Over 40 mouse or human AS Kinases have been reported in peer-reviewed publications making mammalian cells the most commonly studied with AS technology. In many cases, a disease-causing kinase allele may be dominant, allowing simple overexpression of the AS mutant in an appropriate cell line for investigation. In other cases, the WT-kinase gene must be replaced with the AS allele. Although genetic manipulation of mammalian cells is often more difficult than in yeast, they offer the opportunity to address many questions most relevant to human disease. For example, Wong et al. used an AS mutant of the disease-causing fusion protein Bcr-Abl to demonstrate that inhibition of this kinase alone was not sufficient to eliminate all myeloproliferative disorder cell populations (Wong, 2004). This was an important discovery as it revealed the off-target effects of the clinical therapeutic imatinib are, in fact, required for effective treatment of heterogeneous chronic myeloid leukemia cell populations. AS Kinases have also been used to shed light on fundamental processes of mammalian biology and development. An AS mutant of Zap-70, for example, has been used to identify its role in T-cell receptor and CD28 super agonist signaling and to determine which functions are dependent on kinase activity. Thus, the use of AS Kinases in mammalian tissue culture is a powerful method for investigating fundamental biological processes and the molecular basis of human disease.

5. AS KINASES IN LIVING MULTICELLULAR ORGANISMS

Many developmental processes can only be studied in the context of an intact living plant or animal. Furthermore, therapeutic drug targets can only truly be validated by inhibition *in vivo*. Thus, several AS Kinases have been introduced into mice and used to study the *in vivo* biology or pharmacology of kinase-signaling pathways. The *in vivo* use of AS Kinases presents the additional challenges of achieving optimal PK and PD properties for small molecule inhibitors. While the properties of current AS Kinase inhibitors have not been optimized for *in vivo* efficacy, they are nevertheless sufficient for use with several AS Kinases in mice. For example, Wright and coworkers generated mice that are homozygous for an AS mutant of Ret kinase in order to study the role of GDNF signaling in spermatogenesis (Savitt et al., 2012). Taconic is a company that generates mice bearing homozygous AS Kinase alleles, which they refer to as ASKA mice. Although the effectiveness of their ASKA mice is not subject to peer review, they report to have generated functional ASKA mice for the following kinases: Akt1, Gskα, Gsk3β, Ntrk1, Ntrk2, Src, Btk, Map3k5, EphB4, Mapk14, PdgfrB, Met, and Rps6kb1 (http://www.taconic.com). It is likely that such a large number of ASKA mice suggest a strong unmet need in drug target validation. In both drug development enterprises and academic investigations, it is clear that many important questions can only be addressed *in vivo* and that AS Kinase technology may play an important role in furthering these studies.

The methylene position of 1NM-PP1 and the 3-methyl group of 3MB-PP1 are likely to be susceptible to inactivating oxidation by P450 enzymes *in vivo* (Baer, DeLisle, & Allen, 2009; Lourido et al., 2013). This may explain why the inhibitor most commonly used to target AS Kinases in mice is 1NA-PP1, which has neither liability. However, since 1NA-PP1 is not the most potent or selective inhibitor for all AS Kinases, future work to design staralogs and 3-substituted PPs with enhanced PK properties will be of great importance to expanding the use of AS Kinases in mice. An important technical observation is that 1NA-PP1 is much more soluble as the HCl salt. Thus, it may be helpful to convert commercially available 1NA-PP1 to the HCl form in order to achieve the concentration necessary for *in vivo* efficacy.

Protocol for preparation of 1NA-PP1 HCl:

Dissolve 1.16 g of 1NA-PP1 in 50 mL of CH_2Cl_2 and stir at RT until completely dissolved. Add 1.2 equiv. of HCl (2 *M* in diethyl ether) and concentrate *in vacuo*. Wash the resulting off-white solid with cold ether. The HCl form can be distinguished from 1NA-PP1 by ^1H NMR as well as the fact that it has increased solubility (1NA-PP1 HCl but not 1NA-PP1 is soluble at 20 mg/mL in DMSO).

6. SUMMARY

This report provides a practical overview for constructing AS Kinases and their inhibitors as well as a theoretical framework for designing experiments that leverage the power of the AS technique. We outline how to identify the gatekeeper position of a kinase and describe several strategies to rescue the activity of AS Kinases' intolerant of glycine or alanine gatekeeper residues. Next, we explain the logic of AS Kinase inhibitor design and describe the strengths, limitations, and synthetic protocols for several of the most commonly used inhibitors. We highlight examples of how AS Kinases have been used to address questions about basic biological processes in yeast and in mammalian cell lines. Finally, we examined the use of AS Kinases in mice and explained the scope and limitations of current inhibitors *in vivo*.

Recently, Kawashima, Takemoto, Nurse, and Kapoor (2013) described a complementary chemical-genetic approach for studying protein kinases. This technique involves conducting a chemical screen to identify inhibitors of the KOI and then generating a resistant kinase mutant to function as a negative control for off-target inhibitor activity. This strategy has the advantage that the WT kinase is targeted by the inhibitor and thus is the focus of study, while the engineered resistant mutant is only used as a negative control. This eliminates the possibility that processes under study might be perturbed by the reduced activity that is sometimes encountered with AS Kinases. However, this approach is limited in that it remains difficult to identify specific inhibitors for many kinases. In contrast, rescuing the activity of hypomorphic AS Kinases is straightforward using the steps we detail herein should it be necessary. In summary, the AS technique is a facile and general approach for probing kinase-signaling pathways with small molecule inhibitors. The stepwise strategy that we describe in this report allows for the identification of appropriate mutations and inhibitors and enables the application of AS technology to kinases from diverse families and organisms in cells and *in vivo*.

REFERENCES

Alaimo, P. J., & Shokat, K. M. (2001). Recent advances in chemical approaches to the study of biological systems. *Annual Review of Cell and Developmental Biology*, *17*, 405–433.

Azam, M., Seeliger, M. A., Gray, N. S., Kuriyan, J., & Daley, G. Q. (2008). Activation of tyrosine kinases by mutation of the gatekeeper threonine. *Nature Structural & Molecular Biology*, *15*, 1109–1118.

Baer, B. R., DeLisle, R. K., & Allen, A. (2009). Benzylic oxidation of gemfibrozil-1-O-beta-glucuronide by P450 2C8 leads to heme alkylation and irreversible inhibition. *Chemical Research in Toxicology*, *22*, 1298–1309.

Bishop, A. C., Kung, C., Shah, K., Witucki, L., Shokat, K. M., & Liu, Y. (1999). Generation of monospecific nanomolar tyrosine kinase inhibitors via a chemical genetic approach. *Journal of the American Chemical Society*, *121*, 627–631.

Blair, J. A., Rauh, D., Kung, C., Yun, C.-H., Fan, Q.-W., Rode, H., et al. (2007). Structure-guided development of affinity probes for tyrosine kinases using chemical genetics. *Nature Chemical Biology*, *3*, 229–238.

Branford, S., Rudzki, Z., Walsh, S., & Grigg, A. (2002). High frequency of point mutations clustered within the adenosine triphosphate-binding region of BCR/ABL in patients with chronic myeloid leukemia or Ph-positive acute lymphoblastic leukemia who develop imatinib (STI571) resistance. *Blood*, *99*, 3472–3475.

Cipak, L., Zhang, C., Kovacikova, I., Rumpf, C., Miadokova, E., Shokat, K. M., et al. (2011). Generation of a set of conditional analog-sensitive alleles of essential protein kinases in the fission yeast Schizosaccharomyces pombe. *Cell Cycle (Georgetown, TX)*, *10*, 3527–3532.

Garske, A. L., Peters, U., Cortesi, A. T., Perez, J. L., & Shokat, K. M. (2011). Chemical genetic strategy for targeting protein kinases based on covalent complementarity. *Proceedings of the National Academy of Sciences of the United States of America*, *108*, 15046–15052.

Hanke, J. H., Gardner, J. P., Dow, R. L., Changelian, P. S., Brissette, W. H., Weringer, E. J., et al. (1996). Discovery of a novel, potent, and Src family-selective tyrosine kinase inhibitor: Study of Lck- and FynT-dependent T cell activation. *The Journal of Biological Chemistry*, *271*, 695–701.

Hertz, N. T., Berthet, A., Sos, M. L., Thorn, K. S., Burlingame, A. L., Nakamura, K., et al. (2013). A neo-substrate that amplifies catalytic activity of Parkinson's disease-related kinase PINK1. *Cell*, *154*, 737–747.

Holt, L. J., Tuch, B. B., Villén, J., Johnson, A. D., Gygi, S. P., & Morgan, D. O. (2009). Global analysis of Cdk1 substrate phosphorylation sites provides insights into evolution. *Science (New York, N.Y.)*, *325*, 1682–1686.

Hsieh, A. C., Liu, Y., Edlind, M. P., Ingolia, N. T., Janes, M. R., Sher, A., et al. (2012). The translational landscape of mTOR signalling steers cancer initiation and metastasis. *Nature*, *485*, 55–61.

Hunter, T., & Plowman, G. D. (1997). The protein kinases of budding yeast: Six score and more. *Trends in Biochemical Sciences*, *22*, 18–22.

Kawashima, S. A., Takemoto, A., Nurse, P., & Kapoor, T. M. (2013). A chemical biology strategy to analyze rheostat-like protein kinase-dependent regulation. *Chemistry & Biology*, *20*, 262–271.

Kliegman, J. I., Fiedler, D., Ryan, C. J., Xu, Y.-F., Su, X.-Y., Thomas, D., et al. (2013). Chemical genetics of rapamycin-insensitive TORC2 in *S. cerevisiae*. *Cell Reports*, *5*, 1725–1736.

Liu, Y., Bishop, A., Witucki, L., Kraybill, B., Shimizu, E., Tsien, J., et al. (1999). Structural basis for selective inhibition of Src family kinases by PP1. *Chemistry & Biology*, *6*, 671–678.

Liu, Y., Shah, K., Yang, F., Witucki, L., & Shokat, K. M. (1998). A molecular gate which controls unnatural ATP analogue recognition by the tyrosine kinase v-Src. *Bioorganic & Medicinal Chemistry*, *6*, 1219–1226.

Lopez, M. S., Choy, J. W., Peters, U., Sos, M. L., Morgan, D. O., & Shokat, K. M. (2013). Staurosporine-derived inhibitors broaden the scope of analog-sensitive kinase technology. *Journal of the American Chemical Society*, *135*, 18153–18159.

Lourido, S., Zhang, C., Lopez, M. S., Tang, K., Barks, J., Wang, Q., et al. (2013). Optimizing small molecule inhibitors of calcium-dependent protein kinase 1 to prevent infection by *Toxoplasma gondii*. *Journal of Medicinal Chemistry*, *56*, 3068–3077.

Merrick, K. A., Wohlbold, L., Zhang, C., Allen, J. J., Horiuchi, D., Huskey, N. E., et al. (2011). Switching Cdk2 on or off with small molecules to reveal requirements in human cell proliferation. *Molecular Cell*, *42*, 624–636.

Niswender, C. M., Ishihara, R. W., Judge, L. M., Zhang, C., Shokat, K. M., & McKnight, G. S. (2002). Protein engineering of protein kinase A catalytic subunits results in the acquisition of novel inhibitor sensitivity. *The Journal of Biological Chemistry*, *277*, 28916–28922.

Papa, F. R., Zhang, C., Shokat, K., & Walter, P. (2003). Bypassing a kinase activity with an ATP-competitive drug. *Science (New York, N.Y.)*, *302*, 1533–1537.

Savitt, J., Singh, D., Zhang, C., Chen, L.-C., Folmer, J., Shokat, K. M., et al. (2012). The in vivo response of stem and other undifferentiated spermatogonia to the reversible inhibition of glial cell line-derived neurotrophic factor signaling in the adult. *Stem Cells (Dayton, Ohio)*, *30*, 732–740.

Schindler, T., Sicheri, F., Pico, A., Gazit, A., Levitzki, A., & Kuriyan, J. (1999). Crystal structure of Hck in complex with a Src family-selective tyrosine kinase inhibitor. *Molecular Cell*, *3*, 639–648.

Shah, K., & Shokat, K. M. (2002). A chemical genetic screen for direct v-Src substrates reveals ordered assembly of a retrograde signaling pathway. *Chemistry & Biology*, *9*, 35–47.

Shokat, K. M., Bishop, A. C., Ubersax, J. A., Petsch, D. T., Matheos, D. P., Gray, N. S., et al. (2000). A chemical switch for inhibitor-sensitive alleles of any protein kinase. *Nature*, *407*, 395–401.

Wong, S. (2004). Sole BCR-ABL inhibition is insufficient to eliminate all myeloproliferative disorder cell populations. *Proceedings of the National Academy of Sciences of the United States of America*, *101*, 17456–17461.

Zhang, C., Kenski, D. M., Paulson, J. L., Bonshtien, A., Sessa, G., Cross, J. V., et al. (2005). A second-site suppressor strategy for chemical genetic analysis of diverse protein kinases. *Nature Methods*, *2*, 435–441.

Zhang, C., Lopez, M. S., Dar, A. C., Ladow, E., Finkbeiner, S., Yun, C.-H., et al. (2013). Structure-guided inhibitor design expands the scope of analog-sensitive kinase technology. *ACS Chemical Biology*, *8*, 1931–1938.

AUTHOR INDEX

Note: Page numbers followed by "*f*" indicate figures and "*t*" indicate tables.

D

E

F

G

H

I

J

K

T

U

V

W

X

Y

Z

SUBJECT INDEX

Note: Page numbers followed by "*f*" indicate figures and "*t*" indicate tables.

A

B

A

R−OH + (O)P(OADP)(O⁻)(O⁻)

Associative transition state

Dissociative (metaphosphate-like) transition state

(O)P(OR)(O⁻)(O⁻) + ADP

B

Substrate-binding site

Linker

Nucleotide-binding site

C

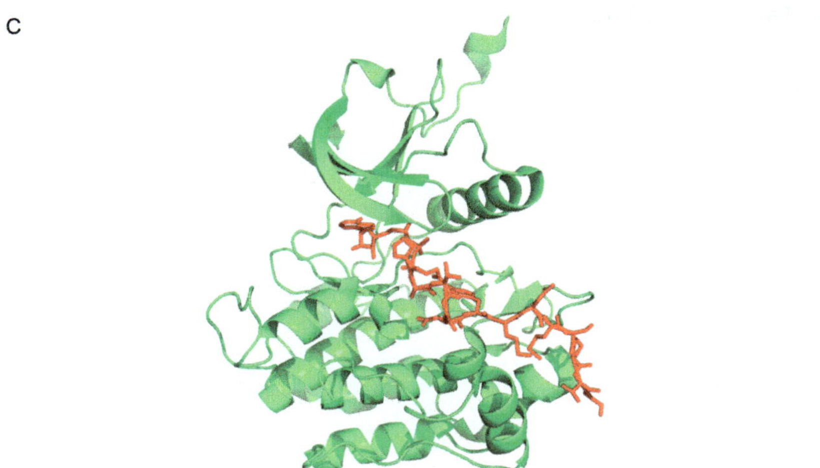

Zhihong Wang and Philip A. Cole, Figure 1.1 See legend on the next page.

Zhihong Wang and Philip A. Cole, Figure 1.1 Protein kinase mechanism and bisubstrate analog inhibitors. (A) Mechanistic scheme of associative and dissociative transition states of phosphoryl transfer. A dissociative state model is proposed for the transfer of the γ-phosphoryl group of ATP to hydroxy group in a kinase–substrate. (B) Designed bisubstrate analog inhibitor for a protein tyrosine kinase. R1 and R2 amino acid sequences are derived from efficient substrate motifs. The linker is predicted to result in a 5–5.7 Å spacer between the substrate-binding site and the nucleotide-binding site, compatible with a dissociative transition state. ATPγS was used as the ATP-mimic analog. (C) The crystal structure of IRK (green ribbon) in complex with the above peptide–ATP conjugate (red stick), determined at 2.7 Å resolution (PDB code: 1GAG; Parang et al., 2001).

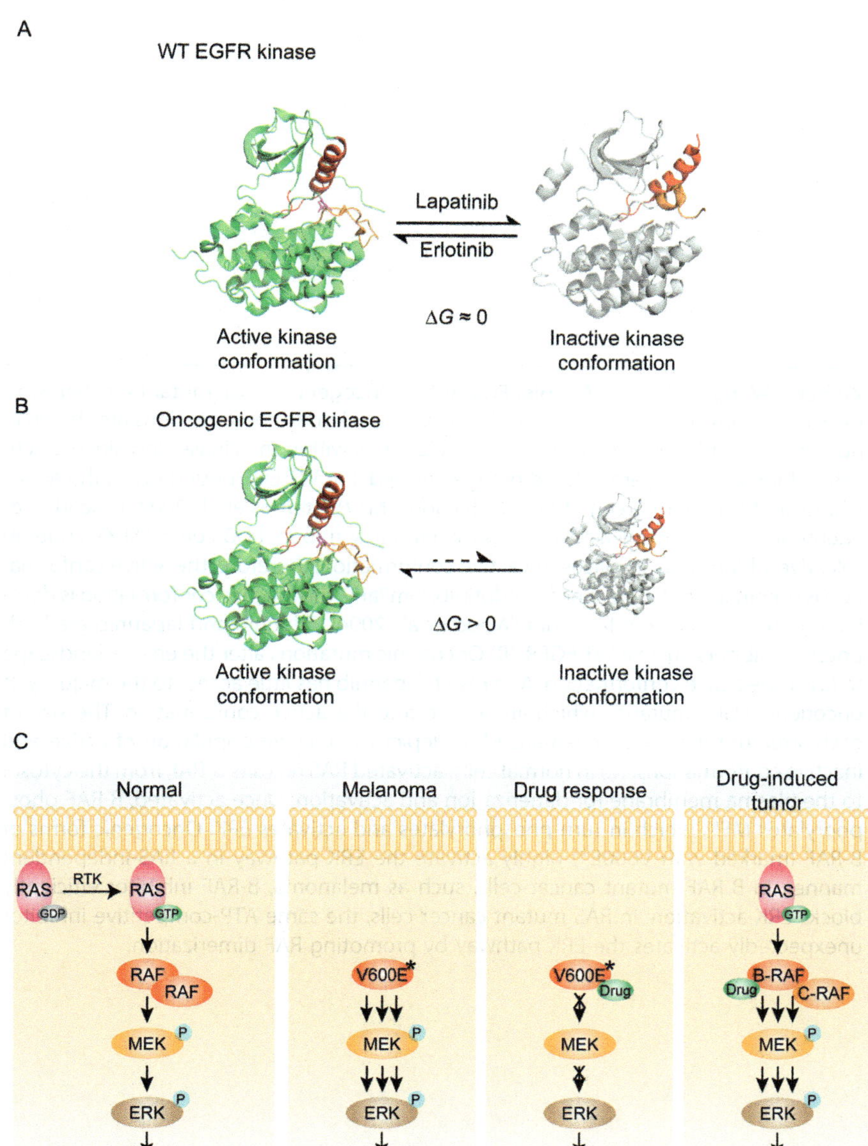

Zhihong Wang and Philip A. Cole, Figure 1.2 See legend on the next page.

Zhihong Wang and Philip A. Cole, Figure 1.2 Oncogenic kinase mutant mechanisms. (A) The crystal structures of active and inactive EGFR kinase conformations are shown in ribbon representation. Two key structural elements within the kinase domain, the activation loop and helix, are colored orange and red. For WT EGF-bound EGFR, the active kinase conformation (green ribbon, PDB code: 1M17) (Stamos et al., 2002) is nearly isoenergetic to the inactive kinase conformation (gray ribbon, PDB code: 1XKK). Erlotinib selectively binds and stabilizes the active conformation; therefore, the active conformation is dominant in the presence of erlotinib. Similarly, the inactive conformation is dominant in the presence of lapatinib (Wood et al., 2004). Erlotinib and lapatinib are both potent inhibitors against WT EGFR. (B) Oncogenic mutations alter the energy landscape to favor the active conformation. As a result, lapatinib has little access to the majority of oncogenic EGFR mutants, which are locked into the active conformation. The size of each structure has been schematized to depict the relative population of active and inactive conformations. (C) In normal cells, activated RAS recruits B-RAF from the cytosol to the plasma membrane for dimerization and activation. Once activated, B-RAF phosphorylates MEK, which in turn phosphorylates and activates ERK. Oncogenic forms of B-RAF (marked with V600E*) highly activate the ERK pathway in a RAS-independent manner. In B-RAF mutant cancer cells, such as melanoma, B-RAF inhibitor efficiently blocks ERK activation. In RAS mutant cancer cells, the same ATP-competitive inhibitor unexpectedly activates the ERK pathway by promoting RAF dimerization.

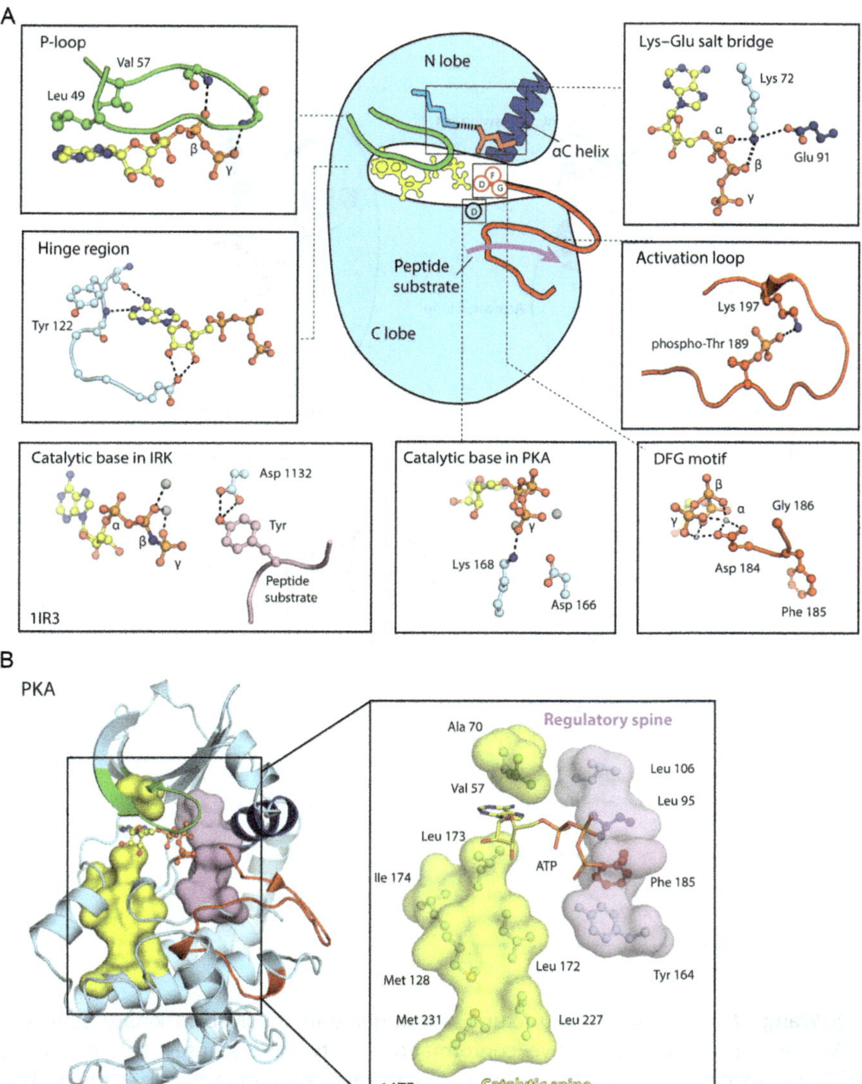

Qi Wang *et al.*, Figure 2.1 The key structural elements in a kinase domain. (A) An overview of catalytic and regulatory elements in a protein kinase. The active conformation is shown in light blue. The αC helix, P-loop, activation loop, and Lys–Glu ion pair are shown in stick representation. The bound ATP is illustrated in yellow. The Asp–Phe–Gly (DFG) motif is shown as red circles. The catalytic base is shown as a black circle. The zoomed-in windows display detailed atomic views of important regulatory elements, with hydrogen bonds indicated by dashed lines. Mg^{2+} or Mn^{2+} is shown as gray spheres. All structural figures are generated using PyMol. (B) The catalytic and regulatory spines of the kinase domain, as defined by Kornev, Taylor, and coworkers. The compact organization of the residues in these spines is disrupted in an inactive conformation of the kinase domain.

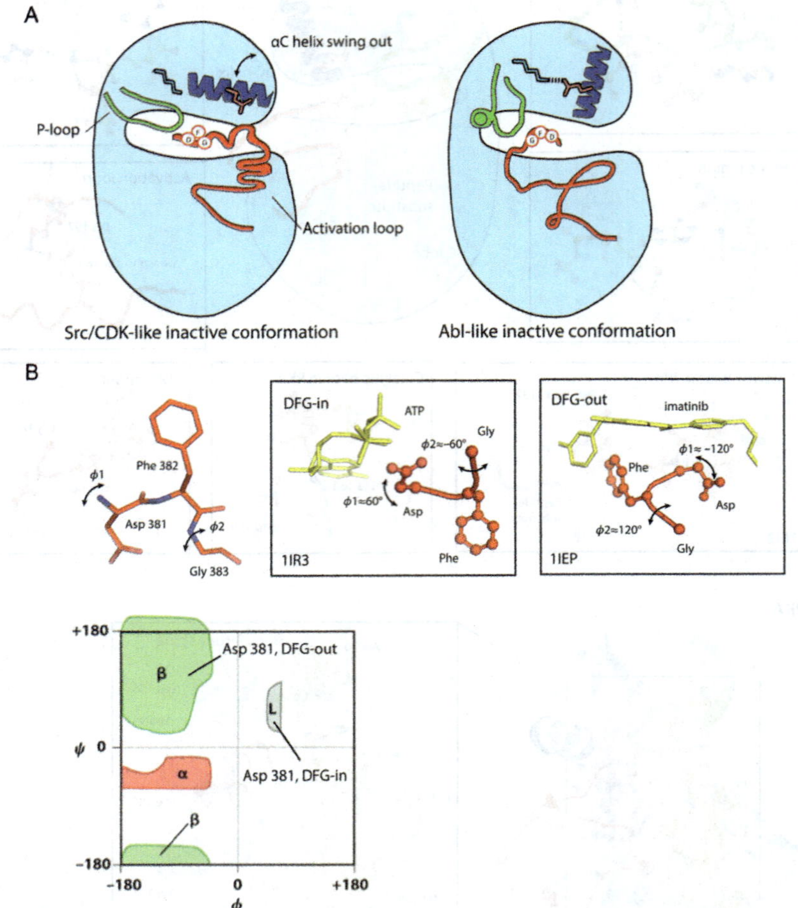

Qi Wang *et al.*, Figure 2.2 Structural rearrangements involved in kinase activation. (A) Two representative inactive conformations of tyrosine kinases. Left panel: Src/CDK-like inactive conformation (PDB: 1QCF); the activation loop stabilizes the αC helix in a distorted conformation. Right panel: Abl-like inactive conformation (PDB: 1IEP); the activation loop mimics substrate binding, and the DFG motif adopts a "DFG-out" conformation that is not compatible with ATP binding. (B) DFG-flip during Abl activation. Upper left: the torsion angle of Asp 381 and Gly 383 in the DFG motif. The Φ value changes about 180° from a "DFG-out" conformation (upper right) to a "DFG-in" conformation (upper middle). Lower panel: Ramachandran plot of peptide backbone torsion angles indicating the orientation of Asp 381 in the two conformations. *Adapted with permission from the Molecules of Life.*

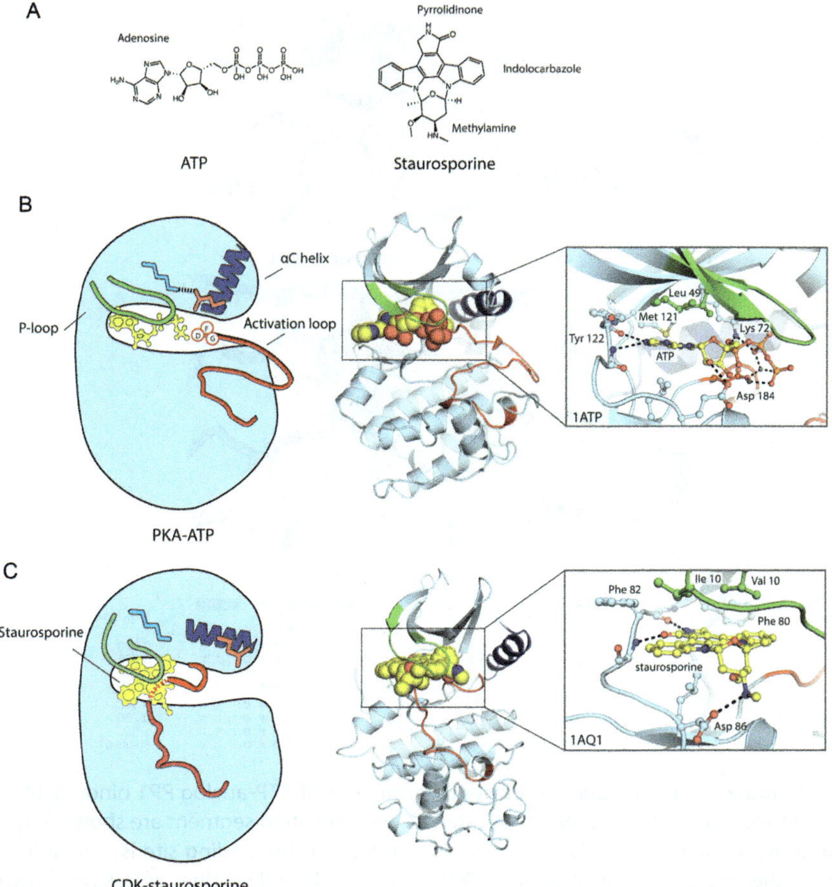

Qi Wang *et al.*, Figure 2.3 ATP and staurosporine. (A) A crystal structure of ATP-bound PKA (PDB: 1ATP). A detailed view of the ATP coordination between the N lobe and C lobe is shown on the right panel. (B) A crystal structure of staurosporine-bound CDK2 (PDB: 1AQ1).

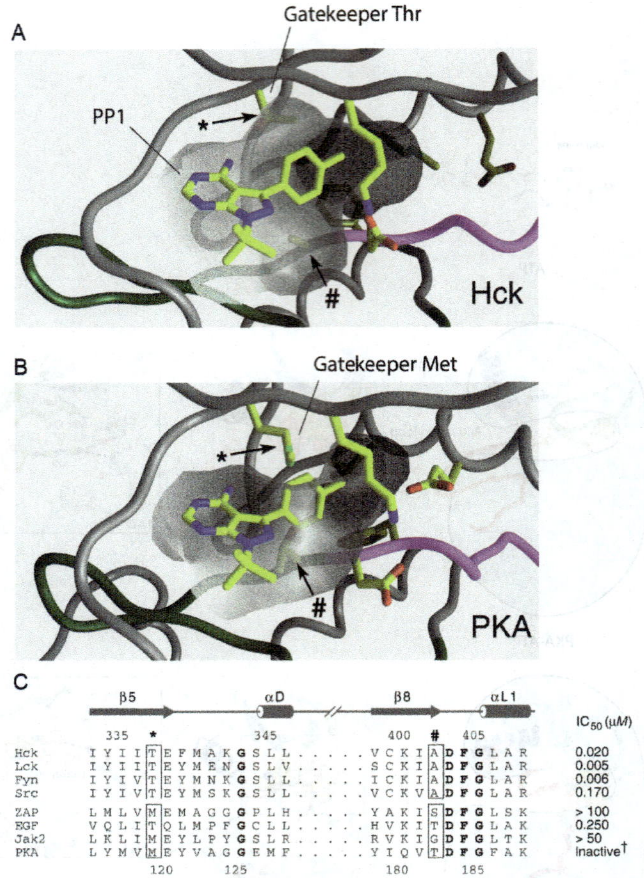

Qi Wang *et al.*, Figure 2.4 Selectivity determinants of ATP-analog PP1 binding. (A) PP1 bound to Hck. The Cα traces of the catalytic and activation segment are shown in green and magenta, respectively. The molecular surface of the binding site is superimposed over the traces in transparent gray. PP1 as well as selected residues of Hck are drawn as stick figures. (B) PP1 modeled into the ATP-binding site of PKA (PDB code 1ATP; Zheng et al., 1993). The model was generated by aligning the pyrazolopyrimidine moiety of PP1 and the adenine ring of ATP. (C) Sequence comparison of Src family tyrosine kinase members with a set of other protein kinases. The inhibition constants (IC50s) for PP1 were taken from Hanke et al. (1996) ([†] PP1 was reported to be essentially inactive against PKA). Thr-338 (*) and Ala 403 (#) are indicated in (A), as are the structurally equivalent residues in PKA, in (B). *Reprinted with permission from Schindler et al. (1999) Molecular Cell.*

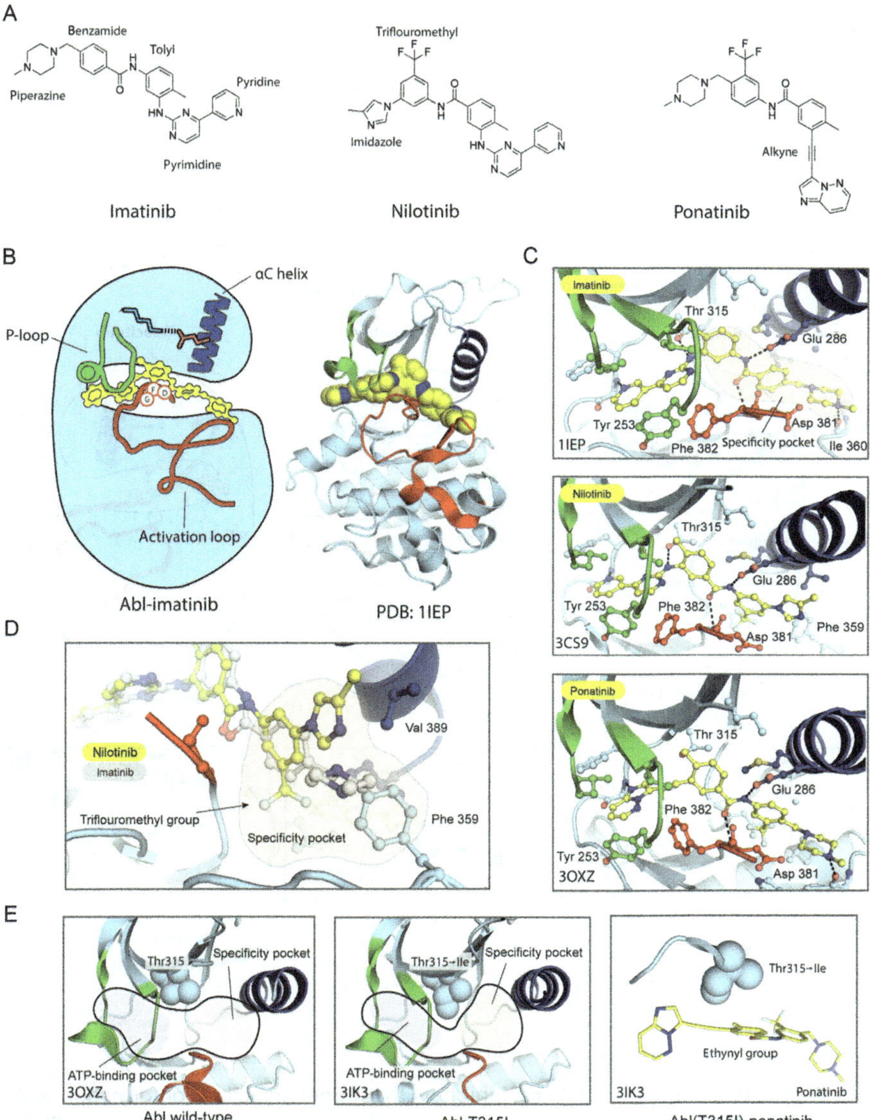

Qi Wang *et al.*, Figure 2.5 Inhibition of c-Abl kinase domain by imatinib and related analogs. (A) Chemical structures of imatinib, nilotinib, and ponatinib. (B) A crystal structure of the Abl kinase domain bound to an imatinib analog (PDB: 1IEP). (C) Detailed views of imatinib (upper panel), nilotinib (middle panel), and ponatinib (lower panel) coordination in Abl kinase domain. The specificity pocket of Abl kinase domain is shaded brown. (D) Structural overlay of nilotinib (yellow) and imatinib (grey) in the Abl specificity pocket (shaded brown). (E) The basis for inhibition of Abl-T315I mutant by ponatinib. Mutations of the gatekeeper threonine restrict access to the specificity pocket (shaded brown) from the ATP binding pocket (shaded purple) in the Abl-T315I crystal structure (middle panel) as compared to the wild-type Abl structure (left panel). Ponatinib is compatible with binding to both pockets of the Abl-T315I mutant.

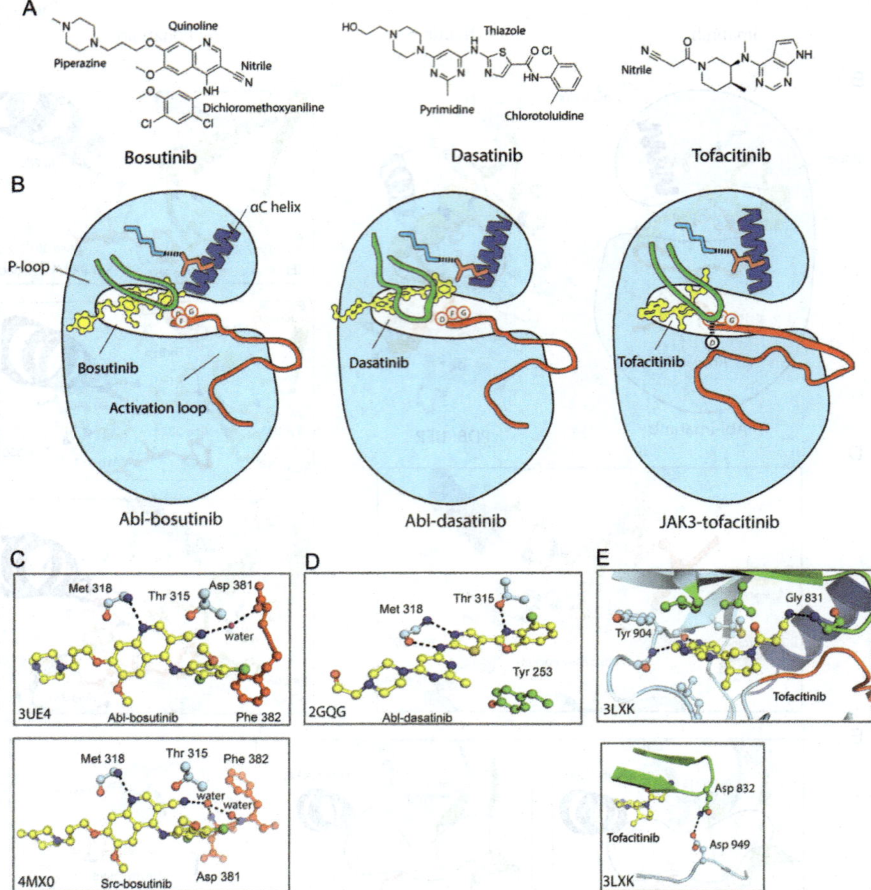

Qi Wang *et al.*, Figure 2.6 Clinical inhibitors that target an active conformation of Abl and JAK3. (A) Chemical structures of bosutinib, dasatinib, and tofacitinib. (B) The bosutinib-bound Abl kinase domain (PDB 3UE4), dasatinib-bound Abl kinase domain (PDB: 2GQG) and tofacitinib-bound JAK3 (PDB 3LXK). (C) Detailed view of the coordination of bosutinib in Abl kinase domain (upper panel) and in Src kinase domain (lower panel). The water molecules are shown as small red spheres. (D) Zoomed-in view of the coordination of dasatinib in the Abl kinase domain. (E) Zoomed-in view of the coordination of tofacitinib in the JAK3 kinase domain (upper panel). The P-loop residue Asn 832 in JAK3 forms hydrogen bonds to the catalytic base, Asp 949 (lower panel).

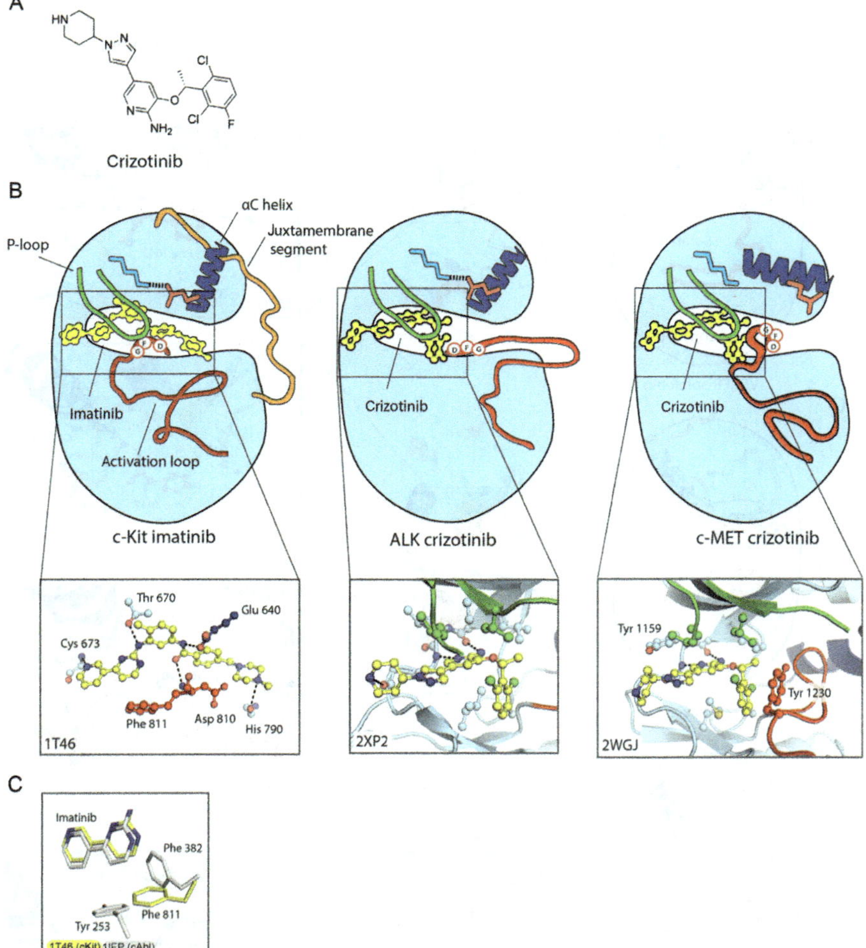

Qi Wang *et al.*, Figure 2.7 Inhibition of c-Kit, ALK, and c-MET by imatinib and crizotinib. (A) Chemical structure of crizotinib. (B) The c-Kit kinase domain bound to imatinib (PDB: 1T46), ALK kinase domain bound to crizotinib (PDB: 2XP2), and c-Met kinase domain bound to crizotinib (PDB 2WGJ). A detailed view of the interactions between the kinase and the inhibitors is shown in the lower panel. (C) Phe 811 in the DFG motif of c-Kit bound to imatinib adopts a different conformation compared to Phe 382 in the DFG motif of c-Abl bound to imatinib.

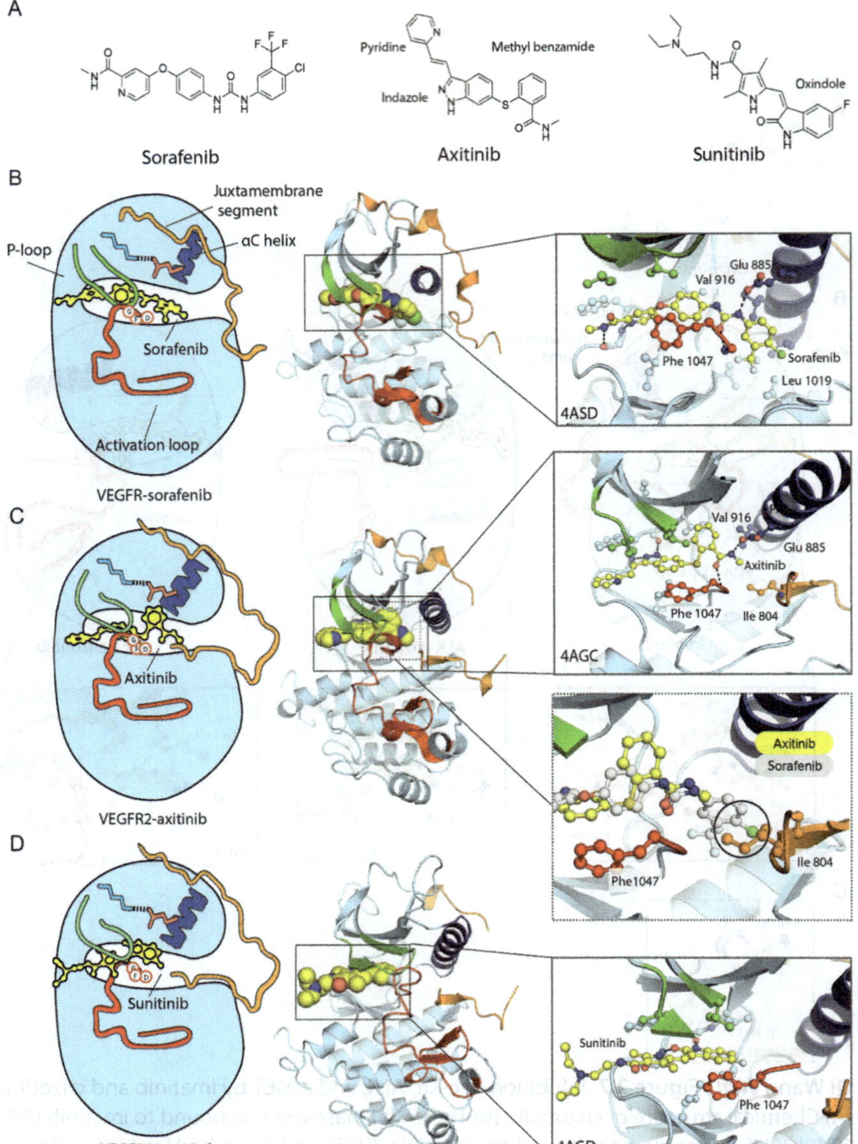

Qi Wang *et al.*, Figure 2.8 Inhibition of VEGFR by sorafenib, axitinib, and sunitinib. (A) Chemical structures of sorafenib, axitinib, and sunitinib. (B) A crystal structure of the VEGFR2 kinase domain bound to sorafenib. The juxtamembrane segment is colored orange. The zoomed-in view of the drug coordination is shown in the right panel. (C) A crystal structure of the VEGFR2 kinase domain bound to axitinib. A comparison of axitinib (yellow) to sorafenib (gray) in the VEGFR kinase domain is shown in the lower right panel. The black circle highlights where the juxtamembrane segment would clash with the phenyl group in sorafenib. (D) A crystal structure of the VEFGR kinase domain bound to sunitinib.

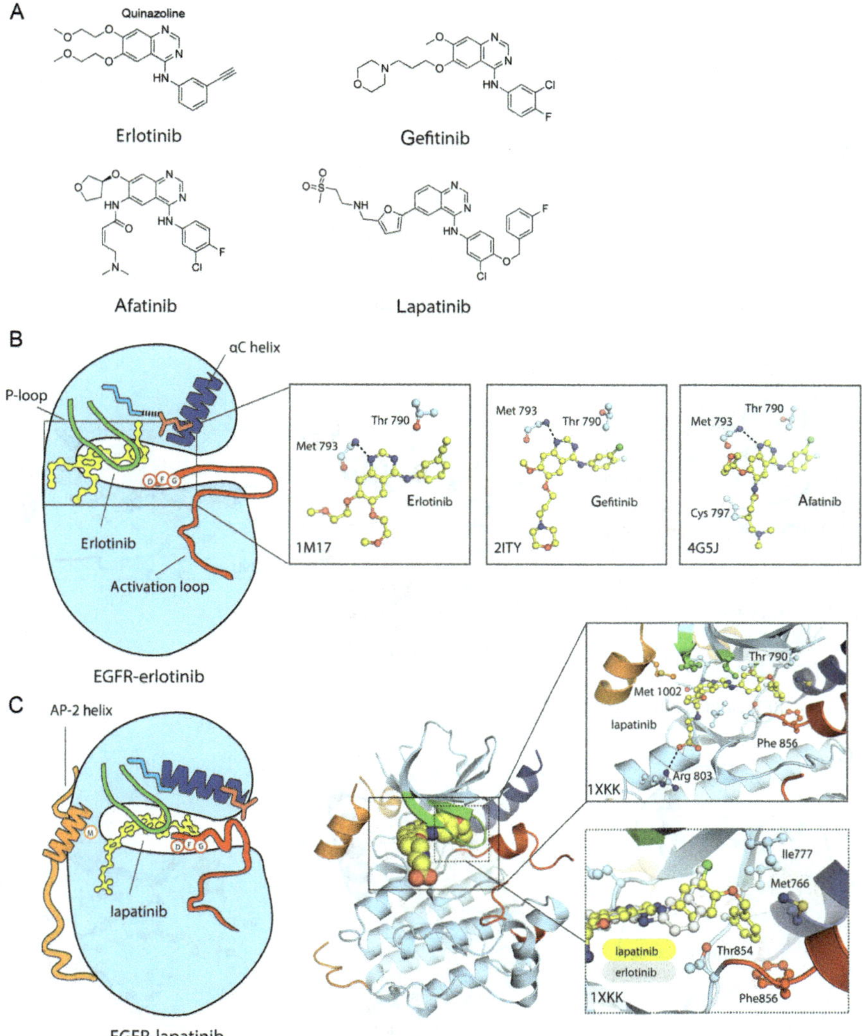

Qi Wang *et al.*, Figure 2.9 Inhibition of EGFR by lapatinib, erlotinib, gefitinib, and afatinib. (A) Chemical structures of erlotinib, gefitinib afatinib, and lapatinib. (B) A crystal structure of the EGFR kinase domain bound to erlotinib. The EGFR kinase domain is in an active conformation. Erlotinib, gefitinb, and afatinib all bind to this active conformation. (C) A crystal structure of the EGFR kinase domain bound to lapatinib. The EGFR kinase domain is in a Src/CDK-like inactive conformation. The C-terminal loop of the kinase domain and the AP-2 helix are colored orange. Met 1002 on the AP-2 helix is shown as an orange circle. A comparison of lapatinib (yellow) and erlotinib (gray) bound to the kinase domain of EGFR is displayed in the zoomed-in figure on the lower right panel.

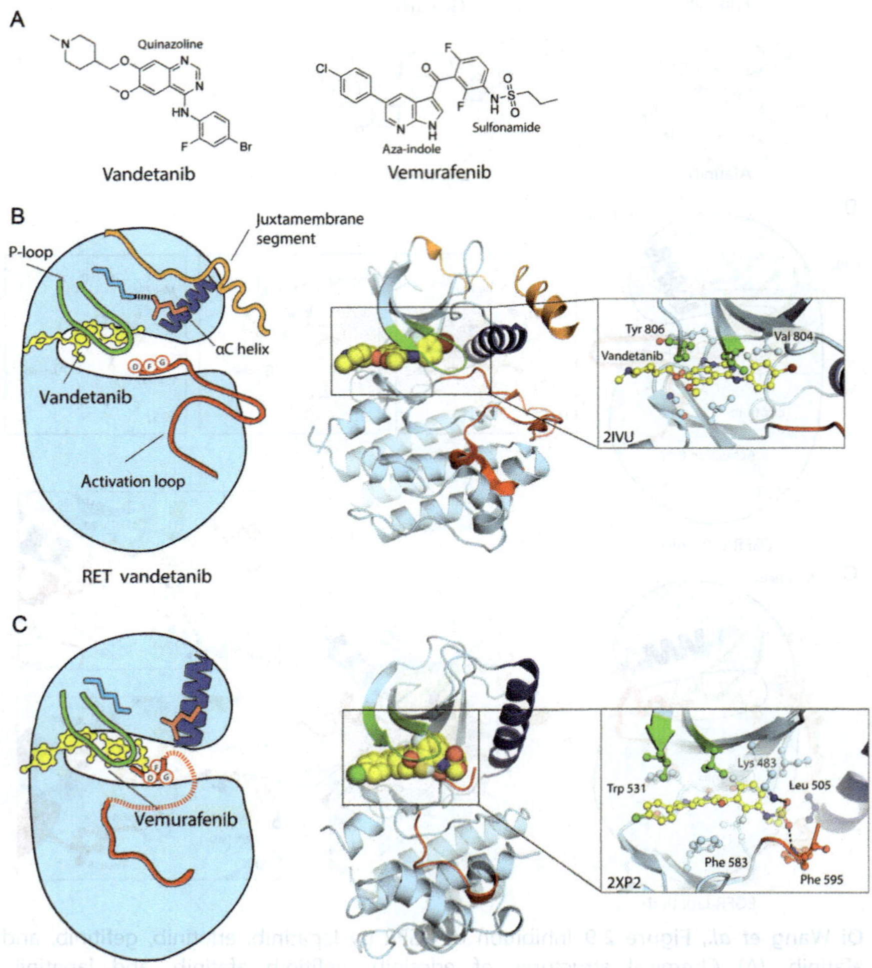

Qi Wang *et al.*, Figure 2.10 Vandetanib and vemurafenib. (A) Chemical structures of vandetanib and vemurafenib. (B) A crystal structure of the vandetanib-bound RET kinase domain. (C) A crystal structure of the vemurafenib-bound B-RAF kinase domain. The activation loop is partially disordered, which is illustrated by the dotted-red line.

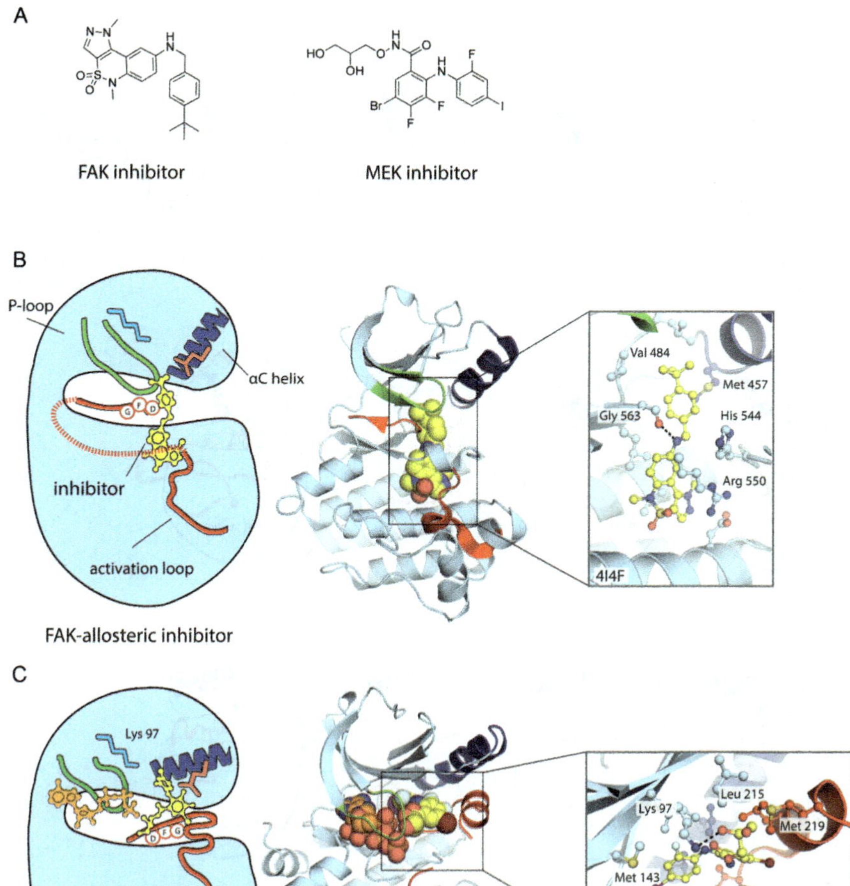

Qi Wang *et al.*, Figure 2.11 Inhibition of protein kinases by non-ATP-competitive compounds. (A) A crystal structure of the FAK kinase domain bound to an allosteric inhibitor. The activation loop is partially disordered, which is shown as a dotted-red line. (B) A crystal structure of the MEK1 kinase domain bound to an allosteric inhibitor and ATP.

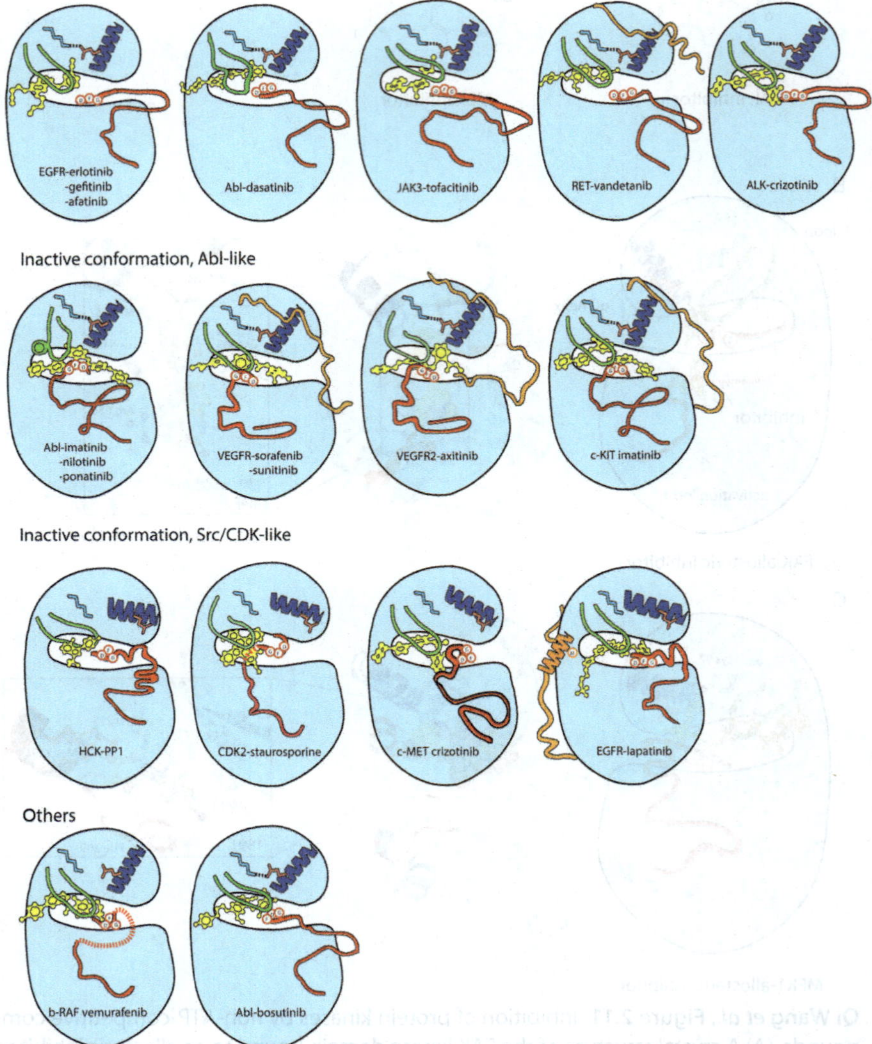

Qi Wang *et al.*, Figure 2.12 Summary of the kinase conformations that are targeted by clinically approved drugs.

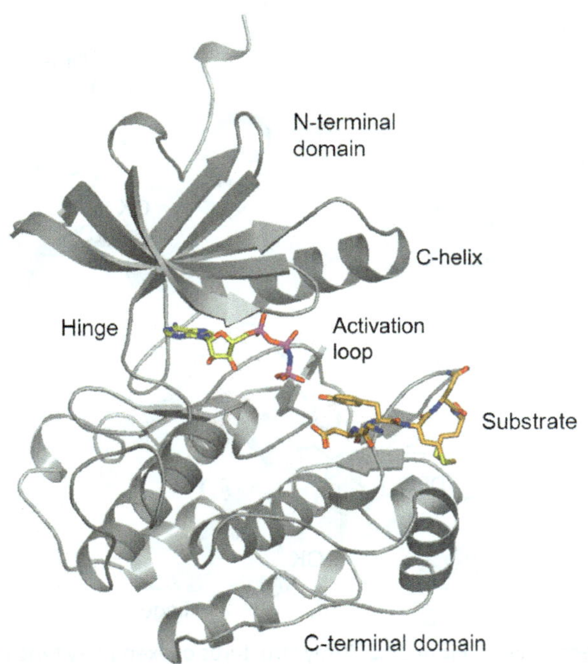

Paul N. Mortenson *et al.*, Figure 3.1 Architecture of a typical protein kinase domain. The protein backbone is shown in cartoon form in gray, and the peptide substrate is shown in orange. The nonhydrolyzable ATP mimetic AMP-PNP is shown in yellow. From PDB structure 1ir3.

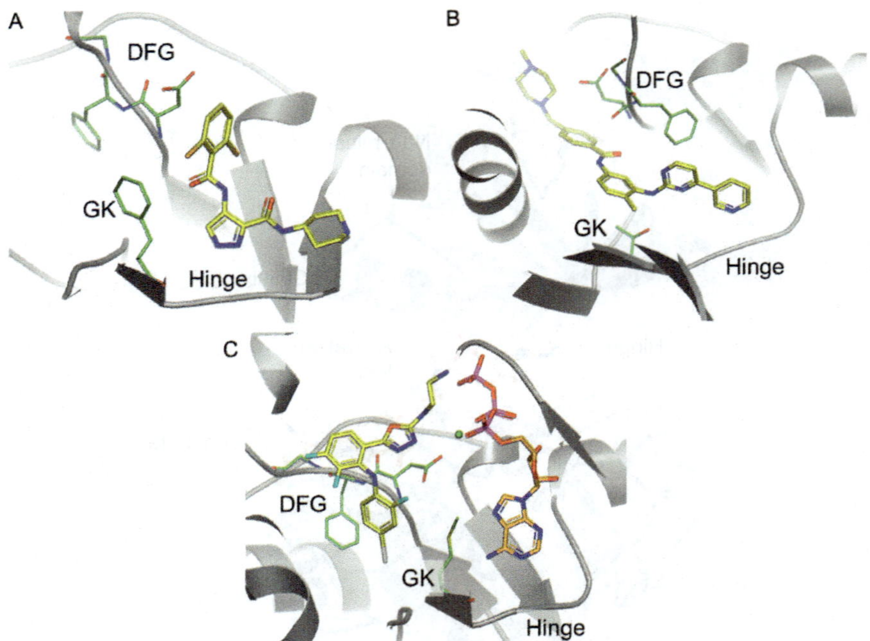

Paul N. Mortenson *et al.*, Figure 3.2 X-ray structures of exemplary type I, II, and III inhibitors. In each case, the ligand is shown in yellow, the protein is shown in cartoon form in gray, and the gatekeeper (GK) and DFG-loop residues are highlighted in green. (A) AT7519, a type I inhibitor, bound to CDK2 (in-house structure). (B) Imatinib, a type II inhibitor, bound to c-Abl (PDB structure 1iep). (C) A type III inhibitor bound to MEK1 (PDB structure 3eqb). ATP is also present in the structure and is shown in orange.

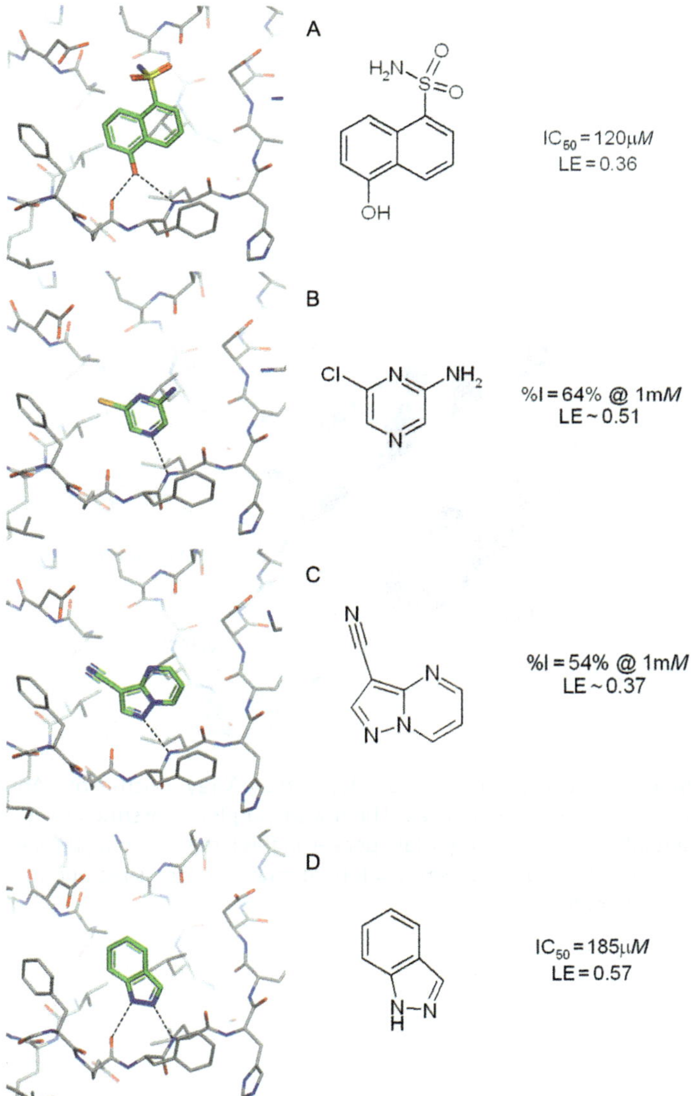

Paul N. Mortenson *et al.*, Figure 3.4 Four fragment hits found in our screening campaign against CDK2, discussed in further detail in the text. The X-ray structures are presented in a consistent frame of reference, with the hinge at the bottom and the gatekeeper (Phe 80) on the left. Hydrogen bonds to the hinge are illustrated by dotted lines.

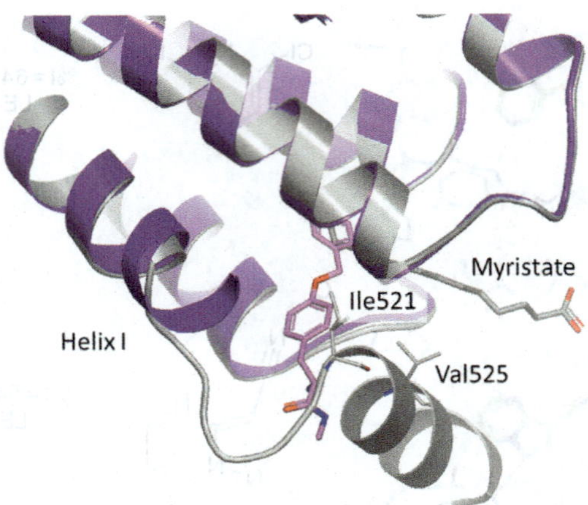

Paul N. Mortenson *et al.*, Figure 3.5 Overlay of (gray) X-ray structure of c-Abl (PDB code 1OPK) with myristate bound and helix I bent with (purple) X-ray structure of c-Abl with a fragment (pink) bound in the myristate pocket. In this latter structure, the second half of helix I is not resolved, but the bent conformation is not accessible due to the clashes between the fragment and Ile521.

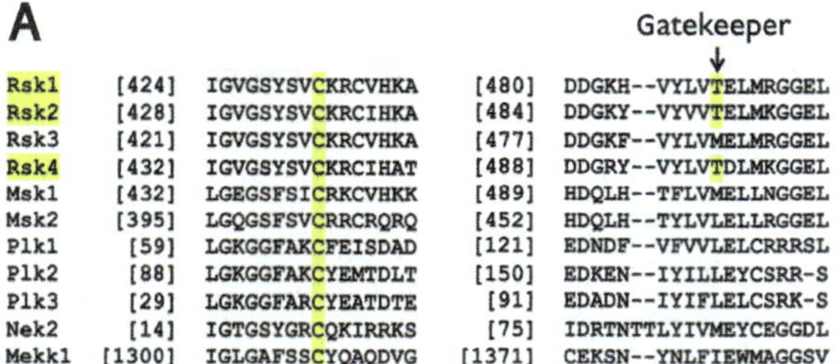

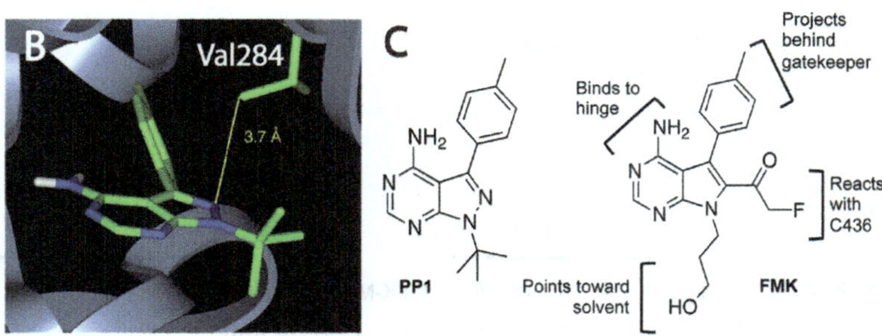

Rand M. Miller and Jack Taunton, Figure 4.1 Structural bioinformatics-based design of cysteine-targeted RSK inhibitors. (A) Sequence alignment reveals two selectivity filters unique to RSK1/2/4: a noncatalytic Cys and a Thr gatekeeper. (B) Multikinase inhibitor PP1 bound to the Src-family kinase HCK (PDB code: 1QCF), with N2 of PP1 proximal to Val284, corresponding to Cys436 in RSK2. (C) Chemical structure of PP1 and the irreversible RSK inhibitor, **FMK**.

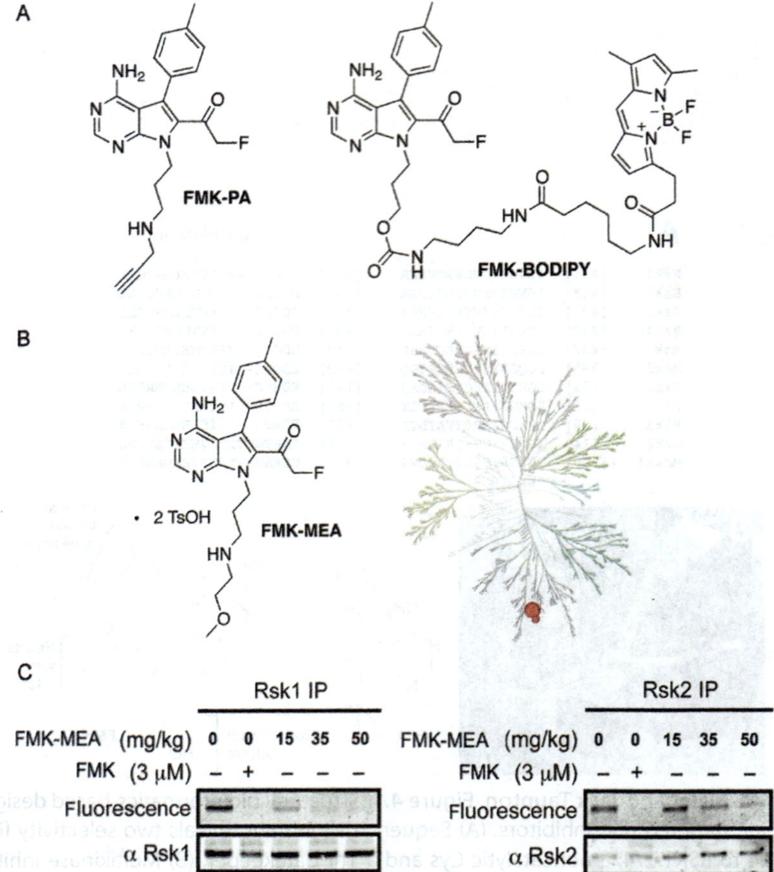

Rand M. Miller and Jack Taunton, Figure 4.2 **FMK** derivatives used to elucidate pathophysiological roles of RSK. (A) Chemical structures of **FMK-PA** and **FMK-BODIPY**. (B) Kinome-wide selectivity of **FMK-MEA**, screened at 1 μ*M* versus 443 kinases (DiscoverX). Only RSK1 and RSK4 CTDs were bound >50% relative to DMSO controls. (C) Using **FMK-BODIPY** as a probe, RSK1/2 occupancy was assessed in cardiac tissue lysates derived from mice treated with the indicated dose of **FMK-MEA.** To visualize **FMK-BODIPY**-labeled RSK1 and RSK2, each kinase was separately immunoprecipitated with isoform-specific antibodies. A dose-dependent reduction in fluorescence indicates RSK occupancy by **FMK-MEA**. *Figure adapted with permission from Lippincott Williams and Wilkins/Wolters Kluwer Health: Circulation, Le et al. (2013).*

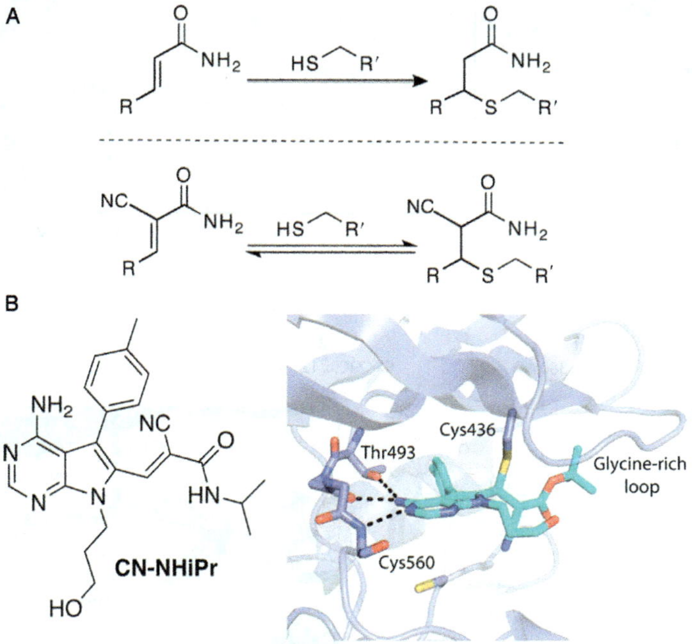

Rand M. Miller and Jack Taunton, Figure 4.3 Reversible targeting of noncatalytic cysteines with cyanoacrylamides. (A) Cyanoacrylamides, unlike acrylamides, form rapidly reversible adducts with thiols. (B) *N*-isopropyl cyanoacrylamide variant of **FMK** is a potent, selective, and reversible RSK inhibitor with slow dissociation kinetics. A cocrystal structure of a *tert*-butyl cyanoacrylate derivative bound to RSK2 shows a network of noncovalent interactions that cooperatively stabilize the covalent complex. Upon unfolding of RSK2, the covalent bond is rapidly reversed. *Adapted with permission from Serafimova et al. (2012). Copyright Nature Publishing Group, 2012.*

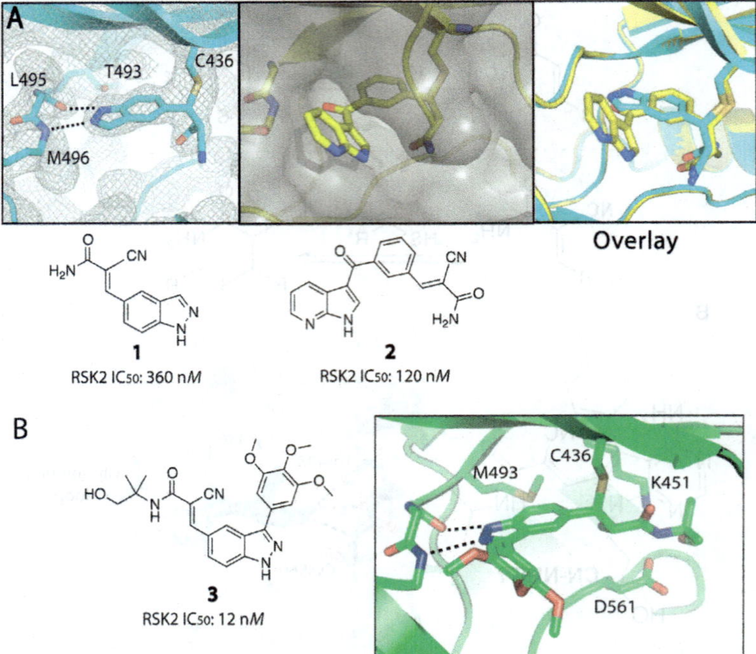

Rand M. Miller and Jack Taunton, Figure 4.4 Cyanoacrylamide fragment screening identifies pan-MSK/RSK inhibitors. (A) Using a RSK2 kinase assay, we screened a panel of 10 cyanoacrylamide fragments (up to 300 μ*M*) and discovered **1** and **2** as the most potent hits. We solved cocrystal structures of **1** and **2** bound RSK2 (PDB codes: 4JG6 and 4JG7). An overlay of both structures suggested the design of 3-aryl indazole variants. (B) Structure-guided optimization led to 3-aryl indazole **3** (RMM-46), which potently and selectively blocks MSK and RSK signaling in cells. *Adapted with permission from Miller et al. (2013). Copyright 2013, American Chemical Society.*

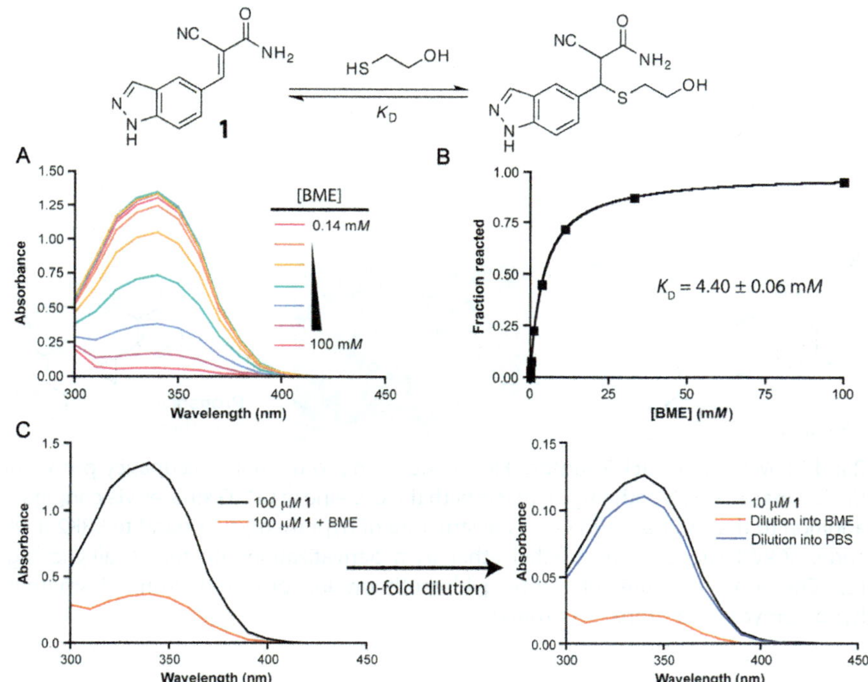

Rand M. Miller and Jack Taunton, Figure 4.5 Characterization of thiol conjugate addition reactions with cyanoacrylamides. (A) Reaction of cyanoacrylamide **1** with BME is accompanied by a decrease in the 340 nm absorption band. (B) The K_d is determined by fitting the titration data to a simple one-site binding model. (C) Dilution of the BME/cyanoacrylamide adduct into buffer lacking BME establishes a new equilibrium favoring the cyanoacrylamide.

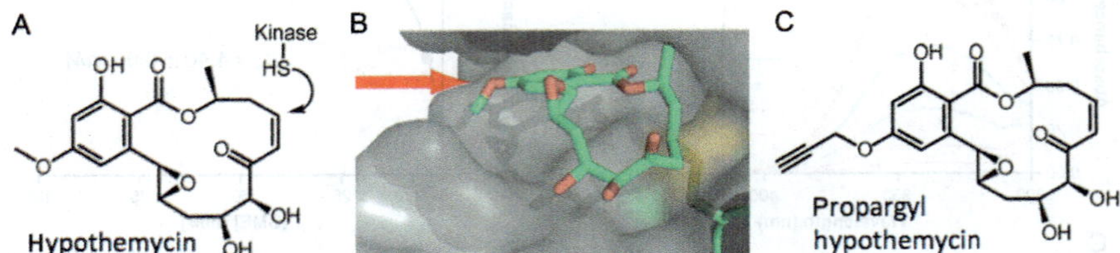

Rand M. Miller and Jack Taunton, Figure 4.6 Semipromiscuous occupancy probe for CDXG kinases. (A) Hypothemycin reacts with the cysteine in CDXG kinases via conjugate addition to the *cis*-enone. (B) Cocrystal structure of hypothemycin bound to ERK2 (PDB code: 3C9W) suggests the methyl ether as a derivatization site for an alkyne tag. (C) Chemical structure of propargyl-hypothemycin, prepared from desmethyl-hypothemycin and propargyl bromide.

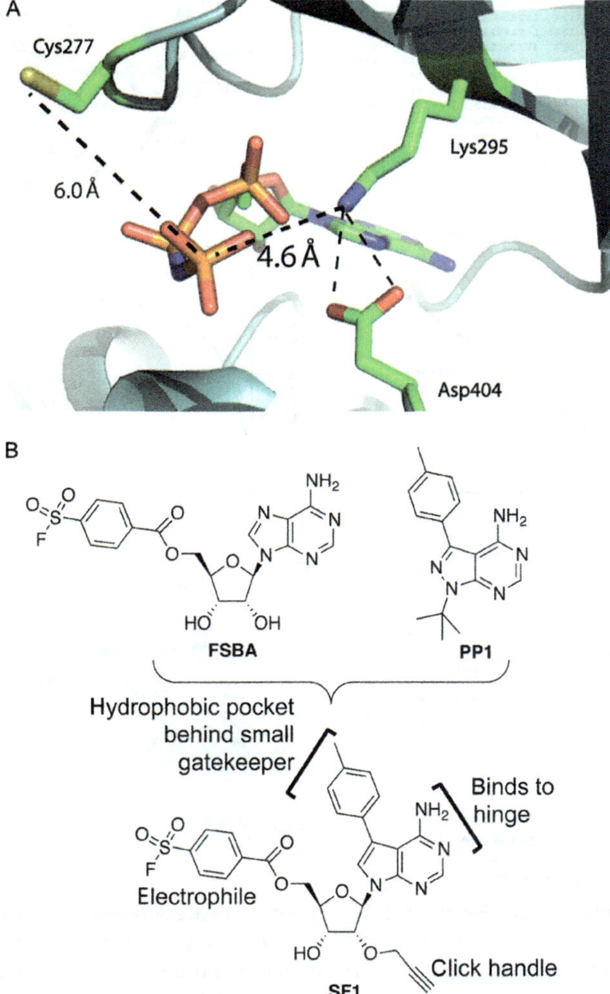

Rand M. Miller and Jack Taunton, Figure 4.7 Design of a semipromiscuous, lysine-targeted kinase probe. (A) Structure of SRC bound to AMP-PNP (PDB code: 2SRC), showing distances from the terminal phosphate to the catalytic Lys295 and nonconserved Cys277. (B) Design of a lysine-targeted sulfonyl fluoride probe, **SF1**, by merging FSBA and PP1. *Adapted with permission from Gushwa et al. (2012). Copyright 2012, American Chemical Society.*

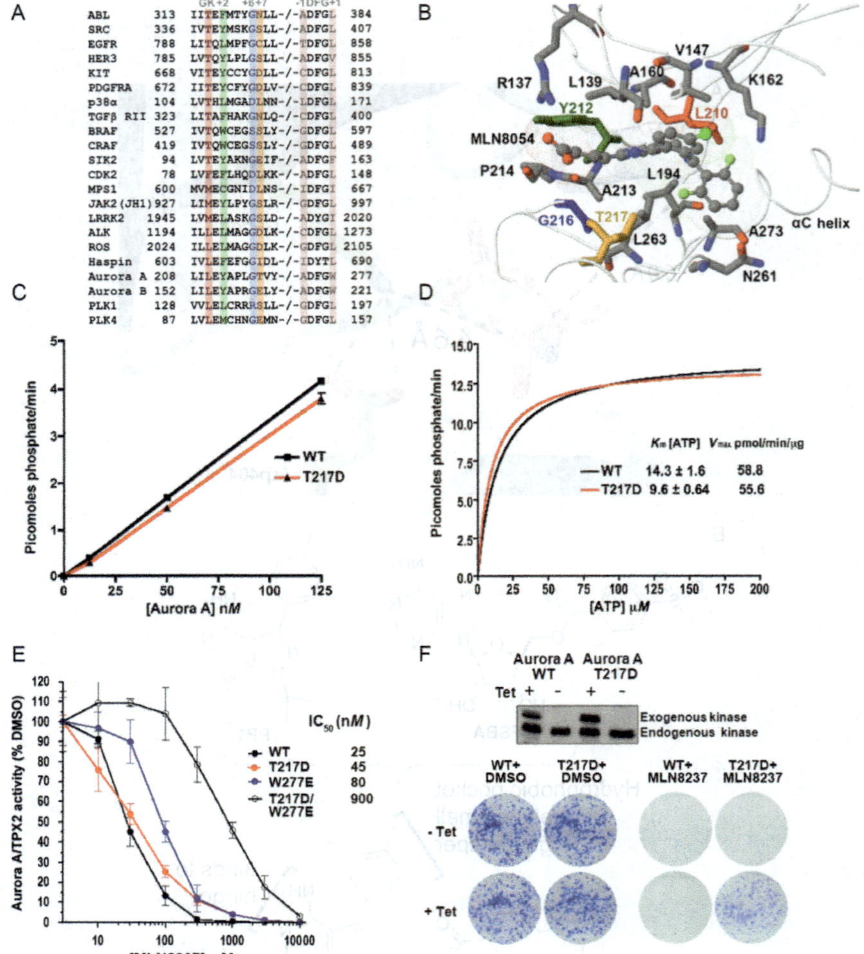

Fiona P. Bailey *et al.*, Figure 5.2 Experimental procedure for evaluating key drug-resistance loci in protein kinases. (A) Alignment of the resistance tetrad and "DFG" region of multiple kinases that have been analyzed using drug-resistance approaches, and for which drug-resistant alleles are readily available. Key side chains highlighted in kinases targeted for drug-resistance approaches are the gatekeeper residue (red), the +2 hydrophobic (green), the +6 (blue), the +7 (orange), and the DFG −1, DFG, and DFG +1 residues (pink). A similar color scheme is employed in (B), demonstrating the binding mode of MLN8054 in Aurora A (PDB ID: 2X81), with the resistance tetrad highlighted. (C) Kinetic analysis of WT and T217D Aurora A mutant (D) K_M[ATP] values for WT and T217D Aurora A. (E) Aurora A/TPX2 drug resistance toward MLN8227 induced by rationally designed mutations *in vitro*. (F) Tet-inducible overexpression of WT or DR (T217D) Aurora A analyzed by Western blotting (top panel) and on-target validation of MLN8237 cytotoxicity using a cellular assay (lower panels). *Panels (C)–(F) are reproduced from Sloane et al. (2010) with permission.*

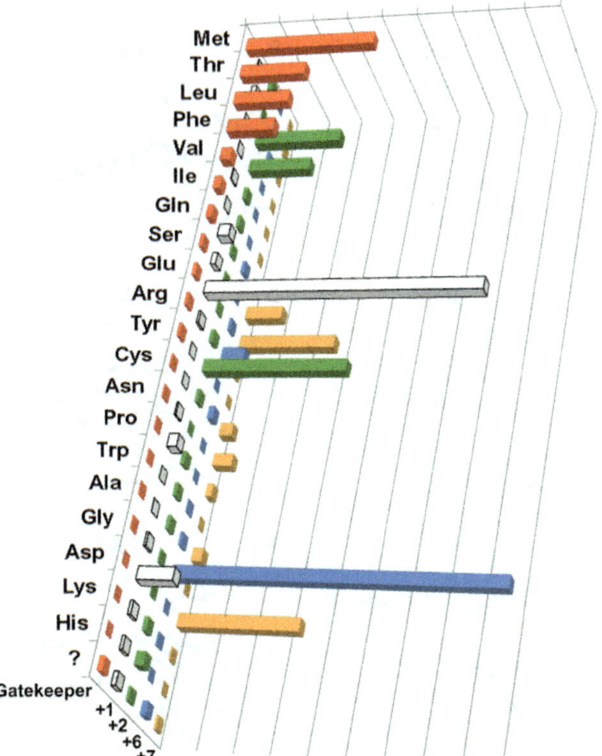

Fiona P. Bailey *et al.*, Figure 5.3 Frequency distribution (%) of amino acids found within the resistance tetrad. Gatekeeper, +1 (commonly an Asp or Glu residue), +2, +6, and +7 amino acid statistics from the human kinome are plotted as percentage of total (Manning et al., 2002). The most common residues at the gatekeeper residue are Met, Thr, Leu, Phe, and Val, whereas Tyr, Leu, Phe, and His predominate at the +2 position. Among benchmark AGC, CAMK, and TKs, the +6 position is very commonly Gly, while Asp, Glu, and Ser prevail at +7. Many amino acids are rare or absent in the resistance tetrad, helping simplify the choice of initial logical mutations for developing drug-resistance alleles.

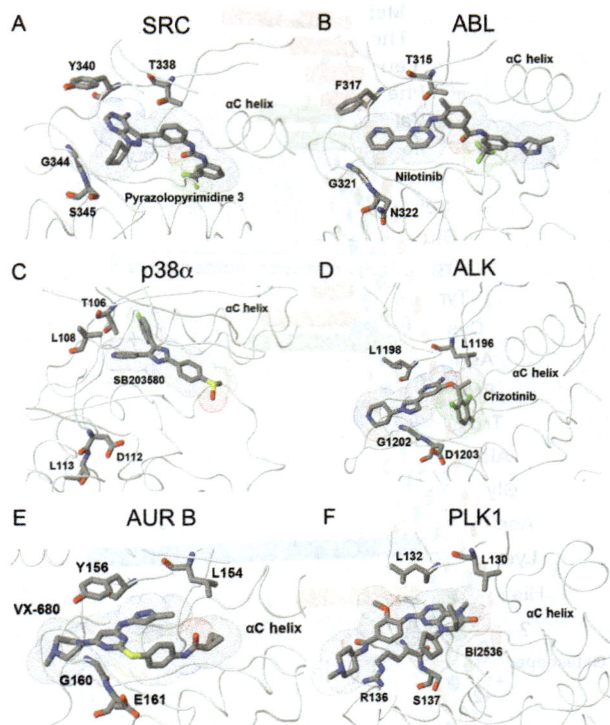

Fiona P. Bailey *et al.*, Figure 5.4 Structural analysis of the resistance tetrad in diverse human kinases. The cocrystal structures of (A) SRC and a PP1 analog (PDB ID: 3EL7), (B) ABL and Nilotinib (PDB ID: 3CS9), (C) p38α and SB203580 (PDB ID: 1A9U), (D) ALK and Crizotinib (PDB ID: 2XPT), (E) Aurora B and VX680 (PDB ID: 4AF3), and (F) Plk1 and BI2536 (PDB ID: 2RKU). Amino acids corresponding to the resistance tetrad are indicated for each kinase.

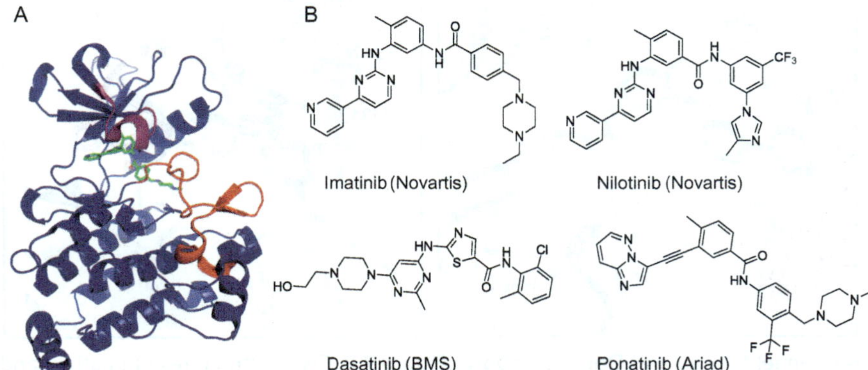

Nathanael S. Gray and Doriano Fabbro, Figure 7.1 Bcr–Abl kinase inhibitors. (A) Crystal structure of the kinase domain of Abl (blue ribbons) with imatinib (green sticks) bound in the ATP pocket. The glycine-rich loop (purple ribbons) which forms the "roof" of the ATP-binding site and the activation loop (red ribbons), both of which undergo conformational rearrangements upon imatinib binding are indicated. (B) Chemical structures of first- (imatinib), second- (nilotinib and dasatinib), and third-generation (ponatinib) FDA approved Bcr–Abl inhibitors.

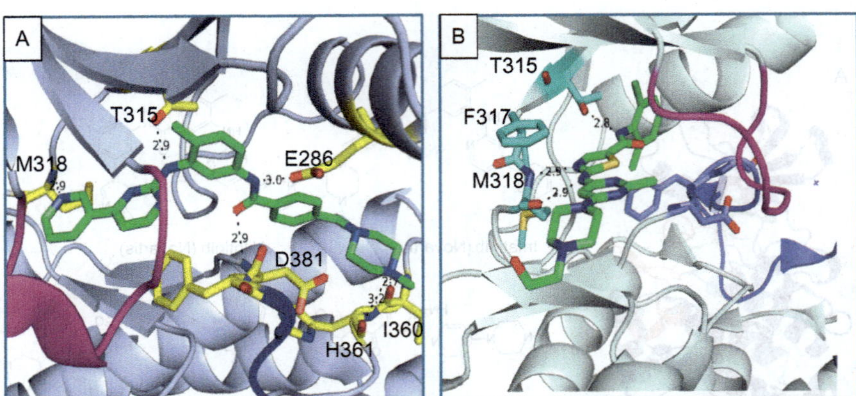

Nathanael S. Gray and Doriano Fabbro, Figure 7.2 Crystal structures of imatinib and dasatinib with Abl highlighting the protein–ligand interactions. (A) Imatinib (green sticks, carbons), nitrogen (blue), oxygen (red) forms five hydrogen-bond interactions with the ATP-binding pocket of Abl as indicated by dotted lines with distances shown in angstroms. The benzamide portion of the inhibitor is situated in a pocket made accessible by the flip of the DFG-motif (yellow sticks). (B) Similar figure for dasatinib. The "DFG-motif" is in not flipped in this structure, thereby closing the pocket accessed by the benzamide group of imatinib. Note both inhibitors make key hydrogen bonds to the "gatekeeper" residue T315 and traverse close to this position suggesting that they require the side-chain of the gatekeeper to be small in order to bind well.

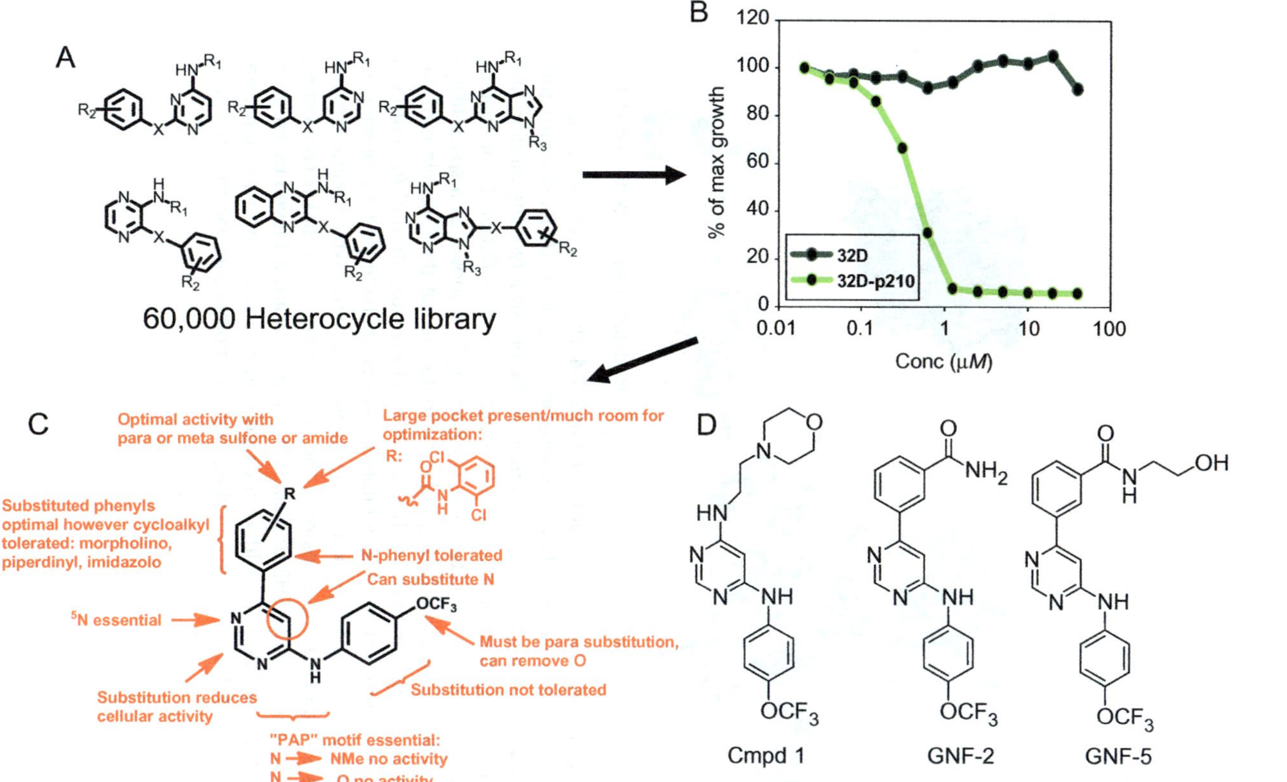

Nathanael S. Gray and Doriano Fabbro, Figure 7.3 Process that lead to the discovery of GNF-2 and -5. (A) Representative heterocyclic structures that were synthesized using solid-phase combinatorial chemistry encoded using directed sorting. (B) The library was screen for differential cytotoxicity between cells addicted to Bcr–Abl kinase activity for survival and proliferation versus isogenic "parental" cells grown under conditions not requiring Bcr–Abl kinase activity. (C) Summary of the salient structure–activity relationships developed by follow-up chemistry performed on the screening "hits." (D) Chemical structures of compound **1** (actual compound identified from the combinatorial library) and of GNF-2 and -5.

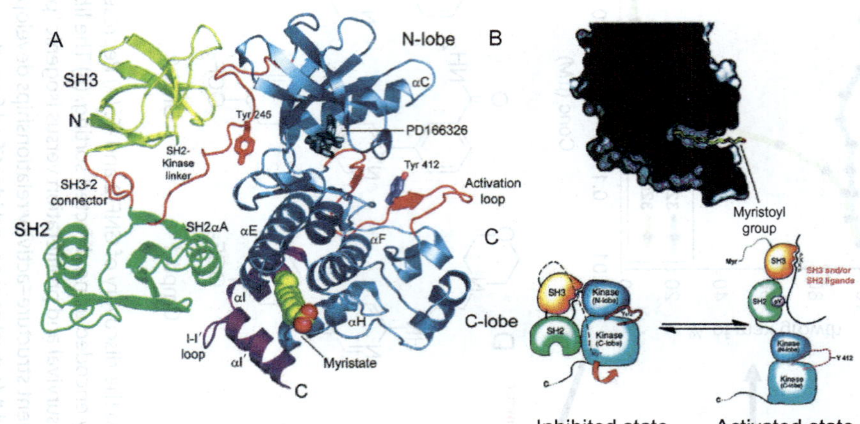

Nathanael S. Gray and Doriano Fabbro, Figure 7.4 Depiction of the proposed "latching" mechanism of Abl enabled by N-terminal myristoylation. (A) Ribbon structure of Abl containing the kinase domain (blue ribbons), the phosphotyrosine-binding SH2 domain (green ribbons), the linker-binding SH3 domain (yellow ribbons), and well as several other key regions as indicated. The myristate lipid is shown as a space-filling model (yellow balls) nestled in a hydrophobic cavity located at the C-terminus of the kinase domain. (B) Proposed model of how N-terminal myristoylation of Abl enforces the autoinhibited conformation. The inhibited state is characterized by binding of the N-terminal myristate into a pocket at the kinase C-terminus which reinforces the binding of the SH2 and SH3 domains onto the kinase domain and linker connecting the kinase N-terminus and the SH2 domain. Activation of Abl is achieved by engagement of the SH2 and SH3 domains with ligands, phosphorylation of the activation loop, and disengagement of the myristate and likely numerous other mechanisms that have not been elucidated.

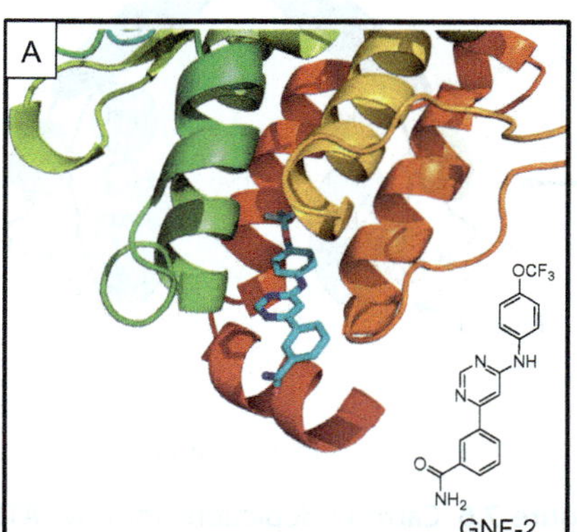

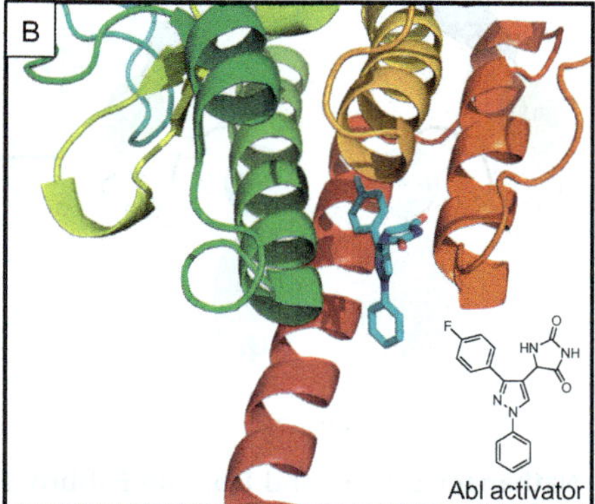

Nathanael S. Gray and Doriano Fabbro, Figure 7.5 Crystal structures of two myristate-site ligands bound to Abl. (A) GNF-5 and (B) Abl-activator compound (carbon atoms of ligands depicted with light blue sticks). The chemical structures of each are shown in the insets. Note the kinking of the red helix (labeled αI and αI′ in Fig. 7.3A) in the GNF-5 costructure which is in sharp contrast to the linear conformation in the Abl-activator costructure.

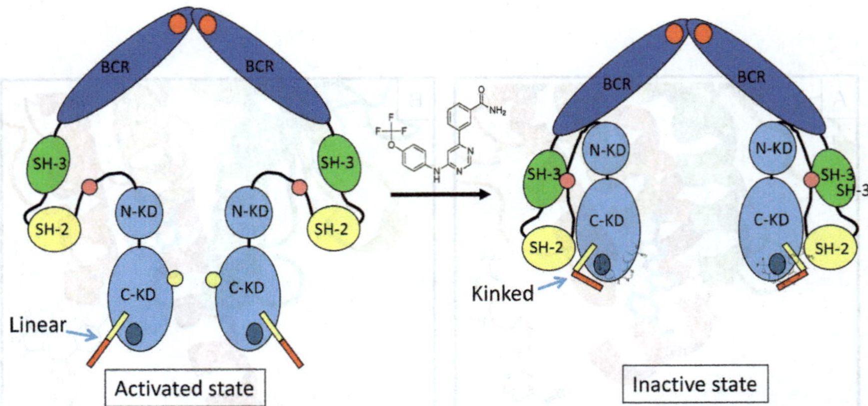

Nathanael S. Gray and Doriano Fabbro, Figure 7.6 Cartoon depictions for how ATP and myristate ligands inhibit Abl kinase. Binding of an ATP site ligand such as imatinib blocks the ATP-binding site, thereby preventing ATP binding but still allows for motion of the SH2 and SH3 domain relative to the kinase domain. Binding of a ligand such as GNF-5 to the myristate pocket induces conformational changes, most notably the dramatic kinking of the αI and αI′ helix (see Fig. 7.7), that favors binding SH2 and SH3 domains. Note that the Bcr-domain, in addition to its established function to oligomerize Abl, is also fused in such a way as to remove the myristate ligand present at the Abl C-terminus in the normal c-Abl. Thus, GNF-5 type compounds exploit a dormant endogenous mechanism of Abl regulation.

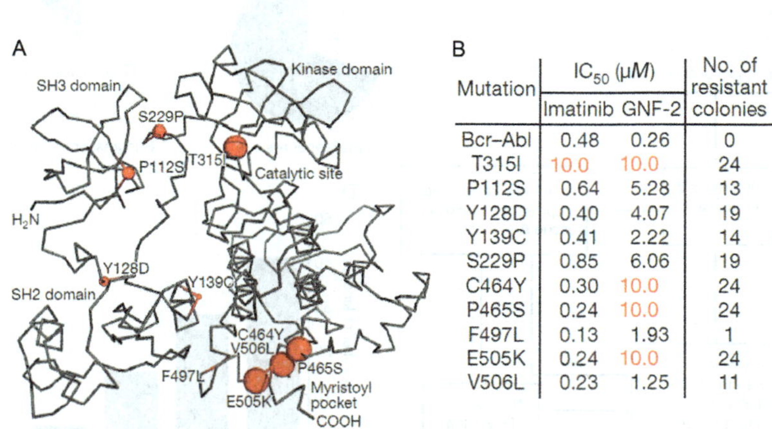

Mutation	IC_{50} (μM)		No. of resistant colonies
	Imatinib	GNF-2	
Bcr–Abl	0.48	0.26	0
T315I	10.0	10.0	24
P112S	0.64	5.28	13
Y128D	0.40	4.07	19
Y139C	0.41	2.22	14
S229P	0.85	6.06	19
C464Y	0.30	10.0	24
P465S	0.24	10.0	24
F497L	0.13	1.93	1
E505K	0.24	10.0	24
V506L	0.23	1.25	11

Nathanael S. Gray and Doriano Fabbro, Figure 7.7 Screen and characterization of GNF-2 resistance mutations in Bcr–Abl. (A) Ribbon diagram of SH1, SH2, and SH3 domains showing location of GNF-5 resistance mutations (red balls, sized in proportion to the number of resistant clones obtained from the screen). The majority of the mutations are clustered in the myristate-binding pocket as expected but also at the gatekeeper residue (T315I). (B) Cell proliferation IC50s in micromolar for imatinib and GNF-2 for Ba/F3 cells engineered to express the Bcr–Abl point mutations recovered from the screen. Also indicated is the number of clones expressing a particular mutation recovered from the screen.

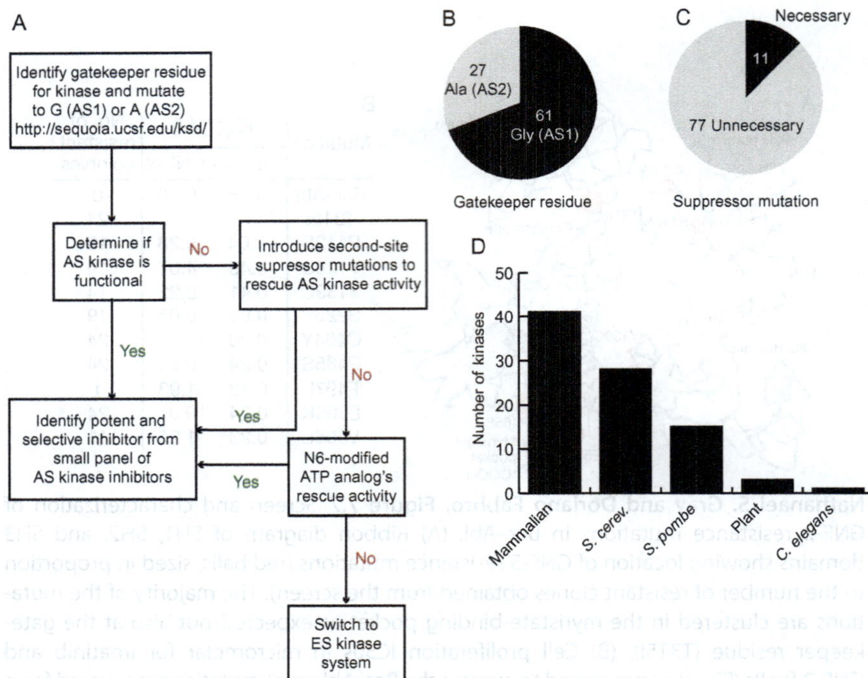

Michael S. Lopez *et al.*, Figure 8.1 Overview of AS-kinase technology. (A) Flowchart outlines systematic approach for identifying functional AS kinases. (B) Distribution of Ala and Gly gatekeeper residues used in reported AS kinases. (C) Fraction of reported AS kinases requiring second-site suppressor mutations. (D) Distribution of reported AS kinases according to species.

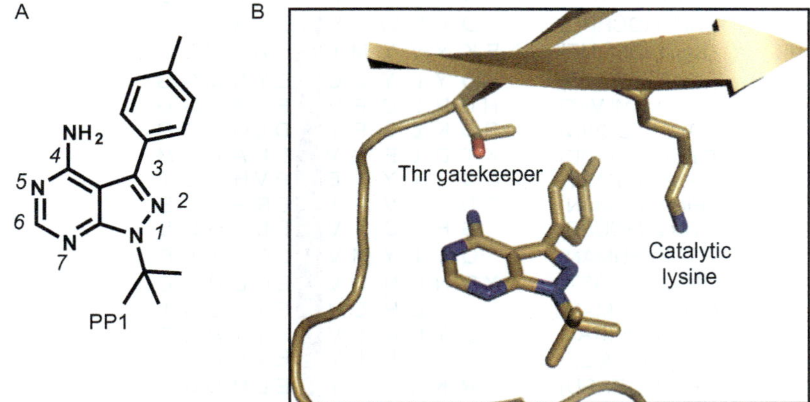

Michael S. Lopez *et al.*, Figure 8.2 Structural basis of PP1 activity. (A) Chemical structure of PP1. (B) X-ray cocrystal structure of PP1 bound to Hck kinase.

Kinase		Gatekeeper (Src T338)	
Src (CHKN)	E P I Y I V	T	E Y M S K G
PhkG2 (RAT)	S S S F M F L V	F	D L M R K G
Cdpk (SOYBN)	D S T A V H L V	M	E L C E G G
Kpk2 (PLAFK)	Q E G F V Y L V	L	E Y L K G G
Cdc5 (YEAST)	S N V Y I L	L	E I C P N G
Cdr1 (SCHPO)	H Q H M Y L A	L	E Y V P D G
Rad53 (YEAST)	D T E S Y Y M V	M	E F V S G G
Kin1 (SCHPO)	N S H Y Y M V	F	E F V D G G
Mek1 (YEAST)	R N N H L Y I F	Q	D L I P G G
Chk1 (SCHPO)	Q W R W V V	L	E F A Q G G
Kgp1 (DROME)	E K Y V Y M L	L	E A C M G G
Kin82 (YEAST)	K D Y L Y L C	M	E Y C M G G
Akt (MLVAT)	H D R L C F V	M	E Y A N G G
Arbk1 (BOVIN)	P D K L S F I	L	D L M N G G
Fused (DROME)	K T D L F V V	T	E F A L - M
Kab7 (YEAST)	D D Y Y Y I E	T	P V H G E T
Pim1 (HUMAN)	P D S F V L I	L	E R P - - E
Clk1 (MOUSE)	G H I C I V	F	E L L G L S
TTK (HUMAN)	Q Y I Y M V	M	E C G N I D
KIN28 (YEAST)	Y D N L N L V	L	E F L P T D
CdkL1 (HUMAN)	L H L V	F	E Y C D H T
Gsk3A (RAT)	L Y L N L V	L	E Y V P E T
Kpro (MAIZE)	S H R L L V	S	E Y V E N G
Phy1 (CERPU)	R K C S I I	M	E L M D G D
Byr2 (SCHPO)	D H L N I F	L	E Y V P G G
Cdc15 (YEAST)	Y E L Y I L	L	E Y C A N G
Ste7 (YEAST)	I N N E I I I L	M	E Y S D C G
Byr1 (SCHPO)	Y K N N I S L C	M	E Y M D C G
MKK1 (YEAST)	E N S S I Y I A	M	E Y M G G R
MOS (CERAE)	S N S L G T I I	M	E F G G N V
Pkn1 (MYXXA)	P R P Y L I	M	E F L D G A
Ran1 (SCHPO)	E D A I Y V V	L	Q Y C P N G
Hal4 (YEAST)	E Y C E V	M	E Y C A G G
Hrr25 (YEAST)	Y N A M V	I	D L L G P S
EphA4 (MOUSE)	K P V M I I	T	E Y M E N G

Michael S. Lopez *et al.*, Figure 8.3 Representative alignment of the hinge region of kinase domains from diverse organisms and kinase families.

Diazo-lactam synthesis

1. MeOH, TEA
2. MeOH, $NaBH_4$

3. DCM, DCC
0 °C → RT

4. EtOH, NaOEt, 60 °C
5. MeCN, H_2O, 70 °C
6. MeCN, TEA, *p*ABSA
0 °C → RT

Biindole synthesis

7. Cu(I), MeOH
pyridine

8. KtOBu, NMP
80 °C

Indolocarbazole formation

9. Pinacolone
$Rh_2(OAc)_4$
120 °C

Br–(CH2)n–NH(Boc)
10. DMF, NaH
11. TFA, CH_2Cl_2
dimethoxybenzene

R =				
n =	2	1	2	2
Star	12	17	18	19

Michael S. Lopez *et al.*, Figure 8.5 Synthetic scheme for the synthesis of staralogs.

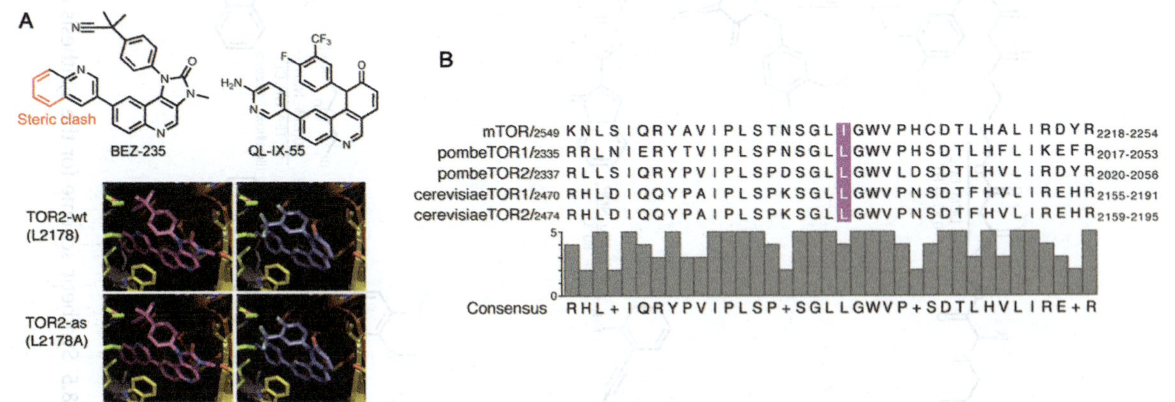

Michael S. Lopez *et al.*, Figure 8.7 (A) Homology model of mTOR based on the structure of PI3Kγ shown with the gatekeeper residue in gray. The known *S. cerevisiae* TOR1/TOR2 inhibitor QL-IX-55 (purple) and BEZ235 (magenta) were oriented based on a typical H-bonding interaction with the backbone carbonyl of valine in the active site at VanDer Waals distances away from other residues that form the ATP-binding pocket. The isoleucine gatekeeper clash with BEZ235 is exacerbated by mutation to leucine and alleviated by mutation to alanine. The smaller QL-IX-55 does not sense this residue. (B) Sequence alignment shows that the gatekeeper residue (in purple) is isoleucine in mTOR and leucine in all other cases. The active site is highly conserved.

Printed by Printforce, the Netherlands